PHYSICS RESEARCH AND TECHNOLOGY

THE BIG BANG

THEORY, ASSUMPTIONS AND PROBLEMS

PHYSICS RESEARCH AND TECHNOLOGY

Additional books in this series can be found on Nova's website under the Series tab.

Additional E-books in this series can be found on Nova's website under the E-book tab.

SPACE SCIENCE, EXPLORATION AND POLICIES

Additional books in this series can be found on Nova's website under the Series tab.

Additional E-books in this series can be found on Nova's website under the E-book tab.

PHYSICS RESEARCH AND TECHNOLOGY

THE BIG BANG

THEORY, ASSUMPTIONS AND PROBLEMS

JASON R. O'CONNELL
AND
ALICE L. HALE
EDITORS

Nova Science Publishers, Inc.
New York

Copyright © 2012 by Nova Science Publishers, Inc.

All rights reserved. No part of this book may be reproduced, stored in a retrieval system or transmitted in any form or by any means: electronic, electrostatic, magnetic, tape, mechanical photocopying, recording or otherwise without the written permission of the Publisher.

For permission to use material from this book please contact us:
Telephone 631-231-7269; Fax 631-231-8175
Web Site: http://www.novapublishers.com

NOTICE TO THE READER

The Publisher has taken reasonable care in the preparation of this book, but makes no expressed or implied warranty of any kind and assumes no responsibility for any errors or omissions. No liability is assumed for incidental or consequential damages in connection with or arising out of information contained in this book. The Publisher shall not be liable for any special, consequential, or exemplary damages resulting, in whole or in part, from the readers' use of, or reliance upon, this material. Any parts of this book based on government reports are so indicated and copyright is claimed for those parts to the extent applicable to compilations of such works.

Independent verification should be sought for any data, advice or recommendations contained in this book. In addition, no responsibility is assumed by the publisher for any injury and/or damage to persons or property arising from any methods, products, instructions, ideas or otherwise contained in this publication.

This publication is designed to provide accurate and authoritative information with regard to the subject matter covered herein. It is sold with the clear understanding that the Publisher is not engaged in rendering legal or any other professional services. If legal or any other expert assistance is required, the services of a competent person should be sought. FROM A DECLARATION OF PARTICIPANTS JOINTLY ADOPTED BY A COMMITTEE OF THE AMERICAN BAR ASSOCIATION AND A COMMITTEE OF PUBLISHERS.

Additional color graphics may be available in the e-book version of this book.

Library of Congress Cataloging-in-Publication Data

The big bang : theory, assumptions, and problems / editors, Jason R. O'Connell and Alice L. Hale.
    p. cm.
 Includes index.
 ISBN 978-1-61324-577-4 (hardcover)
 1. Big bang theory. I. O'Connell, Jason R. II. Hale, Alice L.
 QB991.B54B545 2011
 523.1'8--dc23
                      2011014552

Published by Nova Science Publishers, Inc. † New York

CONTENTS

Preface		vii
Chapter 1	Astrophysical *S*-factors of Proton Radiative Capture in Thermonuclear Reactions in the Stars and the Universe *S. B. Dubovichenko and A. V. Dzhazairov-Kakhramanov*	1
Chapter 2	The Big Bang and What It Was *E. E. Escultura*	61
Chapter 3	Cosmic Structure Formation after the Big Bang *L. M. Chechin*	103
Chapter 4	Temporal Topos and U-Singularities *Goro C. Kato*	173
Chapter 5	Fully Quantum Study of the FRW Model with Radiation and Chaplygin Gas *Sergei P. Maydanyuk and Vladislav S. Olkhovsky*	185
Chapter 6	A Brief Introduction Review on the Problems of the University Origin *V. S. Olkhovsky*	197
Chapter 7	The Quantum Theory of the Big Bang: Effective Theory of Quantum Gravity *Subodha Mishra*	205
Chapter 8	Entropy Growth in the Universe *Marcelo Samuel Berman*	229
Chapter 9	Did The Big Bang Really Take Place? *Ernst Fischer*	243
Chapter 10	De Sitter-Fantappie Universe *E. Benedetto*	259
Chapter 11	Consciousness and Energy *Vikram H. Zaveri*	275

Chapter 12	On the Rotation of the Zero-Energy Expanding Universe *Marcelo Samuel Berman and Fernando de Mello Gomide*	**285**
Index		**311**

PREFACE

In this book, the authors present current research from across the globe in the study of the theory, assumptions and problems of the Big Bang theory. Topics discussed include cosmic structure formation after the Big Bang; temporal topos and u-singularities; the quantum theory of the Big Bang; de Sitter-Fantapppie universe and the astrophysical s-factors of proton radiative capture in thermonuclear reactions in the stars and the universe.

Chapter 1 - In this chapter the authors have considered the possibility to describe the astrophysical S-factors of some reaction with light atomic nuclei on the basis of the potential two-cluster model by taking into account the supermultiplet symmetry of wave functions and splitting the orbital states according to Young's schemes. Two-cluster model constitutes a phenomenological semi-microscopic approach to study many-nucleon systems. Within this model, interaction of the nucleon clusters is described by local potential determined by fit to the scattering data and properties of bound states of these clusters. Many-body character of the problem is taken into account under some approximation, in terms of the allowed or forbidden by the Pauli principle states of the full system of the nucleons. An important feature of the approach is accounting for a dependence of interaction potential between clusters on the orbital Young's scheme, which determines the permutation symmetry of the nucleon system. Photonuclear processes in the p2H, p3H, p7Li, p9Be, p12C systems and corresponding astrophysical S-factors are analyzed on the basis of this approach. Such an approach allows analyzing the structure of inter-cluster interactions, to determine the presence of allowed and forbidden states in the interaction potential and so the number of nodes of the orbital wave function of cluster relative motion. It is shown that the approach allows one to describe quite reasonably experimental data available at low energies, especially for systems with number of nucleons A>4 for the cases in which the phase shifts of cluster-cluster scattering are extracted from the data with minimal errors. In this connection it is very urgent to raise the accuracy of experimental measurements of elastic scattering of light nuclei at astrophysical energies and to perform a more accurate phase shift analysis. The increase in the accuracy will allow making more definite conclusions regarding the mechanisms and conditions of thermonuclear reactions, as well as understanding better their nature in general.

Chapter 2 - The chapter takes as its premise the occurrence of the Big Bang, a colossal explosion that gave rise to the birth of our universe. Then through sequences of mathematical and scientific reasoning called rational thought using the methodology of qualitative modelling the grand unified theory (GUT) is developed on laws of nature which establishes what it was that exploded. This is a new approach. In earlier development both the occurrence of the Big Bang and what caused it were taken as premises.

In its present development GUT builds its pillars – quantum and macro gravity and thermodynamics. Through the discovery of appropriate laws of nature, GUT well defines its fundamental physical concepts such as the superstring, basic constituent of matter; matter and energy; dark matter; and charge and gravity. It also well-defines and explains other physical concepts: primum, photon, matter-anti-matter interaction, atom, cosmic waves, brittle and malleable materials, metal fatigue, turbulence, galaxy, black hole and supernova along with the conversion of dark to visible matter.

The cosmology of our universe is traced from its birth at Cosmic Burst 1.5 billion years from the start of the Big Bang through its evolution to the present as a super...super galaxy 10^{10} billion light years across and a local bubble in the timeless boundless Universe and predicts its destiny as black holes back in dark matter. The existence of universes other than ours is established from available evidences. Then it is deduced from this cosmology what the Big Bang was and what caused it.

Chapter 3 - In the new version of the Universe large-scale structures forming based on the refuse of analyses, only the gravitational instability of cosmological substrate is proposed. It is shown that vacuum is the dominant of non-baryonic matter in the Universe which creates the anti-gravitational instability of baryonic cosmic substrate itself and causes the galaxies' formation.

The growth of a baryonic substance's perturbations that is caused by the non-stationary equation of state of a non-baryonic matter in the very early Universe is searched. It was shown that initially their amplitudes are drastically larger than perturbations of the baryonic matter in the framework of the standard scenario of gravitational instability.

For a deeper understanding of the process of baryonic matter evolution in the expanding Universe, it is necessary to know the physical property of the concrete field that represents the background of the substrate type of dark energy. Besides, it is necessary to explore in detail the influence of such a field on the continuous medium of baryonic matter. These statements were realized for the quintessence field. As the result, the authors describe the quintessence field by two gravitating scalar fields. They give contributions at the total pressure and at the total mass density of baryonic matter. It allowed the authors to show that the evolution of baryonic matter's density perturbations obeys the equation of forced oscillations and admits the resonance case, when amplitude of baryonic matter's density perturbations gets the strong short-time splash. This splash is interpreted as a new macroscopic mechanism of the initial matter density perturbations appearance.

The stochastic equation of a thread's motion in the background of the massive cosmic string whose linear mass density is subjected to stochastic perturbations was investigated. It was shown that stochastic movement of the thread generates, in a cosmological substrate, the long perturbing waves that lead to the galaxies formation, in its turn.

The interval of a weak-oscillating cosmic string with standing flat disturbance waves on it was deduced and the brief discussion of some physical processes occurring in the vicinity of oscillating massive cosmic strings followed. Among them, the dynamics of open "probe" cosmic thread performing the constraint oscillations. The nature of these oscillations, as it was pointed out, is drastically different from the previous well-known mechanisms.

It is shown that domain wall pumps a cosmic string by gravitational energy if it is located in its vicinity. This leads to the tension energy increasing and cosmic string stretching. The discussing process of straighten is the new physical mechanisms of cosmic string stretching.

The rotational galaxy movement produced by vacuum anti-gravitation force was searched. The expression for suitable angular velocity is found and its estimation for the real elliptical galaxies is done. One cosmological consequence of the vacuum angular velocity effect for the Universe description – the Birch effect – is discussed also.

Chapter 4 - Several papers and books by C. Isham, C.Isham-A. Doering, F. Van Oystaeyen, A. Mallios-I. Raptis, C. Mulvey, and Guts and Grinkevich, have been published on the methods of categories and sheaves to study quantum gravity. Needless to say, there are well-written treatises on quantum gravity whose methods are non-categorical and non-sheaf theoretic. This paper may be one of the first papers explaining the methods of sheaves with minimally required background that retains experimental applications.

Temporal topos (t-topos) is related to the topos approach to quantum gravity being developed by Prof. Chris Isham of the Oxford-Imperial research group (with its foundations in the work of F. W. Lawvere). However, in spite of strong influence from papers by Isham, the authors' method of t-topos is much more direct in the following sense. The authors' approach is much closer to the familiar applications of the original algebraic geometric topos where little logic is involved.

The distinguishable aspects of this paper "Temporal Topos and U-Singularity" from other topos theorists' appoaches are the following. For a particle, the authors consider a presheaf associated with the particle. By definition, a presheaf is a contravarinat functor; however, in the t-topos theory, such a presheaf need not be defined for every object of a t-site over which the topos of presheaves are defined. When such an associated presheaf is not defined (or non-reified), the authors say that the presheaf (the particle) is in ur-wave state. Therefore, the duality is already embedded in the authors' t-topos theory. The authors also have the notion of a (micro) decomposition of a presheaf (a particle) to obtain microcosm objects. Another important aspect of their approach is the associated space and time sheaves for a given particle-presheaf. The sheaves associated with space, time, and space-time are treated differently from a particle associated presheaf. Namely, Yoneda Lemma and its embedding are crucial for formulating and capturing the nature of space-time. In this formulation, the space and time sheaves would not exist unless a particle (presheaf) exists. Such a non-locality nature as the EPR type non-locality is also embedded in t-topos. Applications to singularities (a big bang, black holes, and subplanck objects) are formulated in terms of universal mapping properties of direct limit and inverse limit in category theory. Furthermore, the uncertainty principle is formulated through the concept of a micro-morphism in t-site. The authors' t-topos theoretic approach enables us to formulate a light cone in macrocosm and also in microcosm. However, such a light cone in microcosm has non-reified space-time regions because of the uncertainty principle (a miro-morophsim).

Chapter 5 - The authors give an introduction into quantum cosmology with emphasis on its conceptual fully quantum consideration in framework of formalism based on minisuperspace Wheeler–DeWitt equation. After a general review of different approaches of quantum cosmology, the authors study the closed Friedmann-Robertson-Walker model with quantization in presence of the positive cosmological constant, radiation and Chaplygin gas. For analysis of tunneling probability for birth of an asymptotically deSitter, inflationary Universe as a function of the radiation energy they introduce a new definition of a "free" wave propagating inside strong fields. On this basis the authors correct Vilenkin's tunneling boundary condition, define and calculate penetrability and reflection in fully quantum stationary approach (without application of any semiclassical and perturbation theory approximations).

Chapter 6 - In the presented brief introduction review there are briefly analyzed various cosmologic and quantum cosmologic Big-Bang hypothesis and theories. It is made the general conclusion that namely in the *beginning* is arising the general problem of the world origin and a lot of related problems. The schematic descriptions of some problems between them are presented. And it is added the problem of the inevitable choice meta-theoretical dilemma: the beginning of the Universe formation from vacuum ("nothing") is *either* a result of the irrational randomness after passing from other space-time dimensions or from other universe, caused by some unknown process, *or* a result of the creation of the expanding Universe (together with the laws of its functioning) by the supreme intelligent design from *nigilo*. And to the first doctrine there was adjoined in the XXI c the meta-physical doctrine of the parallel other universes with some kind of interaction between them or with an irrational spontaneous passage of the matter from them to our Universe – those hypothetical universes are or the exactly same as ours (in quantum mechanics), or with other space-time dimensions (in some Big-Bang theories), or with other values of the physical constants (in other Big Bang theories).

Chapter 7 - Considering the fact that our universe might have been born out of the *Big Bang* and formed subsequently out of a system of fictitious self-gravitating particles, fermionic in nature, the authors formulate a quantum mechanical theory of the universe which turns out to be an effective theory of quantum gravity. This theory is the special relativized quantized Newton-Cartan theory which is the nonspecial relativistic limit of Einstein's General Theory of Relativity. The authors are able to obtain a compact expression for the radius R_0 of the universe by using a model density distribution $\rho(r)$ for the particles which is singular at the origin. This singularity in $\rho(r)$ can be considered to be consistent with the so called Big Bang theory of the universe. By assuming that Mach's principle holds good in the evolution of the universe and taking the age of the present universe to be $\tau_0 \simeq 14 \times 10^9 yr$, the authors obtain the total mass of the universe ($M \simeq 1.0 \times 10^{23} M_\odot$), $M_\odot$ being the solar mass and the value of ratio of the variation of the universal gravitational constant with time to G in extremely good agreement with the results obtained by others. The authors' theory reproduces the ratio of the number of neutrinos to that of nucleons and density of neutrinos in the present universe correctly. The authors also study quantum mechanically our expanding universe which is made up of gravitationally interacting particles such as particles of luminous matter, dark matter and dark energy as a selfgravitating system using a well-known many-particle Hamiltonian, but only recently shown as representing a soluble sector of quantum gravity. Describing dark energy by a repulsive harmonic potential among the points in the flat 3-space and incorporating Mach's principle to relativize the problem, the authors derive a quantum mechanical relation connecting temperature of the cosmic microwave background radiation, age, and cosmological constant of the universe. When the cosmological constant is zero, they get back Gamow's relation with a much better coefficient. Otherwise, our theory predicts a value of the cosmological constant $2.0 \times 10^{-56} cm^{-2}$ when the present values of cosmic microwave background temperature of 2.728 K and age of the universe 14 billion years are taken as input.

Chapter 8 - In a companion Chapter to this volume, Berman and Gomide proved that, under General Relativity theory, the total energy of a possibly rotating expanding Universe, is zero. Now, it will be shown that if the Universe is a zero-total energy entity and if, this energy is time-invariant, each type of energy contribution, to the total energy of the Universe, divided by Mc^2, yields a constant, during all times; for instance, if the present contribution of the cosmological "constant" Λ , drives the present Universe, it also must have driven alike, in all the lifespan of the Machian Universe (the relative contributions of energy densities of each kind, towards the total density, remain the same during all times). This fact is supported by the recent discovery that the Universe has been accelerating since a long time ago.

The immediate consequence of the above theory, is that all energy densities in the Universe are R^{-2} – dependent.

In consequence, the authors find that the absolute temperature, T, varies like $R^{-\frac{1}{2}}$ lambda, as R^{-2} so that the total entropy of the Universe grows with $R^{\frac{3}{2}}$ and, then, also as $M^{\frac{3}{2}}$

These conclusions are supported by other calculations involving variants of General Relativity theory.

Chapter 9 - The Big Bang model suffers from the occurrence of singularities and infinity conditions, where all known physical laws break down. Besides it is based on improvable new physics. The overwhelming part of the universe should consist of so called dark energy with very strange properties, unlike all we know. Besides not only the energy density should have been infinitely high, this state should have been present in an infinite space. The model takes its justification from the fact that it can explain the cosmological red shift, the microwave background radiation and (with restrictions) the chemical composition of matter.

But based on special and general relativity static spatially curved solutions of the field equations are possible, which allow a stable static homogeneous universe, if only potential energy is introduced as a part of the energy tensor and global Lorentz invariance exists.

But while a static homogeneous universe is stable as a whole, locally it is unstable, leading to the formation of structures, observed as galaxies and stars. There exists a continuous matter cycle, starting from a hot ubiquitous plasma, which locally cools down, first by gravitational loss of momentum, then by emission of electromagnetic radiation. This leads to fragmentation into clusters, galaxies and then into individual stars. The negative potential energy finally stops further collapse into a singularity. Instead this collapsed state may become unstable, leading to the emission of relativistic matter jets, delivering fresh hydrogen to the intergalactic plasma. The Big Bang is replaced by these small bangs.

This scenario is possible within the framework of the confirmed theory of relativity without singularities or new physics.

Chapter 10 - In this chapter the authors analyze the development of de Sitter-Fantappi_e Relativity by analyzing, particularly, its applications to cosmology. The cosmological applications seem to be interesting since they resolve many problems of the standard cosmology. In this scenario the space atness is linked to the observer geometry and it is independent from the presence and distribution of matter-energy. This resolves the atness problem without introducing inationary hypotesis and the global spacetime structure is univocally individuated by the algebraic structure of the physical laws.

Chapter 11 - This article presents further development of the periodic relativity theory (PR), which lead to the derivation of the quantum invariant resulting in a suggestion that the universe originated with a vibration in an ocean of unmanifested fundamental substance called energy. Here the authors propose that the ocean of unmanifest energy without any oscillation is ocean of infinite indivisible motionless pure consciousness beyond space and time. When this infinite indivisible consciousness becomes active, it oscillates by its own power and generates discrete quanta of consciousness which the physicists know as the discrete quanta of energy because of a very low degree of manifestation of the consciousness which makes the quanta appear almost like insentient matter. Thus the one becomes many and the laws of relativity becomes operative. Since the consciousness ever remains indivisible, all the individual consciousness always remain connected with the infinite motionless consciousness like the waves with the ocean. Therefore union of both is possible.

Chapter 12 - The usual expanding metric due to Robertson and Walker, results in two general relativistic equations, which are called Friedman-Robertson-Walker Cosmological equations. The first one expresses conservation of energy, and involves the energy density of the Universe. The second one can be thought of as a definition of cosmic pressure, as the volume-derivative of energy, with negative sign. If rotation is present, both equations are altered. In Berman, the authors find a general relativistic treatment of rotation plus expansion. The results of the authors' published research on the subject imply that not only such rotation is to be expected on General Relativity or even Newtonian theories, but also in other frameworks like Sciama's inertia model, and even in P.A.M. Dirac's large numbers hypothesis, for time-varying finestructure along with other variable "constants".

The constancy of the (zero) energy of Universe, was dealt elsewhere (Berman, 2009c). The authors now go a step further, and show that even when a metric temporal coefficient is added to the usual Robertson-Walker's metric, and thus responds for the rotational state in addition to the expansion, by pseudo-tensor calculation, the zerototal-energy hypothesis is still valid. The Pioneers anomaly is given now a general relativistic full explanation. Another explanation, which is equivalent, is obtained by the theory of time-varying speed of light, which yields the same Pioneers anomalous deceleration.

Chapter 1

ASTROPHYSICAL *S*-FACTORS OF PROTON RADIATIVE CAPTURE IN THERMONUCLEAR REACTIONS IN THE STARS AND THE UNIVERSE

S. B. Dubovichenko and A. V. Dzhazairov-Kakhramanov

V. G. Fessenkov Astrophysical Institute of the National Centre of Space Researches and Technologies, 050020, Almaty, Republic of Kazakhstan,
National Space Agency of the Republic of Kazakhstan

ABSTRACT

In this chapter we have considered the possibility to describe the astrophysical S-factors of some reaction with light atomic nuclei on the basis of the potential two-cluster model by taking into account the supermultiplet symmetry of wave functions and splitting the orbital states according to Young's schemes. Two-cluster model constitutes a phenomenological semi-microscopic approach to study many-nucleon systems. Within this model, interaction of the nucleon clusters is described by local potential determined by fit to the scattering data and properties of bound states of these clusters. Many-body character of the problem is taken into account under some approximation, in terms of the allowed or forbidden by the Pauli principle states of the full system of the nucleons. An important feature of the approach is accounting for a dependence of interaction potential between clusters on the orbital Young's scheme, which determines the permutation symmetry of the nucleon system. Photonuclear processes in the p2H, p3H, p7Li, p9Be, p12C systems and corresponding astrophysical S-factors are analyzed on the basis of this approach. Such an approach allows analyzing the structure of inter-cluster interactions, to determine the presence of allowed and forbidden states in the interaction potential and so the number of nodes of the orbital wave function of cluster relative motion. It is shown that the approach allows one to describe quite reasonably experimental data available at low energies, especially for systems with number of nucleons A>4 for the cases in which the phase shifts of cluster-cluster scattering are extracted from the data with minimal errors. In this connection it is very urgent to raise the accuracy of experimental measurements of elastic scattering of light nuclei at astrophysical energies and to perform a more accurate phase shift analysis. The increase in the accuracy will allow making more definite conclusions regarding the mechanisms and conditions of thermonuclear reactions, as well as understanding better their nature in general.

PACS: 24.10.-i, 25.10.+s, 25.20.-x, 24.50.+g

1. INTRODUCTION

The astrophysical S-factors which determine the reaction rate, i.e. probability of the thermonuclear reactions in the Stars and the Universe at different stages of its evolution very often can not be defined experimentally because of the low energy of the interacting particles in such reactions, even at the modern stage of development of the experimental measurement methods. Currently, only for radiative p^{2}H capture the experimental measurements of the astrophysical S-factor have been made at the energies down to 2.5 keV, i.e. the energy range that can be considered as astrophysical. For the other nuclear systems, which take part in the thermonuclear processes, such measurements are made accurately at best (for p^{3}H system) down to 50 keV.

At the same time, the experimental data on cross-sections of nuclear reactions and their analysis within various theoretical models are the major source of information about nuclear structure, nature and mechanisms of nucleus-nucleus (cluster-cluster) interaction. Such researches in nuclear astrophysics are complicated since in many cases only theoretical predictions can supply deficient experimental information on characteristics of thermonuclear reactions. This difficulty, as it was said, is associated with the low energy of interaction of matter in the stars which ranges from tenth to tens of keV.

Thus, as a rule, it is impossible to measure directly the cross-sections required for astrophysical calculations of nuclear reactions. Usually, the cross-sections are measured at higher energies (about hundreds of keV) and then extrapolated into the energy range representing interest for nuclear astrophysics [1]. However, a simple extrapolation of experimental data into the astrophysical range is not always correct due to the fact that the experimental measurements of thermonuclear cross-sections are carried out at rather high energies (0.2-1.0 MeV). The experimental error band of the determination of the astrophysical S-factor at the energy range of 10-300 keV in different cluster systems can reach up to ±100% extremely lowering the value of such an extrapolation. In this situation the role of theoretical calculations becomes considerably more important.

The calculations carried out on the basis of the chosen theoretical concepts are compared with the existing experimental data (at the energy ranges where these data are available). This allows making certain conclusions about the quality of the physical model used and thus to select concepts and approaches leading to the best agreement with the experiment, which means that they best describe the real situation in the atomic nucleus at these energies. Then on the basis of the chosen model one can make the calculations in the astrophysical energy range, and this is not a simple extrapolation of the experimental data, because of the fact that such an approach has quite a definite microscopic rationale.

The considered potential cluster model (PCM) of atomic nucleus on the basis of calculated potentials of nuclear interaction allows calculating the required nuclear characteristics, such us the cross-sections of different photoreactions and the astrophysical S-

factors, quite easily. The approach used here for the theoretical analysis of such characteristics allows us to obtain results at the lowest energies (down to 1 keV).

The PCM model used here is based on the assumption that the nuclei under consideration consist of two clusters. We have chosen potential cluster model because for many light atomic nuclei the probability of formation of isolated nucleon associations is relatively high. It is confirmed by numerous experimental data and theoretical results obtained over the last fifty years [2].

Thus, the one-channel potential cluster model is a good approximation to the situation really existing in the atomic nucleus in many cases and for various light nuclei. Such a model allows making any calculations of nuclear characteristics in the scattering processes and bound states quite easily, even in the systems where many-body problem solution methods are very cumbersome in the digital implementation or do not lead to certain numerical results at all.

For example, on the basis of measurements of the differential cross-sections of the elastic scattering of nuclear particles [3] and [4] it is possible to perform the phase shift analysis, which at energies lower than 1 MeV usually includes the S-wave only. The data of differential cross-sections usually measured at 10-15 degrees of scattering at the required energy range allow us carrying out the most complete and accurate phase shift analysis and receiving the phase shifts of elastic scattering. Further, we can construct the nuclear potential of the inter-cluster interaction using the received phase shifts of scattering. This potential, in its turn, allows us to make any calculations of nuclear processes, for example, the calculation of the astrophysical S-factor of radiative capture at low energies, which is of interest for nuclear astrophysics.

We considered the astrophysical S-factors on the basis of the PCM which takes into account the supermultiplet symmetry of wave functions (WF) and the splitting of orbital states according to Young's schemes. This approach allows us to analyze the structure of inter-cluster interactions, identifying allowed states (AS) and forbidden states (FS) in the interaction potential, and thus, the number of WF nodes of cluster relative motion [5,6].

2. OUR METHODS

2.1. Astrophysical S-factors

The formula for the astrophysical S-factor of the radiative capture process is of the form [7]

$$S(EJ) = \sigma(EJ)E_{cm}\exp\left(\frac{31.335\,Z_1Z_2\,\sqrt{\mu}}{\sqrt{E_{cm}}}\right), \quad (1)$$

where σ is the total cross-section of the radiative capture (barn), E_{cm} is the center-of-mass energy of particles (keV), μ is the reduced mass of input channel particles at the radiative capture (atomic mass unit) and $Z_{1,2}$ are the particle charges in elementary charge units. The numerical coefficient 31.335 was received on the basis of up-to-date values of fundamental constants, which are given in [8].

The total cross-sections of radiative capture for electric $EJ(L)$ transitions, caused by the orbital part of electric operator, in a cluster model are given, for example, in works [9] or [10] and may be written as

$$\sigma(E) = \sum_{J,J_f} \sigma(EJ, J_f), \qquad (2)$$

$$\sigma(EJ, J_f) = \frac{8\pi Ke^2}{\hbar^2 q^3} \frac{\mu}{(2S_1+1)(2S_2+1)} \frac{J+1}{J[(2J+1)!!]^2} A_J^2(K) \sum_{L_i,J_i} |P_J(EJ, J_f) I_J|^2,$$

where

$$P_J^2(EJ, J_f, J_i) = \delta_{S_i S_f} \left[(2J+1)(2L_i+1)(2J_i+1)(2J_f+1) \right] (L_i 0 J 0 | L_f 0)^2 \begin{Bmatrix} L_i & S & J_i \\ J_f & J & L_f \end{Bmatrix}^2$$

$$A_J(K) = K^J \mu^J \left(\frac{Z_1}{M_1^J} + (-1)^J \frac{Z_2}{M_2^J} \right),$$

$$I_J = \langle L_f J_f | R^J | L_i J_i \rangle. \qquad (3)$$

Here, μ is the reduced mass and q is the wave number of input channel particles; L_f, L_i, J_f, J_i are particle momenta for input (i) and output (f) channels; S_1, S_2 - spins; $M_{1,2}$, $Z_{1,2}$, are masses and charges of input channel particles (1 or 2); K^J, J - the wave number and the momentum of γ-quanta; I_J is the integral taken over wave functions of initial and final states, that is functions of the relative cluster motion with the intercluster distance R. Sometimes, the spectroscopic factor S_{Jf} of the final state is used in the given formulas for cross-sections, but it is equal to one in the potential cluster model that we used, as it was in work [9].

Using the formula from [11] for the magnetic transition $M1(S)$ caused by the spin part of the magnetic operator we can obtain

$$P_1^2(M1, J_f, J_i) = \delta_{S_i S_f} \delta_{L_i L_f} \left[S(S+1)(2S+1)(2J_i+1)(2J_f+1) \right] \begin{Bmatrix} S & L & J_i \\ J_f & 1 & S \end{Bmatrix}^2$$

$$A_1(M1, K) = i \frac{e\hbar}{m_0 c} K \sqrt{3} \left[\mu_1 \frac{m_2}{m} - \mu_2 \frac{m_1}{m} \right], \qquad (4)$$

$$I_1 = \langle \Phi_f | \Phi_i \rangle,$$

where μ_1 and μ_2 are magnetic momenta of proton and ^2H, which are taken from work [12] (μ_H=0.857 and μ_p=2.793).

2.2. Potentials and Functions

Potentials of intercluster interactions with a point-like Coulomb potential are represented as

$$V(R) = V_0\exp(-\alpha R^2) + V_1\exp(-\gamma R) \qquad (5)$$

or

$$V(R) = V_0\exp(-\alpha R^2). \qquad (6)$$

The expansion of WF of relative cluster motion in nonorthogonal Gaussian basis and the independent variation of parameters [10] are used in the variational method (VM)

$$\Phi_L(R) = \frac{\chi_L(R)}{R} = R^L \sum_i C_i \exp(-\beta_i R^2), \qquad (7)$$

where β_i and C_i are the variational expansion parameters and expansion coefficients.

The behavior of the wave function of bound states (BS) at long distances is characterized by the asymptotic constant C_W, having a form [13]

$$\chi_L = \sqrt{2k_0} C_W W_{\eta L}(2k_0 R), \qquad (8)$$

where χ_L is the numerical wave function of the bound state obtained from the solution of the radial Schrödinger equation and normalized to unity; W is the Whittaker function of the bound state which determines the asymptotic behavior of the WF and represents the solution of the same equation without nuclear potential, i.e. long distance solution; k_0 is the wave number determined by the channel bound energy; η is the Coulomb parameter; L is the orbital momentum of the bound state.

The root-mean-square mass radius is represented as

$$R_m^2 = \frac{M_1}{M}\langle r_m^2 \rangle_1 + \frac{M_2}{M}\langle r_m^2 \rangle_2 + \frac{M_1 M_2}{M^2} I_2,$$

where $M_{1,2}$ and $\langle r_m^2 \rangle_{1,2}$ are the masses and squares of mass radii of clusters, $M=M_1+M_2$, I_2 - the integral

$$I_2 = \langle \chi_L(R) | R^2 | \chi_L(R) \rangle$$

of the R inter-cluster distance and the integration is over radial WF $\chi_L(R)$ of cluster relative motion with the orbital momentum L (3).

The root-mean-square charge radius is represented as

$$R_z^2 = \frac{Z_1}{Z}\langle r_z^2\rangle_1 + \frac{Z_2}{Z}\langle r_z^2\rangle_2 + \frac{(Z_2 M_1^2 + Z_1 M_2^2)}{ZM^2}I_2,$$

where $Z_{1,2}$ and $\langle r_z^2\rangle_{1,2}$ are the charges and squares of charge radii of clusters, $Z=Z_1+Z_2$, I_2 - the abovementioned integral.

The wave function $\chi_L(R)$ or $|L_i J_i\rangle$ is the solution of the radial Schrödinger equation of the form

$$\chi''_L(R) + [k^2 - V(R) - V_c(R) - L(L+1)/R^2]\chi_L(R) = 0,$$

where $V(R)$ is the inter-cluster potential represented as (5) or (6) (dim. fm^{-2}); $V_c(R)$ is the Coulomb potential; k is the wave number determined by the energy E of interaction particles $k^2=2\mu E/\hbar^2$; μ is the reduced mass.

2.3. Cluster States Clussification

The states with the minimal spin in the scattering processes of some light atomic nuclei are "mixed" according to orbital Young's schemes, for example the doublet p^{2}H state [5] is mixed according to schemes {3} and {21}. At the same time, the bound forms of these states are "pure" according to Young's schemes, for example, the doublet p^{2}H channel of the ^{3}He nucleus is "pure" according to scheme {3}. The method of splitting of such states according to Young's schemes is suggested in works [2, 5] where in all cases the "mixed" phase shift of scattering can be represented as a half-sum of "pure" phase shifts {f$_1$} and {f$_2$}

$$\delta^{\{f_1\}+\{f_2\}} = 1/2\left(\delta^{\{f_1\}} + \delta^{\{f_2\}}\right) \qquad (9)$$

In this case it is considered that {f$_1$}={21} and {f$_2$}={3}, and the doublet phase shifts, derived from the experiments, are "mixed" in accordance with these two Young's schemes. If we suppose that instead of the "pure" quartet phase shift with the symmetry {21} one can use the "pure" doublet phase shift of p^{2}H scattering with same symmetry, then it is easy to find the "pure" doublet p^{2}H phase shift with {3} symmetry [5] and use it for the construction of the interaction potential "pure" according to Young's schemes. The latter can be used for the description of the characteristics of the bound state. In this case such a potential allows us to consider the bound p^{2}H state of the ^{3}He nucleus. Similar ratios apply to other light nuclear systems as well, and in each specific case we will analyze the AS and FS structure for both the scattering potentials and the interactions of the ground bound states.

2.4. Phase Shift Analysis

Using experimental data of differential cross-sections of scattering, it is possible to find a set of phase shifts $\delta_{S,L}^J$, which can reproduce the behavior of these cross-sections with certain accuracy. Quality of description of experimental data on the basis of a certain theoretical function (functional of several variables) can be estimated by the χ^2 method which is written as

$$\chi^2 = \frac{1}{N}\sum_{i=1}^{N}\left[\frac{\sigma_i^t(\theta) - \sigma_i^e(\theta)}{\Delta\sigma_i^e(\theta)}\right]^2 = \frac{1}{N}\sum_{i=1}^{N}\chi_i^2, \quad (10)$$

where σ^e and σ^t are experimental and theoretical (i.e. calculated for some defined values of phase shifts $\delta_{S,L}^J$ of scattering) cross-sections of elastic scattering of nuclear particles for i-angle of scattering, $\Delta\sigma^e$ – the error of experimental cross-sections at these angles, N – the number of measurements.

The less χ^2 value, the better description of experimental data on the basis of the chosen phase shift of scattering set. Expressions describing the differential cross-sections represent the expansion of some functional $d\sigma(\theta)/d\Omega$ to the numerical series and it is necessary to find such variational parameters of expansion δ_L which are the best for the description of its behavior. Since the expressions for the differential cross-sections are exact, then as L approaches infinity the value of χ^2 must vanish to zero. This criterion is used for choosing a certain set of phase shifts ensuring the minimum of χ^2 which could possibly be the global minimum of a multiparameter variational problem [14].

So, for example, for p^{6}Li system in order to find nuclear phase shifts of scattering using experimental cross-sections, the procedure of minimization of the functional χ^2 as a function of 2L+2 variables, each of which is a phase shift $^{2,4}\delta_L$ of a certain partial wave without spin-orbital splitting, was carried out. To solve this problem we searched for the minimum of χ^2 within a limited range of values for such variables. But it is possible to find a lot of local minima of χ^2 with the value of about one in this range. Choosing the smallest of them allows hoping that this minimum will correspond to the global minimum which is a solution of this variational problem. Then, the value of this minimum should decrease more or less smoothly as the number of partial waves increases. We used these criteria and methods for the phase shift analyses in the p^{6}Li and p^{12}C systems at low energies – the systems important for the astrophysical calculations.

The exact mass values of the particles were taken for all our calculations [12], and the $\hbar^2/m_0$ constant was taken to be 41.4686 MeV fm^2. The Coulomb parameter $\eta = \mu Z_1 Z_2 e^2/(q\hbar^2)$ was represented as $\eta = 3.44476 \cdot 10^{-2} Z_1 Z_2 \mu/q$, where q is the wave number determined by the energy of interacting particles in the input channel (in fm^{-1}), μ - the reduced mass of the particles (atomic mass unit), Z - the particle charges in elementary charge units. The Coulomb potential with $R_c=0$ was represented as V_c(MeV)$=1.439975\, Z_1 Z_2/r$, where r is the distance between the input channel particles (fm).

3. Radiative p²H Capture

The first process under consideration is the radiative capture

p+²H →³He+γ ,

which is a part of hydrogen cycle and gives a considerable contribution to energy efficiency of thermonuclear reactions [15] accounting for burning of the Sun and stars of our Universe. The potential barrier for interacting nuclear particles of the hydrogen cycle is the lowest. Thus, it is the first chain of nuclear reactions which can take place at ultralow energies and star temperatures.

For this chain, the process of the radiative p²H capture is the basic process for the transition from the primary proton fusion

p+p →²H+e⁻+ν_e

to the capture reaction of two ³He nuclei [16], which is one of the final processes

³He+³He →⁴He+2p

in the p-p-chain.

The theoretical and experimental study of the radiative p²H capture in detail is of fundamental interest not only for nuclear astrophysics, but also for nuclear physics of ultralow energies and lightest atomic nuclei [17]. That is why the experimental researches into this reaction are in progress and a short time ago the new experimental data in the range down to 2.5 keV appeared.

3.1. Potentials and Phase Shifts of Scattering

Earlier, the total cross sections of the photoprocesses of lightest ³He and ³H nuclei were considered in the frame of the potential cluster model with forbidden states in our work [6]. E1 transitions resulting from the orbital part of the electric operator $Q_{Jm}(L)$ [10] were taken into account in these calculations of the photodecays of ³He and ³H nuclei into p²H and n²H channels. The values of E2 cross-sections and cross-sections depending on the spin part of the electric operator turned out to be several orders less.

Further, it was assumed that E1 electric transitions in N²H system are possible between ground "pure" (scheme {3}) ²S state of ³H and ³He nuclei and doublet ²P scattering state mixed according to Young's schemes {3}+{21} [17]. On the basis of the approach used it was possible to obtain quite reasonable results describing the experimental data of ³H and ³He nuclei photodecay into the cluster channels [6].

To calculate photonuclear processes in the systems under consideration the nuclear part of the potential of inter-cluster p²H and n²H interactions is represented as (5) with a point-like Coulomb potential, V_0 - the Gaussian attractive part, and V_1 - the exponential repulsive part. The potential of each partial wave was constructed so as to correctly describe the respective partial phase shift of the elastic scattering [18]. Using this concept, the potentials of the p^{2}H interaction of the scattering processes were received. The parameters of such potentials were fully given in works [6, 10, 19], and parameters for doublet scattering states mixed according to Young's schemes are listed in Table 1.

Then, in the doublet channel mixed according to Young's schemes {3} and {21} [5], the "pure" phases (9) with scheme {3} were separated and on their basis the "pure" ^2S potential of the bound state of the ^{3}He nucleus in the p^{2}H channel was constructed [6, 10, 19].

Table 1. The potentials of the p^{2}H [6] interaction in the doublet channel

^{2S+1}L, {f}	V_0 (MeV)	α (fm^{-2})	V_1 (MeV)	γ (fm^{-1})
^2S, {3}	-34.76170133	0.15	---	---
^2P, {3}+{21}	-10.0	0.16	+0.6	0.1
^2S, {3}+{21}	-55.0	0.2	---	---

With kind permission of the European Physical Journal (EPJ)

The calculations of the $E1$ transition [6] show that the best results for the description of the total cross-sections of the ^{3}He nucleus photodecay for the γ-quanta energy range 6-28 MeV, including the maximum value at $E_γ$=10-13 MeV, can be found if we use the potentials with peripheric repulsion of the ^2P-wave of the p^{2}H scattering (table 1) and the "pure" according to Young's schemes ^2S-interaction of the bound state (BS) of the Gaussian form (5) with parameters

V_0 = -34.75 MeV, α = 0.15 fm^{-2}, V_1 = 0,

which were obtained, primarily, on the basis of the correct description of the bound energy (with the accuracy up to few keV) and the charge radius of the ^{3}He nucleus. The calculations of the total cross-sections of the radiative p^{2}H capture and astrophysical S-factors were made with these potentials at the energy range down to 10 keV [6-10]. Though, at that period of time we only knew S-factor experimental data in the range above 150-200 keV [20].

Recently, the new experimental data on the p^{2}H S-factor in the range down to 2.5 keV appeared in [21-23]. That is why, it is interesting to know if it is possible to describe the new data on the basis of the $E1$ and $M1$ transitions in the potential cluster model with the earlier obtained ^2P-interaction of scattering and ^2S-potential of the bound p^{2}H state adjusted in this work.

Our preliminary results have shown that for the S-factor calculation at the energy range of about 1 keV it is necessary to improve the accuracy of finding the bound energy of the p^{2}H system in the ^{3}He nucleus. It must be better than 1-2 keV [6]. The behavior of the tail of the wave function of the bound state should be controlled more strictly at long distances. Then, it is necessary to improve the accuracy of finding Coulomb wave functions which determine the asymptotic behavior of the scattering WF in the ^2P-wave.

The parameters of the "pure" doublet ^2S-potential according to Young's scheme {3} were adjusted using opportunities of a new computer programs based on the finite-difference method (FDM) for a more accurate description of the experimental bound energy of ^{3}He nuclei in p²H channel. This potential (Table 1) has become somewhat deeper than the potential we used in our work [6] and leads to a total agreement between calculated -5.4934230 MeV and experimental -5.4934230 MeV bound energies, which is obtained by using the exact mass values of particles [12]. The difference between potentials given in work [6] and in Table 1 is primarily due to using the exact mass values of particles and more accurate description of the ³He nucleus bound energy in the p²H channel. For these computations the absolute accuracy of searching for the bound energy in our computer program based on the finite-difference method was taken to be at the level of 10^{-8} MeV.

The value of the ³He charge radius with this potential equals 2.28 fm, which is a little higher than the experimental values listed in Table 2 [12, 24, 25]. The experimental radii of proton and deuteron, which are also given in Table 2, are used for these calculations and the latter is larger than the radius of the ³He nucleus. Thus, if the deuteron is present in the ³He nucleus as a cluster, it must be compressed by about 20-30% of its size in free state for a correct description of the ³He radius [10].

Table 2. Experimental masses and charge radii of light nuclei used in these calculations [12, 24, 25]

Nucleus	Radius, (fm)	Mass, (amu)
p	0.8768(69)	1.00727646677
²H	2.1402(28)	2.013553212724
³H	1.63(3); 1.76(4); 1.81(5) The average value is 1.73	3.0155007134
³He	1.976(15); 1.93(3); 1.877(19); 1.935(30) The average value is 1.93	3.0149322473
⁴He	1.671(14)	4.001506179127

The asymptotic constant C_W with Whittaker asymptotics (8) [26] was calculated for controlling behavior of WF of BS at long distances; its value in the range of 5-20 fm equals C_W=2.333(3). The error given here is found by averaging the constant in the range mentioned above. The experimental data known for this constant give the values of 1.76-1.97 [27, 28], which is slightly less than the value obtained here. It is possible to give results of three-body calculations [29], where a good agreement with the experiment [30] for the ratio of asymptotic constants of ²S and ²D waves was obtained and the value of the constant of ²S wave was found to be C_W=1.878.

But in work [13], which is more recent than [27, 28], the value of 2.26(9) is given for the asymptotic constant, and this is in a good agreement with our calculations. One can see from the considerable data that there is a big difference between the experimental results of asymptotic constants received in different periods. These data are in the range from 1.76 to 2.35 with the average value of 2.06.

In the cluster model the value of C_W constant depends significantly on the width of the potential well and it is always possible to find other parameters of ²S-potential of bound state BS, for example:

$V_0 = -48.04680730$ MeV and $\alpha = 0.25$ fm^{-2}, (11)

$V_0 = -41.55562462$ MeV and $\alpha = 0.2$ fm^{-2}, (12)

$V_0 = -31.20426327$ MeV and $\alpha = 0.125$ fm^{-2}, (13)

which give the same value of the bound energy of ^{3}He in p^{2}H channel. The first of them at distances of 5-20 fm leads to asymptotic constant $C_W=1.945(3)$ and charge radius $R_{ch}=2.18$ fm, the second variant gives $C_W=2.095(5)$ and $R_{ch}=2.22$ fm, the third variant - $C_W=2.519(3)$ and $R_{ch}=2.33$ fm.

It can be seen from these results that the potential (11) allows obtaining the charge radius the closest to the experimental values. Further reduction of the potential width could give a more accurate description of its value, but, as it will be shown later, will not allow us to describe the S-factor of the p^{2}H capture. In this sense, the slightly wider potential (12) has the minimal acceptable width of the potential well which leads to asymptotic constant almost equal to its experimental average value 2.06 and gives a possibility to describe quite well the astrophysical S-factor in a wide energy range.

The variational method is used for an additional control of the accuracy of bound energy calculations for the potential from Table 1, which allowed obtaining the bound energy of -5.4934228 MeV by using independent variation of parameters and the grid having dimension 10. The asymptotic constant C_W of the variational WF at distances of 5-20 fm remains at the level of 2.34(1). The variational parameters and expansion coefficients of the radial wave function for this potential having form (7) are listed in Table 3.

Table 3. The variational parameters and expansion coefficients of the radial WF of the bound state of the p^{2}H system for the potential from Table 1. The normalization of the function with these coefficients in the range 0-25 fm equals N=0.999999997

i	β_i	C_i
1	2.682914012452794E-001	-1.139939646617903E-001
2	1.506898472480031E-002	-3.928173077162038E-003
3	8.150892061325998E-003	-2.596386495718163E-004
4	4.699184204753572E-002	-5.359449556198755E-002
5	2.664477374725231E-002	-1.863994304088623E-002
6	4.468761998654231E+001	1.098799639286601E-003
7	8.482112461789261E-002	-1.172712856304303E-001
8	1.541789664414691E-001	-1.925839668633162E-001
9	1.527248552219977E-000	3.969648696293301E-003
10	6.691341326208045E-000	2.097266548250023E-003

With kind permission of the European Physical Journal (EPJ)

The potential (12) was examined within the frame of VM and the same bound energy of -5.4934228 MeV was received. The variational parameters and expansion coefficients of the radial wave function are listed in Table 4. The asymptotic constant at distances of 5-20 fm turned out to be 2.09(1) and the residual error is of the order of 10^{-13}.

For the real bound energy in this potential it is possible to use the value -5.4934229(1) MeV with the calculation error of finding energy by two methods equal to ±0.1 eV, because the variational energy decreases as the dimension of the basis increases and gives the upper limit of the true bound energy, but the finite-difference energy increases as the size of steps decreases and the number of steps increases.

Table 4. The variational parameters and expansion coefficients of the radial WF of the bound state of the p²H system for the potential (12). The normalization of the function with these coefficients at the range 0-25 fm equals N=0.999999998

i	β_i	C_i
1	3.485070088054969E-001	-1.178894628072507E-001
2	1.739943603152822E-002	-6.168137382276252E-003
3	8.973931554450264E-003	-4.319325351926516E-004
4	5.977571392609325E-002	-7.078243409099880E-002
5	3.245586616581442E-002	-2.743665993408441E-002
6	5.8379917320454490E+001	1.102401456221556E-003
7	1.100441373510820E-001	-1.384847981550261E-001
8	2.005318455817479E-001	-2.114723533577409E-001
9	1.995655373133832E-000	3.955231655325594E-003
10	8.741651544040529E-000	2.101576342365150E-003

With kind permission of the European Physical Journal (EPJ).

3.2. Astrophysical S-factor

In our calculations we considered the energy range of the radiative p²H capture from 1 keV to 10 MeV and found the value of 0.165 eV b for the $S(E1)$-factor at 1 keV for the potentials from Table 1. The value found is slightly lower than the known data if we consider the total S-factor without splitting it into S_s and S_p parts resulting from $M1$ and $E1$ transitions. This splitting was made in work [22], where $S_s(0)$=0.109(10) eV b and $S_p(0)$=0.073(7) eV b. At the same time, the authors give the following values S_0=0.166(5) eV b and S_1=0.0071(4) eV b keV^{-1} in the linear interpolation formula

$$S(E_{c.m.}) = S_0 + E_{c.m.} \cdot S_1, \tag{14}$$

and for $S(0)$ leads to the value of 0.166(14) keV b, which was received taking into account all possible errors. The results with the splitting of the S-factor into $M1$ and $E1$ parts are given in one of the first of works [20], where S_s=0.12(3) eV b, S_p=0.127(13) eV b for the total S-factor 0.25(4) eV b.

As it can be seen, there is a visible difference between these results, so, in future we will take as a reference point the total value of S-factor at zero energy which was measured in various works. Furthermore, the new experimental data [23] lead to the value of total $S(0)=0.216(10)$ eV b and this means that contributions of $M1$ and $E1$ will change. The following parameters of linear extrapolation (14) are given in this work $S_0=0.216(6)$ eV b and $S_1=0.0059(4)$ eV b keV^{-1}, that are noticeably differ from the data of work [22].

The known extractions of the S-factor from the experimental data, without splitting to $M1$ and $E1$ parts, at zero energy give the value of 0.165(14) eV b [31]. The previous measurements by the same authors lead to the value 0.121(12) eV b [32], and for theoretical calculations of work [33] the values $S_s=0.105$ eV b, $S_p=0.08-0.0865$ eV b are received for different models.

One can see that the experimental data over the last 10-15 years is very ambiguous. These results allow to come to a conclusion that, most probably, the value of total S-factor at zero energy is in the range 0.11-0.23 eV b. The average of these experimental measurements equals 0.17(6) eV b what is in a good agreement with the value calculated here on the basis of the $E1$ transition only.

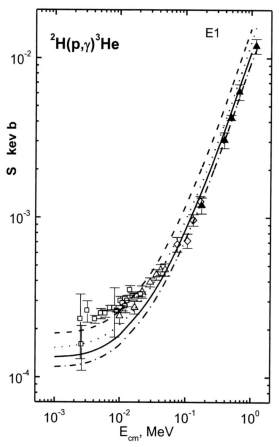

With kind permission of the European Physical Journal (EPJ)

Figure 1a. Astrophysical S-factor of p^{2}H radiative capture in the range 1 keV-1 MeV. Lines: calculations with the potentials mentioned in the text. Triangles denote the experimental data from [20], open rhombs from [21], open triangles from [22], open squares from [23].

Current calculation results for the $S(E1)$-factor of the p^{2}H radiative capture with the potential from table 1 at the energy range from 1 keV to 10 MeV are shown in Figs. 1a and 1b by dotted line. Now the calculated S-factor reproduces experimental data at the energies down to 10-50 keV [22] comparatively well and at lower energies the calculated curve practically falls within the experimental error band of work [23].

Solid lines in Figs. 1a and 1b show the results for potential (12) which describes the behavior of the S-factor somewhat better at energies from 50 keV to 10 MeV and which gives the value of S=0.135 eV b for the energy of 1 keV. At energies of 20-50 keV the calculation curve follows the line of the lower limit of the error band of work [22], and at the energies below 10 keV it falls within the experimental error band of the LUNA project which was received recently [23]. The value of the S-factor at zero energy of this potential is in a good agreement with the data of the S_p from work [20] for the $E1$ transition.

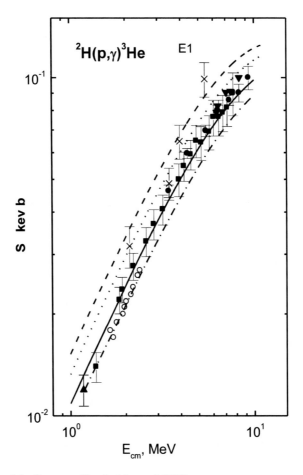

With kind permission of the European Physical Journal (EPJ)

Figure 1b. Astrophysical S-factor of p^{2}H radiative capture in the range 1 MeV-10 MeV. Lines: calculations with the potentials mentioned in the text. Triangles denote the experimental data from [20], squares are from work [34], black points - from [35], crosses - from [36], inverted triangles - from [37], open circles - from [38].

The dashed lines in Figs. 1a and 1b show the results for potential (13) and the dash-dotted lines show those for potential (11). From these calculations one may conclude that the best results are obtained with the BS potential (12) which describes the experimental data in the widest energy range. It represents a sort of a compromise in describing asymptotic constant, charge radius and S-factor of the radiative p^{2}H capture.

As it is seen in Figure 1a, the S-factor at the low energy range near 1-2 keV is practically constant and thus defines the S-factor value at zero energy which turns out to be approximately the same as its value at the energy equal to 1 keV. So, it seems that the difference between the values of the S-factor at energies 0 and 1 keV should be at most 0.005 eV b, and this value can be considered as the calculation error of the S-factor at zero energy and accept its value to be equal to 0.135(5) keV b.

The M1 transition from the S scattering state, which is mixed in accordance with Young's schemes, to the bound state, which is "pure" according to orbital symmetries of the S state of the ^{3}He nucleus, can give a contribution to the astrophysical S-factor at low energies. For our calculations we used the doublet S-potential of the scattering states with the parameters listed in Table 1 and the BS potential (12). The calculation results at the energies 1-100 keV are shown in Figure 2 by the dotted line at the bottom of the figure. The dashed line in Figure 2 shows the calculation results for E1 transition for the potential of the ground state (GS) (12), which is shown in Figure 1a with a solid line. The Total S-factor is shown in Figure 2 with a solid line which demonstrates clearly the small contribution of M1 transition to the S_s-factor at energies above 100 keV and its considerable influence in the energy range of 1-10 keV.

The total S-factor dependence on energy in the range of 2.5-50 keV is in complete accordance with the findings of works [22, 23] and for the S_s-factor of the M1 transition at 1 keV we obtained the value of 0.077 eV b, which leads to the value of 0.212(5) eV b for the total S-factor and which is in a good agreement with the new measurements data from LUNA project [23]. And as it can be seen from Figure 2, at the energies of 1-3 keV the value of the total S-factor is more stable than it was for the E1 transition and we consider it to be absolutely reasonable to write the result as 0.212 eV b with the error of 0.005.

However, it is necessary to note that we are unable to build the scattering S-potential uniquely because of the ambiguities in the results of different phase shift analyses. The other variant of potential with parameters V_0=-35.0 MeV and α=0.1 fm^{-2} [10, 17], which also describes well the S phase shift, leads at these energies to cross-sections of the M1 process several times lower than those of E1.

Such a big ambiguity in parameters of the S-potential of scattering, associated with errors of phase shifts extracted from the experimental data, does not allow us making certain conclusions about the contribution of the M1 process in the p^{2}H radiative capture. If the BS potentials are defined by the bound energy, asymptotic constant and charge radius quite uniquely and the potential description of the scattering phase shifts, which are "pure" in accordance with Young's schemes, is an additional criterion for determination of such parameters, then, for the construction of the scattering potential it is necessary to carry out a more accurate phase shift analysis for the ^2S-wave and to take into account the spin-orbital splitting of ^2P phase shifts at low energies, as it was done for the elastic p^{12}C scattering at energies 0.2-1.2 MeV [39]. This will allow us to adjust the potential parameters used in the calculations of the p^{2}H capture in the potential cluster model, the results of the calculations of which depend strongly on the accuracy of the construction of the interaction potentials in accordance with the scattering phase shifts.

Thus, the S-factor calculations of the p^{2}H radiative capture for the E1 transition at the energy range down to 10 keV, which we carried out about 15 years ago [6], when the experimental data above 150-200 keV were only known, are in a good agreement with the new data of works [21, 22] in the energy range 10-150 keV. And this is true about both the potential from Table 1 and the interaction with parameters from (12). The results for the two considered potentials at the energies lower than 10 keV practically fall within the error band of work [23] and show that the S-factor tends to remain constant at energies 1-3 keV.

In spite of the uncertainty of the M1 contribution to the process, which results from the errors and ambiguity of ^{2}S-phases of scattering, the scattering potential (set forth in Table 1) with mixed Young schemes in the ^{2}S-wave allows obtaining a reasonable value for the astrophysical S_s-factor of the magnetic transition in the range of low energies. At the same time, the value of the total S-factor is in a good agreement with all known experimental measurements at energies from 2.5 keV to 10 MeV.

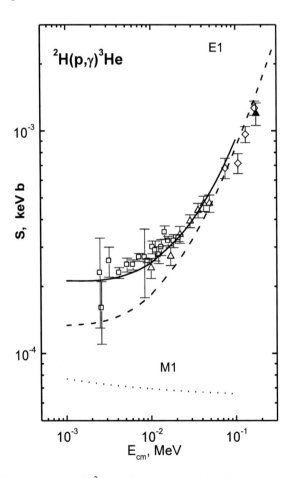

Figure 2. Astrophysical S-factor of p^{2}H radiative capture in the range 1 MeV-0.3 MeV. Lines: calculations with the potentials mentioned in the text. Triangles denote the experimental data from [20], open rhombs from [21], open triangles from [22], open squares from [23].

As a result, the PCM based on the intercluster potentials adjusted for the elastic scattering phase shifts and GS characteristics, for which the FS structure is determined by the classification of BS according to Young's orbital schemes and parameters suggested as early as 15 years ago [16], allows describing correctly the astrophysical *S*-factor for the whole range of energies under consideration.

4. RADIATIVE p³H CAPTURE

Now let's consider the possibility of description of the astrophysical *S*-factor of p³H radiative capture at the energy range down to 1 keV as a continuation of the theoretical investigation of thermonuclear reactions [17] on the basis of the potential cluster model with the splitting of orbital states according to Young's schemes [41]. This reaction probably played a certain role at the prestellar stage of the evolution of the Universe [1] at temperatures of the order of 10^9 K (primordial nucleosynthesis). This reaction can be of some interest for a good understanding of the nature of thermonuclear photoprocesses with lightest atomic nuclei at low energies, both from theoretical and experimental points of view. Thus, experimental researches into this reaction are continued, and quite recently the new data for the total cross-sections of the p³H radiative capture and for the astrophysical *S*-factor at the energy range down to 12 keV (c.m.) were obtained.

4.1. Potentials and Phase Shifts of Scattering

To calculate photonuclear processes in the systems under consideration the nuclear part of the potential of inter-cluster p³H and p³He interactions is represented as (5) with a point-like Coulomb potential. The potential of each partial wave, as for the previously considered p²H system, was constructed so as to describe correctly the respective partial phase shift of the elastic scattering [40].

As a result, the potentials of the p³He interaction for scattering processes which are "pure" in accordance with *T*=1 were received. The parameters of such potentials are fully listed in Table 5 [41, 42]. The singlet *S* phase shift of elastic p³He scattering which is "pure" in accordance with isospin is shown in Figure 3 by the solid line together with the experimental data of works [43-45]. Further it is used for receiving the singlet p³H phase shifts which are "pure" in accordance with *T*=0.

Table 5. The singlet potentials of the p³He system which are "pure" in accordance with *T*=1 isospin [42]

System	^{2S+1}L	V_0 (MeV)	α (fm^{-2})	V_1 (MeV)	γ (fm^{-1})
p³He	^1S	-110.0	0.37	+45.0	0.67
	^1P	-15.0	0.1	---	---

Since there are several variants of the phase shift analyses [40, 43-45] for the $^1P_1$ singlet wave, the parameters of potentials from Table 5 are chosen so as to lead to a kind of a compromise between the different phase shift analyses. The singlet $^1P_1$ phase shift of elastic p³He scattering with $T=1$, used in our calculations of the $E1$ transition to the ground state (GS) of the ⁴He nucleus in the p³H channel with $T=0$, is shown in Figure 4 by the solid line together with the experimental data of works [43-48].

Because of the fact that the p³H system is isospin-mixed the singlet and triplet phase shifts, and consequently the potentials, effectively depend on two values of isospin. The result of mixing in terms of isospin is mixing in terms of Young's schemes. In particular, two orbital schemes {31} and {4} are allowed in the singlet state.

The isospin-mixed singlet S phase shift of the elastic p³H scattering which was calculated from the experimental differential cross-sections and which we used for receiving "pure" p³H phase shifts is shown in Figure 5 by the solid line with the experimental data from works [49-51]. We use the first set of phase shifts of scattering from work [51]. The following parameters were received: $V_0 = -50.0$ MeV, $\alpha = 0.20$ fm^{-2}.

As it was shown in works [10, 41, 42], the isospin-mixed singlet phase shifts of the p³H scattering can be represented as a half-sum of isospin-pure singlet phase shifts

$$\delta^{\{T=1\}+\{T=0\}} = 1/2\delta^{\{T=1\}} + 1/2\delta^{\{T=0\}}, \qquad (15)$$

which is equivalent to the expression

$$\delta^{\{4\}+\{31\}} = 1/2\delta^{\{31\}} + 1/2\delta^{\{4\}} .$$

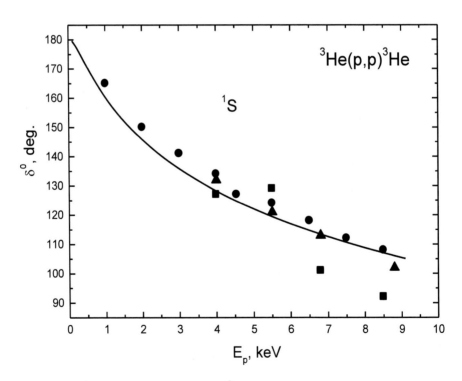

Figure 3. The singlet ^1S phase shift of the elastic p³He scattering. Black points denote the experimental data from [43], squares are from work [44], triangles - from work [45]

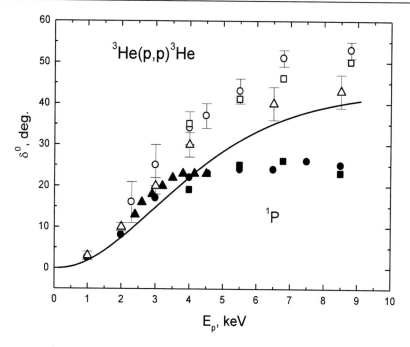

Figure 4. The singlet ^1P phase shift of the elastic p^{3}He scattering. Black points denote the experimental data from [43], squares are from work [44], triangles - from [46], open circles - from [47], open squares - from [45], open triangles - from [48].

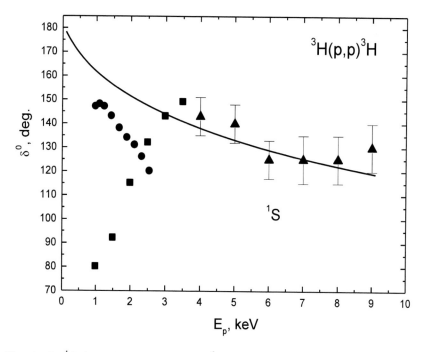

Figure 5. The singlet ^1S phase shift of the elastic p^{3}H scattering. Black points denote the experimental data from [49], squares - from work [50], triangles - from work [51]

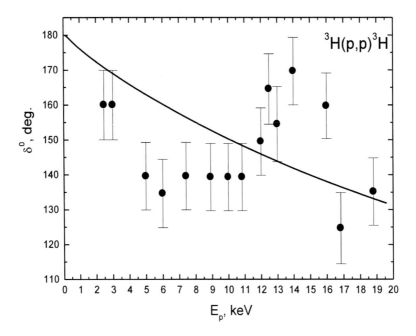

Figure 6. The singlet ^1S phase shift of the elastic p^{3}H scattering "pure" in terms of Young's schemes.

It is usually considered [41] that the "pure" phase shifts correspond to the Young's schemes {31} with $T=1$ for the p^{3}He system and {4} with $T=0$ for the p^{3}H system. The isospin-pure phase shifts of p^{3}H scattering with $T=0$ are constructed on the basis of expression (15) by using known "pure" phase shifts of scattering with $T=1$ for the p^{3}He system [43-45] and "mixed" p^{3}H phase shifts with $T=0$ and 1 [49-51]. They are in turn used for obtaining "pure" p^{3}H interaction potentials [42]. In particular, for the ^1S wave the following parameters were received:

$$V_0 = -63.1 \text{ MeV}, \alpha = 0.17 \text{ fm}^{-2}, V_1 = 0. \tag{16}$$

In Figure 6 the singlet ^1S phase shift of the elastic p^{3}H scattering with "pure" Young's schemes is shown with the black points, and its calculation results with the potential (16) are shown with the solid line. Such "pure" interactions can be used for the calculations of various characteristics of the bound state of the ^{4}He nucleus in the p^{3}H channel and the results will depend on the clusterization rate of this nucleus into the considered channel.

The interactions received in [10, 42] generally give a correct description of the channel bound energy of the p^{3}H system (with an accuracy about few keV) and the root-mean-square radius of the ^{4}He nucleus [42]. The calculations of the differential [41] and total cross-sections of the radiative p^{3}H capture and astrophysical S-factors were done with these potentials at the energy range down to 10 keV [42]. Though, at that time we only knew S-factor experimental data in the range above 700 keV [52].

A short time ago the new experimental data in the energy range from 50 keV to 5 MeV [53] and at 12 and 36 keV [54] appeared. That is why, it was interesting to know if it is possible to describe the new data on the basis of the potential cluster model, with the earlier obtained singlet ^1P potential and adjusted interaction of the ground ^1S state.

Our preliminary results show that for calculations of the S-factor at energies of the order of 1 keV we have to meet the same requirements as for the p^{2}H system (which were discussed in the previous section) and first of all - to raise an accuracy of finding the ^{4}He bound energy in the p^{3}H channel. So, by using the new modified computer programs, we adjusted the parameters of the ground state potential of the p^{3}H system in the ^{4}He nucleus. These potentials differ from the potentials in work [42] by 0.2 MeV and are listed in Table 6.

Table 6. The isospin-pure potentials of the p^{3}H [42] interactions in the singlet channel with T=0. E_{BS} is the calculated energy of the bound state, E_{EXP} - its experimental value [25], the depth of attractive part of the potential (5) V_1=0

System	L	V_0 (MeV)	α (fm^{-2})	E_{BS} (MeV)	E_{EXP} (MeV)
p^{3}H	^1S	-62.906841138	0.17	-19.81381000	-19.813810
	^1P	+8.0	0.03	---	---

Basically, this difference results from the use of the exact mass values of proton and ^{3}H particles [12] in new calculations and more accurate description of the bound state energy of the ^{4}He nucleus. The value -19.813810 MeV was obtained experimentally for the bound state of the ^{4}He nucleus in the p^{3}H channel on the basis of exact mass values [12], and the calculation with such a potential leads to the value of -19.81381000 MeV. The absolute accuracy of searching for the bound energy in our computer program based on the finite-difference method was taken to be at the level of 10^{-8} MeV.

The calculation accuracy of the "tail" of the WF of BS of the p^{3}H system was verified using asymptotic constant C_W with Whittaker asymptotics (8) [13,55], and its value in the range of 5-10 fm turned out to be C_W=4.52(1). The experimental data known for this constant in the p^{3}H channel give the value of 5.16(13) [13]. In the same work for the asymptotic constant of the n^{3}He system there was obtained almost the same value 5.1(4).

At the same time, in works [55] the value of 4.1 is given for the constant of the n^{3}He system and the value of 4.0 for p^{3}H. If we take the average of the data of works [13] and [55], it agrees well with our results. As it is seen, there is a big difference between the experimental results of asymptotic constants. For the n^{3}He system the asymptotic constant is in the range of 4.1-5.5, and for the p^{3}H channel it seems to be in the range from 4.0 to 5.3.

For the ^{4}He charge radius the value of 1.78 fm was obtained with the radius of tritium being 1.73 fm [24] and that of proton - 0.877 fm [12], while the experimental value of the ^{4}He charge radius is 1.671(14) fm [25] (see Table 2).

The variational method with the expansion of the WF of the relative cluster motion in nonorthogonal Gaussian basis is used for an additional control of the accuracy of bound energy calculations for the S-potential from Table 6, which allowed obtaining the bound energy of -19.81380998 MeV by using independent variation of parameters and the grid with dimension 10. The asymptotic constant C_W (8) of the variational WF at distances of 5-10 fm remains at the level of 4.52(2). The variational parameters and expansion coefficients of the radial wave function having form (7) are listed in Table 7.

Table 7. The variational parameters and expansion factors of the radial WF of the bound state of the p^{3}H system for the S-potential from Table 6. The normalization of the function with these factors in the range 0-25 fm equals N=0.9999999998

i	β_i	C_i
1	3.775399682294165E-002	-3.553662130779118E-003
2	7.390030511120065E-002	-4.689092850709087E-002
3	1.377393687979590E-001	-1.893147614352133E-001
4	2.427238748079469E-001	-3.619752356073335E-001
5	4.021993911220914E-001	-1.988757841748206E-001
6	1.780153251456691E+000	5.556224701527299E-003
7	5.459871888661887E+000	3.092889292994009E-003
8	1.9213177238092050E+001	1.819890982631486E-003
9	8.4161171211980260E+001	1.040709526875803E-003
10	5.60393988031844500E+002	5.559240350868498E-004

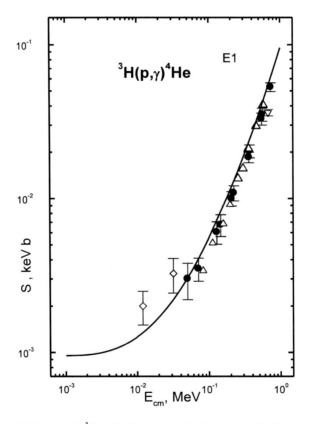

Figure 7. Astrophysical S-factor of p^{3}H radiative capture in the range 1 keV - 1 MeV. Line: calculation with the potential mentioned in the text. Black dots represent the conversion of the total capture cross-sections from work [53] given in work [54], open triangles - from [57], open rhombs - from [54], inverted open triangle - from [52].

For the real bound energy in this potential it is possible to use the value -19.81380999(1) MeV with the calculation error of finding energy by the two methods used equals to ±0.01

eV, because, as we mentioned in the previous section, the variational energy decreases as the dimension of the basis increases and gives the upper limit of the true bound energy, but the finite-difference energy increases as the size of steps decreases and the number of steps increases.

It can be seen from the given results that the simple two-cluster p^{3}H model with the classification of orbital states according to Young's schemes allows obtaining quite reasonable values for such characteristics of the bound state of the ^{4}He nucleus as charge radii and asymptotic constants. Thus, these results are indicative of a comparatively high clusterization rate of the nucleus into the p^{3}H channel. Therefore, the model used leads to reasonable results of the astrophysical S-factor calculations.

4.2. Astrophysical S-factor

Earlier, the total cross-sections and astrophysical S-factor of the process of p^{3}H radiative capture were considered on the basis of the potential cluster model in work [42]. It was assumed that the transitions with isospin changes $\Delta T=1$ [56] give the main contribution to the $E1$ cross-sections of the ^{4}He photodecay into the p^{3}H channel or to the radiative p^{3}H capture. Thus, we should use the $^1P_1$ scattering potential of the isospin-pure singlet state of the p^{3}He system with $T=1$ and the ^1S potential of the isospin-pure ground state of the ^{4}He nucleus in the p^{3}H channel with $T=0$ [42].

Using this concept we have carried out the calculations of the $E1$ transition with the adjusted potential of the ^{4}He ground state from Table 6. The calculation results for the astrophysical S-factor at the energies down to 1 keV are shown in Figs. 7 and 8 by solid lines. These results are almost the same as the previous ones which we obtained in work [42] at the energy range down to 10 keV. The new experimental data are taken from works [53, 54] and the data which we used additionally are from [57].

As one can see from the figures, the calculation results, which we obtained about 15 years ago, reproduced the new S-factor data obtained in work [53] at the energy range from 50 keV to 5 MeV (c.m.) very well. The ambiguity of these data is visibly less at the energy range over 1 MeV when compared with the previous results of works [52, 58-60], and they give a more accurate overall behavior of the S-factor at low energies, practically coinciding with the previous data [57] at the energy range 80-600 keV.

At the energy of 1 keV the value of the S-factor turned out to be equal to 0.95 eV b, and the calculation results of its value at the energy range lower than 50 keV lie slightly below the new data [54], where for the $S(0)$ the value of 2.0(2) eV b was received. Note that a simple extrapolation of existing experimental data to 1 keV of the last three points of works [53,57] leads to its value of about 0.6(3) eV b, which is three times less than the results of work [54].

As it is seen in Figure 7 the S-factor at the lowest energies, approximately in the range of 1-3 keV, is practically constant. Thus, the S-factor value at zero energy can be defined almost the same as its value at the energy of 1 keV. As a result, the difference between the values of the S-factor at energies 0 and 1 keV seems to be at most 0.05 eV b, and this value can be considered as the calculation error of the S-factor at zero energy taking 0.96(5) eV b for the calculated value.

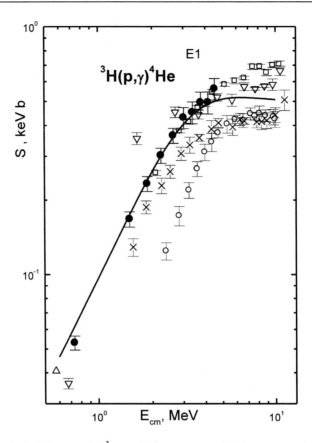

Figure 8. Astrophysical S-factor of p^{3}H radiative capture in the range 1 MeV - 10 MeV. Line: calculation with the potential mentioned in the text. Black dots is the conversion of the total capture cross-sections from work [53] given in work [54], open triangles - from [57], open circles - from [58], open squares - from [59], crosses (×) from [60], inverted open triangles - from [52].

Thus, on the basis of only E1 process, we have managed to predict the general behavior of the S-factor of the p^{3}H capture at the energy range from 50 keV to 700 keV, because our calculations in the energy range down to 10 keV were done about 15 years ago [42], when only experimental data higher than 700 keV were known.

The results of the calculations at the energy range from 50 keV to 5 MeV are in a good agreement with the new data of the S-factor from work [53] (black dots in Figs. 7, 8).

5. RADIATIVE P^{6}LI CAPTURE

For the updating of the existing experimental data the new measurements of differential cross-sections of the elastic p^{6}Li scattering at the energy range from 350 keV to 1.15 MeV (l.s. - laboratory system) were done [3]. The experimental data [3], which we are going to consider in this section, are received at five energy values: 593 keV for 13 angles of scattering in the range 57^0-172^0, 746.7 and 866.8 keV for 11 angles in the range 45^0-170^0 and at energies 976.5 and 1136.6 keV for 15 angles in the range 30^0-170^0.

On the basis of measurements from works [3] and differential cross-sections of the elastic scattering at the energy 500 keV from earlier work [61] we have carried out the phase shift analysis and received $^{2,4}S$ and ^2P-phase shifts of scattering. The p^{6}Li interaction potentials for $L=0$ and 1 at low energies and without taking into account the spin-orbital splitting were constructed according to the phase shifts obtained, and then the calculations of the astrophysical S-factor at the energy above 10 keV were made.

Although the p^{6}Li reaction of radiative capture may be of some interest for the nuclear astrophysics [1, 62], it is not well enough studied experimentally. There are a comparatively small number of measurements of the total cross-sections and calculations of the S-factor [9], and they were performed only in the energy range from 35 keV to 1.2 MeV. Nevertheless, it would be interesting to consider the possibility of its description in the frame of the potential cluster model taking into account the classification of the bound states according to the orbital Young's schemes at the astrophysical energy range where the experimental data exist.

5.1. Phase Shift Analysis

For scattering processes in the particle system with a spin 1/2+1, without taking into account the spin-orbital splitting, the cross-section of the elastic scattering is represented as [63]

$$\frac{d\sigma(\theta)}{d\Omega} = \frac{2}{6}\frac{d\sigma_d(\theta)}{d\Omega} + \frac{4}{6}\frac{d\sigma_q(\theta)}{d\Omega}, \qquad (17)$$

where d and q are related to the doublet (with total spin 1/2) and quartet (with total spin 3/2) states of the p^{6}Li scattering and

$$\frac{d\sigma_d(\theta)}{d\Omega} = |f_d(\theta)|^2, \quad \frac{d\sigma_q(\theta)}{d\Omega} = |f_q(\theta)|^2. \qquad (18)$$

The scattering amplitudes are written as

$$f_{d,q}(\theta) = f_c(\theta) + f_{d,q}^N(\theta), \qquad (19)$$

where

$$f_c(\theta) = -\left(\frac{\eta}{2k\sin^2(\theta/2)}\right)\exp\{i\eta\ln[\sin^{-2}(\theta/2)] + 2i\sigma_0\},$$

$$f_d^N(\theta) = \frac{1}{2ik}\sum_L (2L+1)\exp(2i\sigma_L)[S_L^d - 1]P_L(\cos\theta),$$

$$f_q^N(\theta) = \frac{1}{2ik}\sum_L (2L+1)\exp(2i\sigma_L)[S_L^q - 1]P_L(\cos\theta), \qquad (20)$$

and $S_L^{d,q} = \eta_L^{d,q} \exp(2i\delta_L^{d,q}(k))$ is the scattering matrix in the doublet or quartet spin state [63].

It is possible to use simple formulae (17-20) for the calculations of cross-sections of the elastic scattering because the spin-orbital phase shift splitting at low energies is quite insignificant. It is confirmed by the results of the phase shift analysis given in work [64] where the authors take into account the spin-orbital splitting of the scattering phase shifts.

Earlier the phase shift analysis of the differential cross-sections and the excitation functions of the elastic p^{6}Li scattering was made in work [64], but this analysis did not include the doublet ^2P-wave. Our phase shift analysis is based on the differential cross-sections given only in works [3] and [61]. The calculations are made for lower energies having the importance for the nuclear astrophysics and take into account all partial waves, including the doublet ^2P-wave.

The first energy we considered is 500 keV from work [61] and it leads to ^2S and ^4S-phase shifts of scattering which are listed in Table 8 and quite reasonably describe the experimental results with small average value of χ^2=0.15. The effort to take into account the doublet ^2P and quartet ^4P-phase shifts leads to the low values of these shifts. The error of the differential cross-sections of these data was taken to be 10%.

The next five energies are the new results of measurements of the differential cross-sections taken in work [3]. The first of them is equal to 593 keV and leads to the $^{2,4}S$-phase shifts which differ slightly from those for the previous energy; they have the same value of χ^2 and are listed in Table 8. The phase shifts of $^{2,4}P$-waves vanish to zero.

Table 8. Results of the phase shift analysis of the elastic p^{6}Li scattering

№	E, keV	^2S, deg.	^4S, deg.	^2P, deg.	^4P, deg.	χ^2
1	500	176.2	178.7	---	---	0.15
2	593	174.2	178.8	---	---	0.15
3	746.4	170.1	180.0	---	---	0.23
	746.4	172.5	179.9	1.7	0.0	0.16
4	866.8	157.8	180.0	---	---	0.39
	866.8	170.2	174.9	3.9	0.0	0.22
	866.8	169.6	175.0	3.5	0.1	0.23
5	976.5	160.0	178.5	---	---	0.12
	976.5	167.0	174.5	1.1	0.0	0.12
	976.5	166.9	174.5	1.1	0.0	0.12
6	1136.3	144.9	180.0	---	---	0.58
	1136.3	164.7	171.1	5.8	0.0	0.32
	1136.3	166.4	169.9	5.5	0.1	0.32

The second energy 746.7 keV leads to the $^{2,4}S$-phase shifts (see Table 8) which allow us to describe the cross-sections with χ^2=0.23. In spite of the small value of χ^2, the attempt to take into account $^{2,4}P$-phase shifts was made. At the beginning we supposed that the quartet ^4P-phase shift is negligible as it followed from the results of work [63] where their account begins from 1.0-1.5 MeV only. The results of our analysis taking into account the ^2P-phase shift only are shown in Figure 9 and are listed in Table 8. It can be seen that the small doublet ^2P-phase shift slightly changes the doublet ^2S-phase shift increasing its value and reducing the value of χ^2 to 0.16. The effort to take into account the ^4P-phase shift too leads to the negligible values (less than 0.1^0). This fact is absolutely in conformity with the results of work [63] and it will be demonstrated on the example of the next energy of 866.8 keV.

The results of the phase shift analysis at the energy of 866.8 keV taking into account the $^{2,4}S$-waves only are given in Table 8 at $\chi^2=0.39$. They suggest that the value of the ^2S-phase shift falls sharply in comparison with the previous energy. However, if ^2P-wave is taken into account, its value increases noticeably (Figure 10 and Table 8) while the χ^2 value decreases almost twofold. The effort to take into account the quartet ^4P-phase shift leads to the value not more than 0.1^0 (Table 8) which means that its contribution at this energy range is very small. Any change in this phase shift resulting in its increase leads to the increase in χ^2, even at the different values of other phase shifts. For this energy and for all considered energies from work [3] it is impossible to find some non-zero doublet and quartet phase shifts with the value of χ^2 approaching its minimum.

The next considered energy equals 976.5 keV and, if $^{2,4}P$-waves are not taken into account, it leads to the values of ^2S and ^4S-phase shifts listed in Table 8. The further consideration of the ^2P-wave noticeably increases the value of the ^2S-phase shift when the ^4P-wave equals zero, as it seen in Figure 11 and Table 8 at $\chi^2=0.12$. If we include the quartet ^4P-wave in the analysis, then it also vanishes to zero as χ^2 decreases.

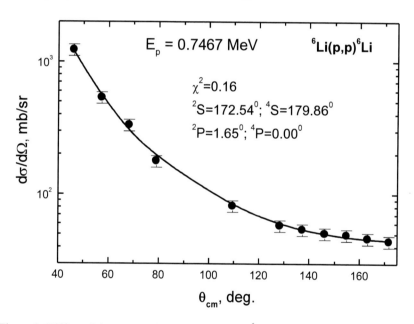

Figure 9. Differential cross-sections of the elastic p^{6}Li scattering at energy 746.7 keV.

The last energy from work [3] is equal to 1.1363 MeV and it leads to a comparatively small value of $\chi^2=0.58$ even if we take into account only $^{2,4}S$-waves in the analysis (see Table 8). However the account of the ^2P-wave noticeably decreases this value, and the calculation results for the differential cross-sections are shown in Figure 12 and are listed in Table 8. The attempt to take into account the ^4P-phase shift leads to its negligible values in this case as well (see Table 8).

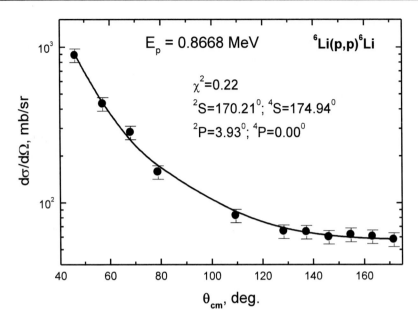

Figure 10. Differential cross-sections of the elastic p^{6}Li scattering at energy 866.8 keV.

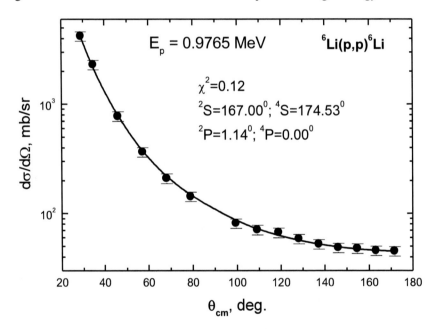

Figure 11. Differential cross-sections of the elastic p^{6}Li scattering at energy 976.5 keV.

Thus, all the experimental data of work [3] have a structure which does not require the presence of the quartet ^4P-waves in this energy range, i.e. their value equals or lower than 0.1^0. This fact, in general, is in agreement with the results of work [64], but the doublet ^2P-phase shift almost comes up to 6^0.

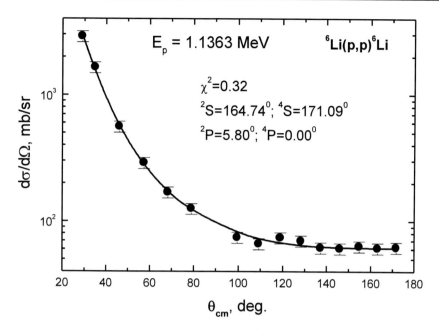

Figure 12. Differential cross-sections of the elastic p^{6}Li scattering at energy 1136.3 keV.

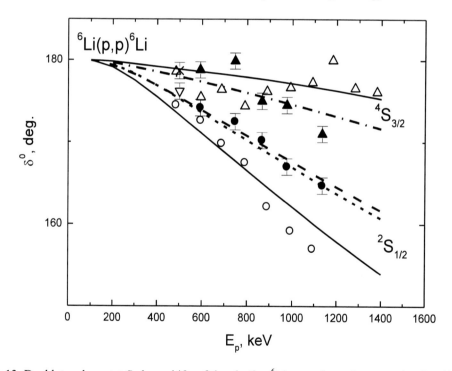

Figure 13. Doublet and quartet S-phase shifts of the elastic p^{6}Li scattering at low energies. Doublet and quartet S-phase shifts taking into account ^2P-wave, when ^4P-phase shift was taken to be zero. Dots - ^2S, triangles - ^4S-phases obtained on the basis of data [3]. For comparison the results of phase shift analysis [64] are represented by open triangles and open circles. Lines: calculation results for different potentials.

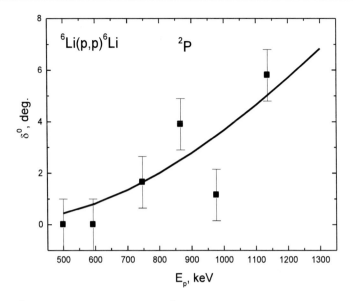

Figure 14. Doublet ^2P-phase shifts of the elastic p^{6}Li scattering at low energies. Squares - results of our phase shift analysis at $^4P=0$. Lines - the result of calculation with received potentials.

The general pattern of the ^2S and ^4S-phase shifts of scattering is shown in Figure 13 and the doublet ^2P-phase shifts are shown in Figure 14. In spite of the large data spread for the ^4S-phase shifts, the doublet ^2S-phase shift tends to decrease, but significantly slower than it could be expected from the results of analysis [64], where the ^2P-wave was not taken into account. If we do not take into account the doublet ^2P-wave in our analysis, then the values of the ^2S-phase shift are very close to the results of the phase shift analysis of work [64].

Errors of the phase shift analysis which are shown in Figure 13 are due to the ambiguity of the phase shift analysis - it is possible to obtain slightly different values of the phase shifts of scattering with approximately the same value of χ^2. This ambiguity is estimated as 1^0-1.5^0 and is shown for the S and ^2P-phase shifts in Figs. 13 and 14.

5.2. Potential Description of the Phase Shifts of Scattering

To calculate the partial inter-cluster p^{6}Li interactions according to the existing phase shifts of scattering we use common Gaussian potential with a point-like Coulomb component, which can be represented as (6). The following parameters for the description of the results of the phase shift analysis of work [64] were received:

^2S - $V_0 = -110$ MeV, $\alpha = 0.15$ fm^{-2},

^4S - $V_0 = -190$ MeV, $\alpha = 0.2$ fm^{-2}.

They include two forbidden bound states which correspond to the Young's schemes {52} and {7} [10, 65]. The calculation results of the phase shifts for these potentials are shown in Figure 13 by the solid lines with the results of the phase shift analysis [64] shown by blank circles and blank triangles.

For the description of our results of the phase shift of scattering the next potential parameters are preferable:

^2S - V_0 = -126 MeV, α = 0.15 fm^{-2}

^4S - V_0 = -142 MeV, α = 0.15 fm^{-2}.

They also include two forbidden bound states which corresponded to the schemes {52} and {7}. The phase shifts for these potentials are shown in Figure 13 by the dashed and dot-dashed lines in comparison with the results of our phase shift analysis given by black points and triangles.

The potential of the doublet ^2P-wave of scattering can be represented, for example, by the next parameters:

^2P - V_0 = -68.0 MeV, α = 0.1 fm^{-2}.

The results of the calculation of the phase shifts with this potential are shown in Figure 14 by the solid line. The potential has one forbidden bound state with the scheme {61} and the allowed state with the Young's schemes {43} and {421}.

Such potential gives the wrong bound energy of the ^{7}Be nucleus in the p^{6}Li channel because the allowed state is mixed in terms of the above mentioned symmetries, but only the scheme {43} corresponds to the ground bound state [65]. But even if we use the methods of receiving the "pure" phase shifts given in [65, 66], it is not possible to obtain the "pure" in Young's schemes potential of the ground state. It seams that it is due to the absence of the spin-orbital splitting and the small probability of clusterization of the ^{7}Be nucleus into the p^{6}Li channel.

That is why the "pure" according to orbital symmetries $^2P_{3/2}$-potential of the ground state of the ^{7}Be nucleus with Young's scheme {43} should be constructed so as to describe primarily the channel energy - the bound energy of the ground state of the nucleus with $J=3/2^-$ as the p^{6}Li system and its root-mean-square radius. Then the parameters of the "pure" $^2P_{3/2}^{\{43\}}$-potential can be represented as

$$^2P_{3/2} - V_P = -252.914744 \text{ MeV}, \alpha_P = 0.25 \text{ fm}^{-2}. \tag{21}$$

Such potential gives the bound energy of the allowed state with scheme {43} equal to -5.605800 MeV, while the experimental value is equal to -5.6058 MeV [67] and has one forbidden state corresponding to Young's scheme {61}. The root-mean-square charge radius is equal to 2.63 fm what is generally in agreement with the data of [67], and the C_W constant (8) is equal to 2.66(1) within the range of 5-13 fm.

For the parameters of the $^2P_{1/2}^{\{43\}}$-potential of the first forbidden state of ^{7}Be nucleus with $J=1/2^-$ the next values are obtained

$$^2P_{1/2} - V_P = -251.029127 \text{ MeV}, \alpha_P = 0.25 \text{ fm}^{-2}. \tag{22}$$

This potential leads to the bound energy -5.176700 MeV while its experimental value is equal to -5.1767 MeV [67] and it contains the forbidden state with scheme {61}. The asymptotic constant (8) is equal to 2.53(1) within the range of 5-13 fm and the charge radius is equal to 2.64 fm. The absolute accuracy of searching for the bound energy in our new computer programs was taken to be at the level of 10^{-7} MeV.

The obtained potential parameters of the bound states a little bit differ from our previous results [65]. This is because we used in these calculations the exact mass values of particles and more accurate description of the experimental values of the energy levels.

The variational method for the energy of the ground state gives the value -5.605797 MeV and hence the average energy for this potential is equal to -5.6057985(15) MeV, i.e. the accuracy of its determination equals ±1.5 eV. The asymptotic constant at the distances of 5-13 fm turned out to be comparatively stable and equal to 2.67(2) and the charge radius is in agreement with the calculation results based on the finite-difference method. The variational wave function (7) for the ground state of the ^{7}Be nucleus in the p^{6}Li channel with potential (21) is listed in Table 9, and the residual error is not more than 10^{-12}.

Table 9. The variational parameters and expansion coefficients of the radial WF of the bound state of the p^{6}Li system for the *P*-potential (21). The normalization of the function with these coefficients in the range 0-25 fm equals N=0.9999999999999895

i	β_i	C_i
1	2.477181344627947E-002	1.315463702527344E-003
2	5.874061769072439E-002	1.819913407984276E-002
3	1.277190608958812E-001	9.837541674753882E-002
4	2.556552559403827E-001	3.090018297080802E-001
5	6.962545656024610E-001	-1.195304944694753
6	87.215179556255360	3.237908749007494E-003
7	20.660304078047520	5.006096657700867E-003
8	1.037788131786810	-6.280751485496025E-001
9	2.768782138965186	1.282309968994793E-002
10	6.753591325944827	8.152343478073063E-003

For the energy of the first excited state with the use of the VM we received the value of -5.176697 MeV and hence the average energy is equal to -5.1766985(15) MeV, with the same accuracy as it was in the case of the GS. The asymptotic constant at distances of 5-13 fm turned out to be of the level of 2.53(2), the residual error being not more than 10^{-12} and the charge radius being almost the same as for the GS. The parameters of the excited state of the WF of the ^{7}Be nucleus with potential (22) are listed in Table 10.

It should be noted that on the basis of the obtained results of the phase shift analysis for the doublet ^2P-phase shift of scattering shown in Figure 14, it is impossible to construct a unique ^2P-potential. The results of the phase shift analysis at higher energies are required and they have to take into account ^2P-wave and spin-orbital phase shift splitting.

The same concerns the ^4S-potential, and only the ^2S-interaction is obtained quite uniquely. This interaction with the above ^2P-potentials of the bound states can be used in future, for example, for the calculations of the astrophysical S-factor with the $E1$ transition from the doublet ^2S-wave of scattering to the ground and first excited doublet bound ^2P-states of the ^{7}Be nucleus.

Table 10. The variational parameters and expansion coefficients of the radial WF of the first excited bound state of the p⁶Li system for the *P*-potential (22). The normalization of the function with these coefficients in the range 0-25 fm equals N=0.9999999999999462

i	β_i	C_i
1	2.337027900191992E-002	1.218101547601343E-003
2	5.560733180673633E-002	1.653319276756672E-002
3	1.214721917930904E-001	9.009619752334307E-002
4	2.474544878067495E-001	3.003291466882630E-001
5	7.132725465249825E-001	-1.332325501226168
6	84.896023494945160	3.273725679869025E-003
7	1.162854732120233	-5.340018423135894E-001
8	1.574203000936825	9.367648737801053E-002
9	5.779896847077723	1.033713941440747E-002
10	19.422905786572090	5.314592946045428E-003

5.3. Astrophysical *S*-factor

The $E1$ transitions from ^2S and ^2D-states of scattering to the ground $^2P_{3/2}$ and the first excited $^2P_{1/2}$ bound states of the ^{7}Be nucleus were taken into account when the astrophysical *S*-factor was considered. The calculation of the wave function of the ^2D-wave without spin-orbital splitting was made on the basis of the ^2S-potential but with the orbital momentum L=2.

When the calculations were made it turned out that the given above ^2S-potential of scattering with the depth of 110 MeV and based on the phase shift analysis [64] led to the astrophysical *S*-factor significantly lower than it had to be. At the same time the doublet ^2S-potential with the depth of 126 MeV, which was obtained in our calculations, gives quite correct description of the general behavior of the experimental *S*-factor. The results received are shown in Figure 15. The results of the transitions from ^2S and ^2D-waves of scattering to the ground state of the ^{7}Be nucleus are shown by the dashed line, the dotted line is for the transitions to the first excited state and the solid line is the total *S*-factor.

The calculated *S*-factor at 10 keV is equal to $S(3/2^-)$=76 eV b and $S(1/2^-)$=38 eV b while the total value is equal to 114 eV b. The behavior of the $S(1/2^-)$-factor well describes the experimental data (circles in Figure 15) for the transition to the first excited state of the ^{7}Be nucleus at low energies.

For comparison of the calculated *S*-factor at zero energy (10 keV) we will give the known results for the total $S(0)$: 79(18) eV b [70], 105 eV b (at 10 keV) [69] and 106 eV b [71]. In work [72] for the *S*-factor of transitions to the ground state the value of 39 eV b is given and for the transition to the first excited state the value of 26 eV b, so the total *S*-factor is equal to 65 eV b. As it seen the difference between these data is comparatively large, and our results are in agreement with them in general.

Besides, a small change in the depth of the ^2S-potential of scattering, for example if we take 124 MeV, which practically does not affect the behavior of the calculated phase shifts shown in Figure 13 by the short dashes, influences the *S*-factor significantly and leads to the value of 105 eV b at 10 keV. The total *S*-factor with this potential is shown in Figure 15 by the dot-dashed line which lies within the experimental error band at the energies below 1 MeV.

It should be mentioned that if we use the potentials without the forbidden states in S and P-waves or with another number of FS, then the value of the calculated S-factor turns out to be from 3 to 100 times lower that the values obtained above. For example, the ^2S-potential with one forbidden state and parameters 25 MeV and 0.15 fm^{-2}, which gives a good description of the phase shifts of scattering and the given above potential of the ground state, leads to the S-factor of about 1 eV b.

Thus, the doublet ^2S-phase shifts obtained in our phase shift analysis, which takes into account the doublet ^2P-phase shift, lead to the potential which allowed to describe the experimental S-factor at the energies down to 1 MeV, in distinction from the interaction constructed on the basis of the analysis results [63]. The potential cluster model used and the potentials given above allow in general obtaining quite reasonable results for the description of the process of radiative p^{6}Li capture at the astrophysical energy range, as it was in the case of lighter nuclei [73].

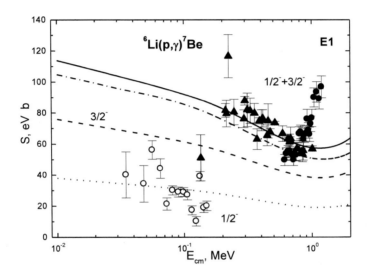

Figure 15. Astrophysical S-factor of p^{6}Li radiative capture. Black points, triangles and circles are the experimental data from works [68] given in work [69]. The result for transitions from ^2S and ^2D-waves of scattering to the ground state of the ^{7}Be nucleus is shown by the dashed line and for transitions to the first excited state - by the dotted line. The solid line shows the total S-factor. Dash-dot line is result for other variant of the scattering potential.

6. RADIATIVE P^{12}C CAPTURE

In this section we will consider the p^{12}C system and the process of proton radiative capture by the ^{12}C nucleus at astrophysical energies. The new measurement of differential cross-sections of the elastic p^{12}C scattering at energies from 200 keV up to 1.1 MeV (center-of-mass system) within the range of 10^0-170^0 with 10% errors was carried out in works [4].

Further, the standard phase shift analysis made and the potential of S-state of p^{12}C system was reconstructed in this paper on the basis of these measurements [39], and then the astrophysical S-factor at the energies down to 20 keV was considered in the frame of potential cluster model.

Before we start describing the results obtained, we would like to note that this process is the first thermonuclear reaction of the CNO-cycle which took place at a later stage of stellar evolution when partial hydrogen burning occurred. As the hydrogen is burned, the core of the star starts contracting which results in the increase in pressure and temperature in the star and triggers along with the proton-proton cycle the next chain of thermonuclear processes, called CNO-cycle.

6.1. Phase Shift Analysis

While examining scattering in the particle system with a total spin 1/2, i.e. where one of the particles has spin 0, and the second has spin 1/2, it is necessary to take into account spin-orbital splitting of phase shifts. This sort of scattering takes place in the nuclear systems such as N^{4}He, ^{3}H^{4}He, p^{12}C etc. The differential cross-section of the elastic scattering of nuclear particles is represented as [63]

$$\frac{d\sigma(\theta)}{d\Omega} = |A(\theta)|^2 + |B(\theta)|^2.$$

The connection between the differential cross-sections of the elastic scattering and the phase shifts is given in work [63].

Earlier, the phase shift analysis of excitation functions of the p^{12}C scattering, measured in [74] at energies 400-1300 keV (l.s.) and angles from 106^0 to 169^0, was carried out in work [75] where it was found that the S-phase must be in the range of 153^0-154^0 at the energy E_{lab}=900 keV.

For the same experimental data we have received 152.7^0. The cross-sections were extracted from the excitation functions [75] at energies 866-900 keV. Results of our calculations σ_t in comparison with experimental data σ_e are given in Table 11. Partial χ^2_i for each point with 10% errors in experimental cross-sections are given in the last column of the table and the value of 0.11 is received for the average of χ^2.

The values 155^0-157^0 of the S-phase shift are found in work [75] at the energy 751 keV (l.s.). The received results for this energy are listed in Table 12. We took the cross-section data from excitation functions at energies in the range of 749-754 keV and obtained the value 156.8^0 for the S-phase shift at the average value of χ^2=0.30.

Table 11. Comparison of theoretical and experimental cross-sections of the elastic p^{12}C scattering at the energy 900 keV

θ^0	σ_e, (mb)	σ_t, (mb)	χ^2_i
106	341	341.5	1.90E-04
127	280	282.1	5.76E-03
148	241	251.2	1.80E-01
169	250	237.5	2.50E-01

Table 12. Comparison of theoretical and experimental cross-sections of the elastic p^{12}C scattering at the energy 750 keV

θ^0	σ_e, (mb)	σ_t, (mb)	χ^2_i
106	428	428.3	3.44E-05
127	334	342.8	6.91E-02
148	282	299.1	3.66E-01
169	307	279.9	7.82E-01

Table 13. Results of the phase shift analysis of the elastic p^{12}C scattering at low energies taking into account S-wave only

E_{cm}, (keV)	$S_{1/2}$, (deg.)	χ^2
213	2.0	1.35
317	2.5	0.31
371	7.2	0.51
409	36.2	0.98
422	58.0	3.75
434	107.8	0.78
478	153.3	2.56
689	156.3	2.79
900	153.6	2.55
1110	149.9	1.77

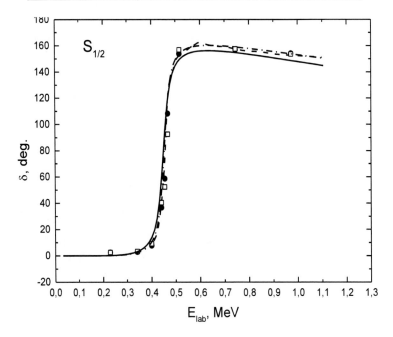

Figure 16. ^{2}S-phase shift of the elastic p^{12}C scattering at low energies. Black points - results of the phase shift analysis for the S-phase shift taking into account the S-wave only; open squares - results of the phase shift analysis for the S-phase shift taking into account S and P-waves; dotted line - results of work [75]; other lines - results calculated with different potentials.

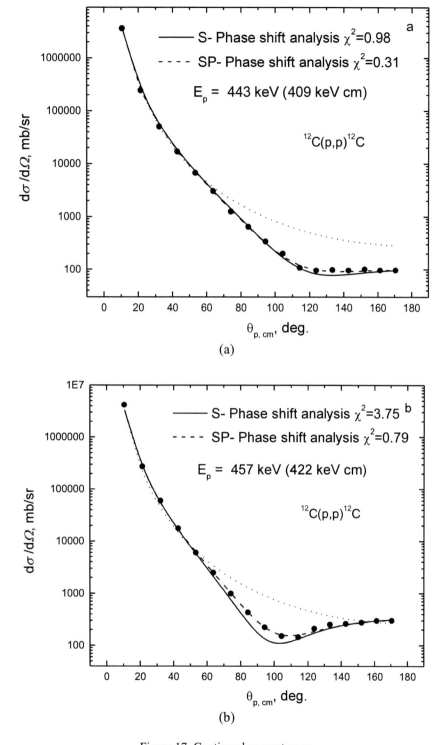

Figure 17. Continued on next page.

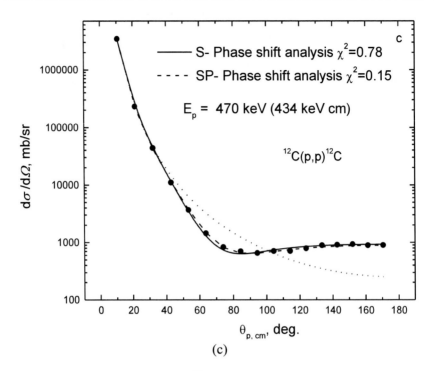

Figure 17 a,b,c. Differential cross-sections of p^{12}C scattering. Solid line is the phase shift analysis which takes into account the *S*-wave only; dotted line - the Rutherford scattering; dashed line - the phase shift analysis where *S* and *P*-waves are taken into account; black points - experimental data [4].

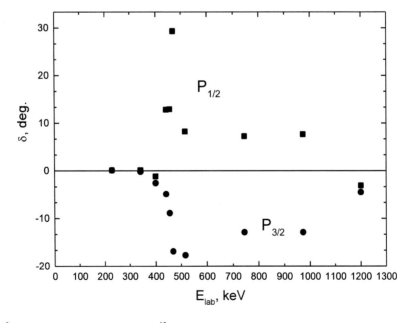

Figure 18. ^2P-phase shifts of the elastic p^{12}C scattering. Black points ($P_{3/2}$) and squares ($P_{1/2}$) are the results of the phase shift analysis for *P*-phase shifts taking into account *S* and *P*-waves.

All these reference results are in a good agreement with each other and the phase shift analysis [39] of the new experimental data of the differential cross-section of p^{12}C scattering was carried out using our program at energies in the range of 230-1200 keV (l.s.) [4]. The results of our phase shift analysis are given in Table 13 and are shown in Figure 16 in comparison with the values of work [75] which are shown by the dashed line.

Figure 17a,b,c shows the differential cross-sections in the resonance region at 457 keV (l.s.), the calculation results of these cross-sections on the basis of Rutherford formula - (dotted line), the cross-sections on the basis of our phase shift analysis which takes into account S-phase shift only (solid line). The cross-sections which take into account the S and P-waves of scattering in the phase shift analysis are shown by the dashed line.

One can see from the figures that it is impossible to describe the cross-section in the resonance region on the basis of S-wave only. In this case, the P-wave shown in Figure 18 starts playing a considerable role and its consideration improves the experiment description. At the energy 457 keV (l.s.) the value of χ^2 can be improved from 3.75 to 0.79 (Figure 17b) by taking into account P-wave, what considerably affects the quality of description of differential cross-sections.

It is seen in Figure 18, that at low energies $P_{1/2}$-phase shift lies above $P_{3/2}$ but at the energy about 1.2 MeV they intersect and then $P_{3/2}$-phase shift lies above in the region of negative angles [76,77]. The value of the S-phase shift practically doesn't change when the P-wave is taken into account (see Figure 16 - blank blocks). Consideration of the D-wave in the phase shift analysis leads to the value 1-1.5 degrees in the resonance region and practically doesn't affect the behavior of calculated differential cross-sections.

6.2. Astrophysical *S*-factor

The radiative p^{12}C capture at low energies is the part of the CNO thermonuclear cycle and gives a considerable contribution into the energy output of thermonuclear reactions [15, 66]. The existing experimental data of the astrophysical S-factor [9] indicates the presence of the narrow resonance with the width of about 32 keV at the energy 0.422 MeV (center-of-mass system), which leads to the two-three order increase in the S-factor.

It is interesting to find out if there is a possibility to describe the resonance S-factor on the basis of the PCM with FS and with the classification of orbital states according to Young's schemes. The phase shift analysis of the new experimental data [4] of differential cross-sections of the elastic p^{12}C scattering at astrophysical energies [39], which we have shown above, allows constructing the potentials of the p^{12}C interaction for the phase shift analysis of the elastic scattering.

The $E1(L)$ transition resulting from the orbital part of electric $Q_{JM}(L)$ operator [10] is taken into account in present calculations of the process of radiative p^{12}C capture. The cross-sections of $E2(L)$ and $MJ(L)$ processes and the cross-sections depending on the spin part $EJ(S)$, $M2(S)$ turned out to be a few orders less. The electrical $E1(L)$ transition in the p^{12}C→γ^{13}N process is possible between the doublet $^2S_{1/2}$ and $^2D_{3/2}$-states of scattering and the ground bound $^2P_{1/2}$-state of the ^{13}N nucleus in the p^{12}C channel.

Let's examine the classification of orbital states according to Young's schemes in the p^{12}C system for the purposes of construction of the interaction potential. The possible orbital Young's schemes in the $N=n_1+n_2$ particle system can be defined as the direct external product of orbital schemes of each subsystem, in our case it gives $\{1\}\times\{444\}=\{544\}$ and $\{4441\}$ [78]. The first of them is consistent only with the orbital momentum $L=0$ and is forbidden, because s-shell cannot contain more than four nucleons. The second scheme is allowed with orbital momenta 1 and 3 [78], the first of which corresponds to the ground bound state of the ^{13}N nucleus with $J=1/2^-$. Therefore, in the potential of the ^2S-wave there must be a forbidden bound state and ^2P-wave should have only one allowed state at the energy of -1.9435 MeV [79].

For the calculations of photonuclear processes the nuclear part of the inter-cluster $p^{12}C$ interaction is represented as (6) with the point-like Coulomb component. The potential of $^2S_{1/2}$-wave is constructed so as to describe correctly the corresponding partial phase shift of the elastic scattering, which has a well defined resonance at 0.457 MeV (l.s.).

Using the results of the phase shift analysis [39] the $^2S_{1/2}$-potential of the $p^{12}C$ interaction with FS at energy E_{FS}=-25.5 MeV was obtained together with parameters:

V_S = -67.75 MeV, α_S = 0.125 fm^{-2}.

The results of calculation of $^2S_{1/2}$-phase shift with this potential are shown in Figure 16 by the solid line.

The potential of the bound $^2P_{1/2}$-state has to reproduce correctly the bound energy of the ^{13}N nucleus in the $p^{12}C$ channel -1.9435 MeV [79] and reasonably describe the root-mean-square radius, which probably does not differ significantly from the radius of the ^{14}N nucleus equal to 2.560(11) fm [79]. As a result the following parameters were received:

V_{GS} = -81.698725 MeV, α_{GS} = 0.22 fm^{-2}. (23)

The potential gives the bound energy equal to -1.943500 MeV and the root-mean-square radius R_{ch}=2.54 fm. We use the following values for the radii of proton and ^{12}C: 0.8768(69) fm [12] and 2.472(15) fm [80]. The asymptotic constant C_W with Whittaker asymptotics (8) was calculated for controlling behavior of WF of BS at long distances; its value in the range of 5-20 fm equals 1.96(1).

The results of calculations of the S-factor of the radiative $p^{12}C$ capture with the abovementioned potentials of $^2P_{1/2}$ and $^2S_{1/2}$-waves at energies from 20 keV to 1.0 MeV are shown in Figure 19 by the solid line. The value 3.0 keV b of the S-factor is received at energy 25 keV. The extrapolation of the S-factor experimental measurements gives: 1.45(20) keV b and 1.54^{+15}_{-10} keV b [79]. The ^2S-potential given here is not the only one which can describe the resonance behavior of the S-phase shift at energies lower than 1 MeV.

Thus, it is always possible to find other combinations of bound and scattering state potentials with FS which lead to the similar results for the $^2S_{1/2}$-phase shift and describe well the value and location of the maximum of the S-factor, for example:

V_{GS} = -121.788933 MeV, α_{GS} = 0.35 fm^{-2},

R_{ch} = 2.49 fm, C_W = 1.50(1), E_{GS} = -1.943500 MeV,

V_S = -102.05 MeV, α_S = 0.195 fm^{-2}, E_{FS} = -12.8 MeV. (24)

They lead to a sharper fall of the S-factor at energies near the resonance. The phase shift of potential (24) and the behavior of its S-factor are shown in Figs. 16 and 19 by dot-dashed lines. The S-factor value for this combination of the potentials at 25 keV is equal to 1.85 keV b, what generally agrees with the values given in review [79].

A narrower bound state potential with the same potential of scattering (24)

V_{GS} = -144.492278 MeV, α_{GS} = 0.425 fm^{-2},

R_{ch} = 2.47 fm, C_W = 1.36(1), E_{GS} = -1.943500 MeV, (25)

leads to a small decrease in the S-factor at the resonance energy, as it is shown in Figure 19 by the dashed line, and gives the value $S(25)$=1.52 keV b which is in a good agreement with data [79]. At the same time, in the range of 20-30 keV the S-factor value is practically constant and one can consider it as the S-factor value at zero energy with an error of about 0.02 keV b. As it can be seen from the above results, the asymptotic constant and charge radius of the nucleus became smaller as the width of the BS potentials decreases, and potential (25) gives their smallest values.

The variational method was used for an additional control of the accuracy of bound energy calculations, which allowed to obtain the bound energy of -1.943498 MeV for the first variant of the potential (23) by using an independent variation of parameters and the grid having dimension 10. The asymptotic constant C_W of the variational WF at distances of 5-20 fm remains at the level of 1.97(2). Its variational parameters are listed in Table 14. The charge radius does not differ from the value obtained in FDM calculations.

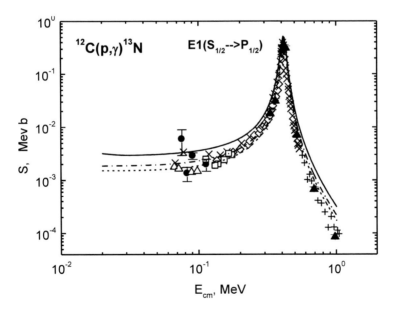

Figure 19. Astrophysical S-factor of p^{12}C radiative capture at low energies. The experimental data specified as ×, •, □, +, ◊ and ∆ are taken from review [9], triangles are from [81]. Lines: calculations with different potentials.

Table 14. The variational parameters and expansion coefficients of the radial WF of the p^{12}C system for the first variant (23) of the BS potential

i	β$_i$	C$_i$
1	4.310731038130567E-001	-2.059674967002619E-001
2	1.110252143696502E-002	-1.539976053334172E-004
3	4.617318488940146E-003	-2.292772895754105E-006
4	5.244199809745243E-002	-1.240687319547592E-002
5	2.431248255158095E-002	-1.909626327101099E-003
6	8.481652230536312 E-000	5.823965673819461E-003
7	1.121588023402944E-001	-5.725546189065398E-002
8	2.309223399000618E-001	-1.886468874357471E-001
9	2.297327380843046 E-000	1.244238759439573E-002
10	3.7567721497435540 E+001	3.435757447077250E-003

Table 15. The variational parameters and expansion coefficients of the radial WF of the p^{12}C system for the second variant (24) of the BS potential

i	β$_i$	C$_i$
1	1.393662782203888E-002	3.536427343510346E-004
2	1.041704259743847E-001	3.075071412877344E-002
3	4.068236340341411E-001	3.364496084003433E-001
4	3.517787678267637E-002	4.039427231852849E-003
5	2.074448420678197E-001	1.284484754736406E-001
6	7.360025091178769E-001	2.785322894825304E-001
7	3.551046173695889E-000	-1.636661944722212E-002
8	1.5131407009411240E+001	-9.289494991217288E-003
9	9.726024028584802E-001	-1.594107798542716E-002
10	6.634603967502104E-002	8.648073851532037E-003

Table 16. The variational parameters and expansion coefficients of the radial WF of the p^{12}C system for the third variant (25) of the BS potential

i	β$_i$	C$_i$
1	1.271482702554672E-002	2.219877609724907E-004
2	9.284155511162226E-002	2.240043561912315E-002
3	3.485413978134982E-001	2.407314126671507E-001
4	3.088717918378341E-002	2.494885124596691E-003
5	1.815363020074388E-001	8.792233462610707E-002
6	5.918532693855678E-001	3.652121068403727E-001
7	3.909887088341156E+000	-1.906081640167417E-002
8	1.6356080812096500 E+001	-1.111922033874987E-002
9	9.358886757095011E-001	2.314583156796476E-001
10	5.673177540516311E-002	5.956470542991426E-003

For the real bound energy in this potential it is possible to use the value -1.943499(1) MeV, i.e. the calculation error of finding bound energy is on the level of ±1 eV, because the variational energy decreases as the dimension of the basis increases and gives the upper limit of the true bound energy, but the finite-difference energy increases as the size of steps decreases and the number of steps increases.

The variant (24) of the BS potential, which was examined within the frame of the variational method, leads to the bound energy of -1.943498 MeV with the residual error of the order of 3×10^{-14}, the radius is equal to 2.49 fm and the asymptotic constant at distances of 5-17 fm is equal to 1.50(2). The variational parameters and expansion coefficients of the radial wave function are listed in Table 15.

The third variant (25) of the BS potential, which was examined within the frame of variational method, leads to the bound energy of -1.943499 MeV with the residual error of 6×10^{-14}, the radius being the same as in the FDM calculations and the asymptotic constant at distances of 5-17 fm being equal to 1.36(2). The variational parameters of the radial wave function are listed in Table 16.

Thus, the given above pairs of the potentials with FS for the $^2S_{1/2}$-wave and the bound state, which gives the correct bound energy, lead to the joint description of the resonance in the S-factor and the resonance in the ^2S-phase of scattering. The best results of the description of the scattering process characteristics of the radiative p^{12}C capture at low energies and of the S-factor at 25 keV are obtained with the third variant (25) of the interaction potential.

At the same time, if we use the potentials of the $^2S_{1/2}$-wave with small depth and without forbidden states, for example, with parameters:

$V_S = -15.87$ MeV, $\alpha_S = 0.1$ fm^{-2},

$V_S = -18.95$ MeV, $\alpha_S = 0.125$ fm^{-2}, (26)

$V_S = -21.91$ MeV, $\alpha_S = 0.15$ fm^{-2},

then we can't obtain the correct description of the maximum of the S-factor of the radiative capture. It is impossible to describe the absolute value of the S-factor which for all variants of the scattering potentials (26) and the BS potentials is 2-3 times as much as the experimental maximum. At the same time, for all given depthless potentials of the form (26) the resonance behavior of the $^2S_{1/2}$-phase shift of scattering is well described. As the width of the $^2S_{1/2}$-potential decreases, i.e. the α value increases, the value of the S-factor maximum grows up, e.g. for the last variant of the $^2S_{1/2}$-scattering potential its value is approximately three times as much as the experimental value [83].

It should be noted that in all calculations the cross-section of the $E1$ electrical process due to transition from the doublet $^2D_{3/2}$-state of scattering to the ground bound $^2P_{1/2}$-state of the ^{13}N nucleus is 4-5 orders less than the cross-section of the transition from $^2S_{1/2}$-state of scattering. Thus, the main contribution to the calculated S-factor of the p^{12}C→^{13}Nγ process is made by the $E1$ transition from the ^2S-wave of scattering to the ground state of the ^{13}N nucleus. The mass of proton was taken to be 1 in all calculations for the p^{12}C system.

Thus, it is possible to combine the description of the astrophysical S-factor and the $^2S_{1/2}$-phase shift in the resonance energy range 0.457 MeV (l.s.) on the basis of the PCM and the deep $^2S_{1/2}$-potential with the FS, and to receive the reasonable values for the charge radius and asymptotic constant. The depthless potentials of scattering do not lead to the joint description of the S-factor and the ^2S-phase shift of scattering at any considered combinations of $p^{12}C$ interactions [82, 83].

7. RADIATIVE p^{7}Li CAPTURE

The reaction of radiative capture

$$p + {}^7Li \to {}^8Be + \gamma$$

at ultralow energies resulting in formation of unstable ^8Be nucleus which then decays into two α-particles may take place along with the weak process

$$^7Be + e^- \to {}^7Li + \gamma + \nu_e,$$

as one of the final reactions of the proton-proton chain [1]. Therefore the in-depth study of this reaction, in particular of the form and energy dependence of the astrophysical S-factor, is of a certain interest for the nuclear astrophysics.

To calculate the astrophysical S-factor of radiative p^{7}Li capture in the potential cluster model [5, 10] which we usually use for such calculations [73, 82] it is necessary to know partial potentials of p^{7}Li interaction in the continuous and discrete spectra. We will again assume that such potentials should follow the classification of cluster states by orbital symmetries [5] as it was assumed in our earlier works [17, 83] and previous chapters of the book for other nuclear systems.

We would like to remind that in the approach used the potentials of scattering processes are usually constructed on the basis of elastic scattering phase shifts obtained from experimental data while the interactions in bound states are determined by the requirement to reproduce the main characteristics of the bound state of the nucleus assuming that it is mainly due to cluster channel consisting of the input particles of the reaction under consideration.

For example, in the radiative capture process the $^2H^4He$ particles colliding at low energies form ^6Li nucleus in the ground state and the remaining energy is released as a γ quantum. Since there is no restructuring in such reactions we can consider potentials of one and the same nuclear system of particles; that is the $^2H^4He$ system in continuous and discrete spectra. In the latter case it is assumed that the ground state of the ^6Li nucleus is very likely caused by the cluster $^2H^4He$ configuration. Such approach leads to quite reasonable results of the description of the astrophysical S-factors of this and some other reactions of radiative capture [84].

It seems that in this case the ^8Be nucleus does not consist of cluster p^{7}Li system and most probably is determined by the $^4H^4He$ configuration into which it decays. However, it is possible that the ^8Be nucleus is in the bound state of the p^{7}Li channel for a while just after of the reaction of the radiative p^{7}Li capture and only after this it changes to the state defined by the unbound $^4H^4He$ system. Such an assumption makes it possible to consider the ^8Be nucleus as the cluster p^{7}Li system and use PCM methods, at least at the initial stage of its formation in the reaction $p+{}^7Li \to {}^8Be+\gamma$ [85].

7.1. Classification of the Orbital States

First, we would like to note that the p^{7}Li system has the $T_z = 0$ isospin projection and it is possible for two values of total isospin $T = 1$ and 0 [86], therefore p^{7}Li channel is mixed by isospin as p^{3}H system [73], even though as it will be shown later both of isospin states ($T = 1,0$), in contrast to p^{3}H system, in the triplet spin state correspond to the allowed Young's scheme {431} [10]. In this case the cluster channels p^{7}Be and n^{7}Li with $T_z = \pm 1$ and $T = 1$ are pure by isospin in a complete analogy with the p^{3}He and n^{3}H systems [41].

The spin-isospin schemes of the ^{8}Be nucleus for the p^{7}Li channel are the product of spin and isospin parts of the WF. Particularly, for any of these momenta we will have {44} scheme at the ground state of the ^{8}Be nucleus with the momentum equal to zero, scheme {53} for a certain state with momentum equal to one and for the state with momentum equal to 2 - {62} symmetry form.

If the scheme {7} is used for the ^{7}Li nucleus then possible Young's schemes of p^{7}Li system turn out to be forbidden, because of the rule that there can not be more than four cells in a row [66, 78], and they correspond to forbidden states with configurations {8} and {71} and relative motion momenta $L = 0$ and 1, which is determined by Elliot rule [78]. The p^{7}Li system contains forbidden states with the scheme {53} in P_1-wave and {44} in S_1-wave and allowed state with the configuration {431} at $L = 1$ when the scheme {43} is accepted for the ^{7}Li nucleus.

Thus, the p^{7}Li potentials in the different partial waves should have the forbidden bound state {44} in the S_1-wave and forbidden and allowed bound levels in the P_1-wave with schemes {53} and {431}, respectively. The considered classification is true for any isospin state of the p^{7}Li system ($T = 0$ or 1) in triplet spin channel. Allowed symmetries are absent for spin $S = 2$ and all Young's schemes listed above correspond to forbidden states.

Probably, as it was in a previous case for the p^{6}Li system, it is more correctly to consider both allowed schemes {7} and {43} for bound states of the ^{7}Li because of the fact that they are present in FS and AS in the ^{3}H^{4}He configuration of this nucleus [85]. Then the level classification will be slightly different, the number of forbidden states will increase and an extra forbidden state will appear in each partial wave. Such more complete scheme of FS and AS states, per se, is a sum of the first and the second cases considered above.

7.2. Potential Description of the Phase Shifts of Scattering

Because of the isospin mixing the phase shifts of the p^{7}Li elastic scattering are represented as a half sum of the isospin pure phase shifts [41] in complete analogy with the p^{3}H system considered above [73]. The phase shifts with $T = 1,0$ mixed by isospin are usually determined as a result of the phase shift analysis of the experimental data of differential cross-sections of the elastic scattering or excitation function. The pure phase shifts with isospin $T = 1$ are determined from the phase shift analysis of the p^{7}Be or n^{7}Li elastic scattering. As a result it is possible to find pure p^{7}Li phase shifts of scattering with $T = 0$ and construct the interaction model using these results which have to correspond to the potential of the bound state of the p^{7}Li system in the ^{8}Be nucleus [86]. Just the same method of phase shift separation was used for the p^{3}H system [41] and its absolute validity was shown [10, 42].

However, we failed to find experimental data of differential cross-sections or phase shifts of the p^{7}Be or n^{7}Li elastic scattering at astrophysical energies, so here we will consider only isospin-mixed potentials of the elastic scattering processes in the p^{7}Li system and pure potentials of the bound state with $T = 0$ which are constructed on the base of description BS characteristics and are chosen in the Gaussian form with point-like Coulomb term (6).

The phase shifts of the p^{7}Li elastic scattering received from the phase shift analysis of the experimental data of excitation functions [87] taking into account spin-orbital splitting at the energies up to 2.5 MeV are given in the work [88]. These phases, which are equal to zero at energies down from 600-700 keV, we will use later for the intercluster potential construction for the p^{7}Li elastic scattering in S_1- and P_1-waves. Since we will consider the low and astrophysical energy range only, then we will limit the energy range from 0 keV to 700 keV. Practically zero phase shifts at these energies are received with the potential of the form (6) and parameters:

$V_0 = -147$ MeV and $\alpha = 0.15$ fm^{-2}.

Such potential contains two FS as it follows from the state classification given above. Of course, S_1-phase shift at about zero one can obtain from the other variants of potential parameters with two FS. In this regard it is not possible to fix its parameters unambiguously and the other combinations of V_0 and α are possible. However, this potential, as the potential given above, should have comparatively large width which gives small phase shift change when the energy changes in the range from 0 to 700 keV.

There is an over-threshold level in the P_1-wave with the energy 17.640 MeV and $J^PT = 1^+1$ or 0.441 MeV (l.s.) which is above the threshold of the cluster p^{7}Li channel in the ^{8}Be nucleus, with the bound energy of this channel being -17.2551 MeV [86]. The 0.441 MeV level has very small width of only 12.2(5) keV [86] for the p^{7}Li → ^{8}Beγ radiative capture reaction and p^{7}Li elastic scattering. Such a narrow level leads to the sharp rise of the P_1-phase shift of elastic scattering which is mixed by spin states $^5P_1$ and $^3P_1$ [86] for the total moment $J = 1$. The phase shift, which is shown with points in Figure 20, can be described by the Gaussian potential (6) with parameters:

$V_0 = -5862.43$ MeV and $\alpha = 3.5$ fm^{-2}.

This potential, mixed in isospin $T = 0$ and 1, has two FS and the calculation results of the P_1-phase shift of the elastic scattering are shown in Figure 20 with a solid line. The potential parameters which describe the P_1-phase shift are fixed quite unambiguously at the interval of sharp increase obtained from the experimental data and the potential itself should have very small width.

Since, later we will consider astrophysical S-factor only at the energies from 0 to 700 keV, it can be deemed that both of the potentials received above give a good description of the results of the phase shift analysis for two considered partial waves in this energy range.

The following parameters of the potential of the bound P_0-state of the p^{7}Li system corresponding to the ground state of the ^{8}B nucleus in the examined cluster channel are obtained:

$V_0 = -433.937674$ MeV and $\alpha = 0.2$ fm^{-2}.

The bound energy -17.255100 MeV with the accuracy of 10^{-6} MeV, the root-mean-square radius equal to 2.5 fm and the asymptotic constant, calculated with the help of Whittaker functions (8), equal to C_W = 12.4(1) were obtained with such a potential. The error of the constant is estimated by its averaging in the range 6-10 fm where the asymptotic constant is practically stable. In addition to the allowed BS corresponding to the ground state of the ^{8}Be nucleus such P-potential has two FS in total correspondence with the classification of orbital cluster states.

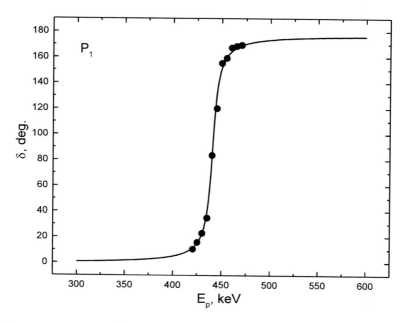

Figure 20. $^5P_1$-phase mixed with $^3P_1$-phase of the elastic p^{7}Li scattering at low energies. Points - phase shifts received from the experimental data in work [89]. Line - calculations with the Gaussian potential based on parameters given in the text.

Table 17. The variational parameters and expansion coefficients of the radial wave function of the form (7) for ground state of the ^{8}Be nucleus in the p^{7}Li channel in nonorthogonal Gaussian basis [90]. The normalization coefficient of the wave function in the range 0-25 fm equals N=1.000000000000001

i	β_i	C_i
1	1.140370098659333E-001	-9.035361688615057E-002
2	5.441057961629589E-002	-5.552214961281388E-003
3	2.200385338662954E-001	-4.776382639167991E-001
4	5.657244883872561E-001	3.790054587274382
5	9.613849915820404E-001	-2.409004172680931
6	1.216602174819119	-3.280156202364487
7	4.797601726001004	2.475815245412750E-002
8	14.137444509612200	1.070215776034501E-002
9	45.160915627598030	6.119172187062497E-003
10	191.081716320368200	3.950399055271339E-003

It seems that the root-mean-square radius of the ^{8}Be nucleus in the cluster p^{7}Li channel should not differ a lot from the ^{7}Li radius which equals 2.35(10) fm [89], since the nucleus is in a strongly bound (~ -17 MeV) i.e. compact state. Moreover, at such bound energy the ^{7}Li nucleus itself can be in deformed, compressed form as it is for deuteron in the ^{3}He nucleus [17]. Therefore, the value of the root-mean-square radius for the p^{7}Li channel in the GS of the ^{8}Be nucleus received above has quite a reasonable value.

The variational method with the expansion of the cluster wave function of the p^{7}Li system in nonorthogonal Gaussian basis (17) is used for an additional control of the accuracy of bound energy calculations and the energy -17.255098 MeV with N=10 order of matrix were obtained for this potential which differ from the given above finite-difference value by 2 eV only. Residuals [90] are of the order of 10^{-11}, asymptotic constant at the range 5-10 fm equals 12.3(2), the charge radius does not differ from previous results. Expansion parameters of the received variational GS radial wave function of the ^{8}Be nucleus in the p^{7}Li cluster channel are listed in Table 17.

We want to remind that the variational energy decreases as the dimension of the basis increases and gives the upper limit of the true bound energy, but the finite-difference energy increases as the size of steps decreases and the number of steps increases [90], therefore it is possible to use the average value -17.255099(1) MeV for the real bound energy in this potential. Thus, the calculation error of the bound energy of the ^{8}Be nucleus in the cluster p^{7}Li channel using two different methods is about ±1 eV.

7.3. Astrophysical S-factor

While considering electromagnetic transitions we will take into account the E1 process from the $^3S_1$-wave of scattering to the ground bound state of the ^{8}Be nucleus in the cluster p^{7}Li channel with $J^PT = 0^+0$ and the M1 transition from the $^3P_1$-wave of scattering (see Figure 20) also to the P_0(GS) of the nucleus. Cross-sections of the E1 transition from the $^3D_1$-wave of scattering (with potential for the $^3S_1$-wave at L = 2) to the GS of the ^{8}Be nucleus are by 2-4 orders lower than from the $^3S_1$-wave transition at the energy range 0-700 keV. Further on we will consider only S-factor for the transition to the ground state of the ^{8}Be nucleus i.e. the reaction: ^{7}Li$(p,\gamma_0)^8$Be. One of the last experimental measurements of the S-factor of this reaction in the energy range from 100 keV to 1.5 MeV was made in the work [91].

Expressions given above are used for the S-factor calculations. Values: μ_p=2.792847 and $\mu(^7$Li$)$=3.256427 are accepted for magnetic momentum of proton and ^{7}Li nucleus. The calculation results for the S-factor with the given above potentials at the energy range 5-800 keV (l.s.) are shown in Figure 21. The E1 transition is shown by the dashed line, dotted line - M1 process, solid line - the sum of these processes. In the considered reaction the M1 transition like the E1 transition in the p^{3}H system [41] goes with change of the isospin ΔT=1, since the ground state of the ^{8}Be nucleus has T=0 and resonance isospin in the P_1-wave of scattering equals 1.

The value of 0.50 keV b was obtained for the astrophysical S-factor at 5 keV (center-of-mass system) for the transition to GS of the ^{8}Be nucleus, where the E1 process gives the value of 0.48 keV b, which is in a good agreement with the data from [91]. The calculated and experimental S-factor values at the energy range 5-300 keV (l.s.) are given in Table 18. As it can be seen from Figure 21 and Table 18, the value of the theoretical S-factor at the energy range 30-200 keV is almost constant and approximately equal to 0.41-0.43 keV b, which agrees with data of the work [91] for the energy range 100-200 keV practically within the experimental errors.

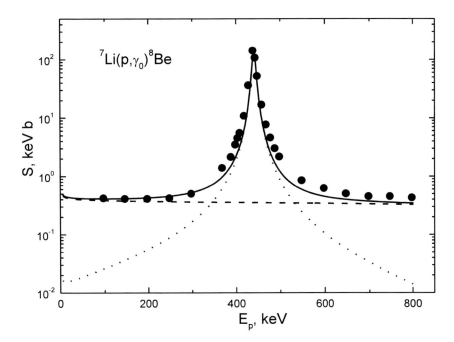

Figure 21. Astrophysical S-factor of the reaction of p^{7}Li radiative capture. Dots: experimental data from work [91]. Lines: calculation results for different electromagnetic transitions with the potentials mentioned in the text.

Let's compare some extrapolation results of different experimental data to zero energy. The value 0.25(5) keV b was obtained in work [92] and the value 0.40(3) keV b in work [93] on the basis of data [91]. Then in work [94] on the basis of new measurements of the total cross-sections of ^{7}Li(p,γ_0)^{8}Be reaction at the energy range 40-100 keV the value 0.50(7) keV b was suggested which is in a good agreement with the obtained above value at the energy 5 keV.

Table 18. Calculated astrophysical S-factor of the reaction of p^{7}Li radiative capture at low energies and its comparison with the experimental data [91]

E_{lab}, keV	S_{exp}, keV b [91]	S_{E1}, keV b	S_{M1}, keV b	S_{E1+M1}, keV b
5,7	---	0,48	0,02	0,50
29,7	---	0,41	0,02	0,43
60,6	---	0,39	0,02	0,41
98,3	0,41(3)	0,39	0,03	0,42
198,3	0.40(2)	0,37	0,06	0,43
298,6	0.49(2)	0,36	0,16	0,52

It is interesting to look at the chronology of different works for determination of the astrophysical S-factor of the ^{7}Li(p,γ_0)^{8}Be reaction. It was believed in 1992 that its value equals 0.25(5) keV b [92], the value 0.40(3) keV b [93] was obtained in 1997 on the basis of measurements made in 1995 [91] and the measurements in 1999 at lower energies led to the value of 0.50(7) keV b [94]. This chronology demonstrates well the constant increase in the value obtained for the astrophysical S-factor of ^{7}Li(p,γ_0)^{8}Be reaction (two fold increase) as the energy of experimental measurements decreased.

Thus, E1 and M1 transitions from $^3S_1$ and $^{3-5}P_1$-wave of scattering to the ground bound state in the p^{7}Li channel of the ^{8}Be nucleus were considered in the potential cluster model. It is possible to completely describe present day experimental data for the astrophysical S-factor at the energies up to 800 keV taking into account certain assumptions concerning the channel restructuring in the ^{8}Be nucleus and to obtain its value for zero (5 keV) energy, which is in a good agreement with the latest experimental measurements.

8. RADIATIVE P^{9}BE CAPTURE

Let's consider p^{9}Be → ^{10}Bγ reaction in the astrophysical energy range in the potential cluster model with splitting of orbital states according to Young's schemes and, in some cases, with forbidden states. We should note that we managed to find only one work devoted to a detailed experimental measurement of cross-sections and astrophysical S-factor for this reaction [95] at low energies. We will use the results of this work later for comparison with our model calculations.

The knowledge of p^{9}Be interaction potentials in continuous and discrete spectrum is required for the calculation of the astrophysical S-factor of radiative p^{9}Be capture in PCM [10] which we usually used for the analysis of similar reactions [10, 17, 41, 42, 73]. Again we will consider that such potentials should correspond to the classification of cluster states according to orbital symmetries [5], as it was assumed earlier for other light nuclear systems.

8.1. Classification of the Orbital States

First, we define the orbital Young's schemes of the ^{9}Be nucleus considering it, for example, in the p^{8}Li or n^{8}Be channel. If we assume that it is possible to use schemes {44}+{1} in the 8+1 system, then two possible symmetries - {54}+{441} are obtained for this system. The first of them is forbidden because it contains five cells in one matrix row [78]. We note, promptly, that the given classification of orbital states according to Young's schemes has qualitative character, because there is no table of Young's schemes products for the system with A = 9,10 particles, though they exist for all systems with A < 9 [96] and they are used for the analysis of the number of allowed and forbidden states in wave functions of different cluster systems [10].

Further, if the scheme {54} is used for the ^{9}Be nucleus then possible Young's schemes of p^{9}Be system turn out to be forbidden, because of the rule that it can not be more than four cells in one row [78] and they are corresponded to forbidden states with configurations {64}, {55} and relative motion momenta L = 0 and 1, which is determined by Elliot rule [78].

Another forbidden scheme {541} is present in this product as in the examined case too and correspond to $L = 1$.

When the scheme {441} is accepted for the ^{9}Be nucleus, the p^{9}Be system contains forbidden sates with {541} scheme in the P-wave and {442} in the S-wave, and AS with {4411} configuration with $L = 1,3$. Thus the p^{9}Be potentials in different partial waves should have forbidden bound {442} state in the S-wave and forbidden and allowed bound levels in the P-wave with {541} and {4411} Young's schemes, respectively.

We can examine the case when for the ^{9}Be nucleus both possible orbital Young's schemes {54} and {441} are used. We used the same approach quite successfully earlier for consideration of the p^{6}Li [97] and p^{7}Li [85] systems. Then the level classification will be slightly different - the number of forbidden states will increase and an additional forbidden bound level will appear in every partial wave with $L = 0$ and 1.

Such a more complete scheme of states, which we will use later, equals the sum of the first and the second cases considered above and there are two FS in S- and P-waves with allowed states in the P-wave. One of them - the $^5P_3$-state - can correspond to the ground state of the ^{10}Be nucleus in the p^{9}Be channel.

8.2. Potential Description of the Phase Shifts of Scattering

The considered p^{9}Be channel in the ^{10}Be nucleus has the isospin projection equal to $T_z = 0$, which is possible with two values of the total isospin $T = 1$ and 0 [86], so the p^{9}Be system, as well as p^{3}H [41], turn out to be isospin-mixed. In this case the cluster p^{9}Be and n^{9}Be channels with $T_z = \pm 1$ and $T = 1$ are isospin-pure in a complete analogy with p^{3}He and n^{3}H systems [41]. Phase shifts of the elastic p^{9}Be scattering are represented as a half-sum of pure in isospin phase shifts (15) because this system is isospin-mixed.

The isospin-mixed phase shifts with $T = 1,0$ are derived from the phase shift analysis of experimental data, which is usually the differential cross-sections of the elastic p^{9}Be scattering. Pure phase shifts with isospin $T = 1$ are obtained from the phase shift analysis of the p^{9}Be or n^{9}Be elastic scattering. As a result one can find pure $T = 0$ phase shifts for the p^{9}Be scattering and construct the interaction potential which should correspond to the potential of the bound state of the p^{9}Be system in the ^{10}Be nucleus [86].

However, we have not found the data on phase shifts for n^{9}Be, p^{9}B and p^{9}Be elastic scattering at astrophysical energies, so we will consider only isospin-mixed potentials of the scattering processes in the p^{9}Be system and pure potentials for bound states with $T = 0$ which are usually constructed on the basis of the description of BS characteristics - bound energy, charge radius, asymptotic constant. Exactly the same approach we used earlier for the p^{6}Li and p^{7}Li systems and the potential is selected in a simple Gaussian form with a point-like Coulomb term (6).

Since we don't have the phases of p^{9}Be elastic scattering obtained from the phase shift analysis of experimental data, we will only rely on the purely qualitative views about their behavior as an energy function. It is known in particular that there is $J = 1^-$ over-threshold level with $T = 0+1$, energy 0.319(5) MeV (l.s.) and 133 keV width [89, 98]. This resonance state can be formed by the $^3S_1$ configuration in the p^{9}Be channel of the ^{10}B nucleus because $J(^9\text{Be}) = 3/2^-$ and $J(\text{p}) = 1/2^+$. The presence of such a level leads to the resonance of the phase shift which equals 90^0 at this energy.

However, the resonance of the S-factor measured in [95] is observed at the energy of 299 keV (l.s.) which is listed in Table 1 and Figure 4 of work [95]. At the same time the value of 0.380(30) MeV (l.s.) with the width of 330(30) keV is given for the resonance energy in Table 2 of work [95]. Both of these values do not correspond to the well known data [86, 98]. That is why an additional analysis of the experimental results was carried out in later works [99, 100] and the values of 328-329 keV (l.s.) with the width of 155-161 keV were obtained for the energy of this level, which slightly differs from data of [86, 98].

Since there is a large difference between various data, we slightly varied parameters of this potential to receive the best description of the S-factor resonance location given in work [95]. As a result, the potential of the $^3S_1$-wave of scattering was obtained, which leads to the resonance of the 90^0 phase shift at 333 keV (l.s.) and has the following parameters:

$V_0 = -69.5$ MeV and $\alpha = 0.058$ fm^{-2}.

The triplet $^3S_1$-phase shift of this potential is shown in Figure 22 by a solid line and has a resonance nature, while the potential itself contains two FS in accordance with the classification given above.

If we use the expression for calculation of the level width using δ phase shift of scattering

$\Gamma_{lab.} = 2(d\delta/dE_{lab.})^{-1}$,

then the width of such resonance approximately equals 150(3) keV (l.s.), which is in a complete agreement with the results of works [99, 100].

Further we assume that the $^5S_2$-phase shift almost vanishes to zero at the energy range up to 600 keV which will be considered here because there are no such levels at ^{10}B spectra comparable to this partial wave at these energies [86, 98]. Practically zero phase shift is obtained with the Gaussian potential and parameters

$V_0 = -283.5$ MeV and $\alpha = 0.3$ fm^{-2}.

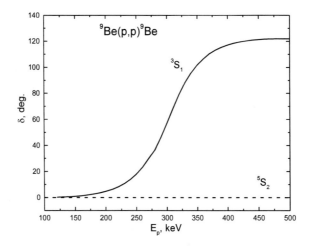

Figure 22. S-phase shifts of the elastic p^{9}Be scattering at low energies. Solid line - results obtained with the Gaussian potentials parameters of which are given in the text.

It contains two FS as it follows from the classification of orbital states given earlier. The phase shift of scattering is shown in Figure 22 by the dashed line. Of course, it is possible to obtain the $^5S_2$-phase shift in the vicinity of zero by using some other variants of potential parameters with two FS. In this regard it is not possible to fix its parameters definitely and other combinations of V_0 and α are possible. However, further calculations of the E1-transition from the elastic $^5S_2$-wave to the bound $^5P_3$-state have shown the weak dependence of the S-factor of the radiative p^{9}Be capture on the parameters of this potential.

The following parameters of the potential of the bound $^5P_3$-state of the p^{9}Be system corresponding to the ground state of the ^{10}B nucleus in the cluster channel under consideration are obtained:

$V_0 = -719.565645$ MeV and $\alpha = 0.4$ fm^{-2}.

With this potential we have obtained the bound energy -6.585900 MeV with the accuracy of 10^{-6} MeV, the root-mean-square radius equal to 2.58 fm while the experimental value is 2.58(10) fm [86,98] and the asymptotic constant calculated by Whittaker functions equal to C_W=2.94(1). Values R_p=0.8768(69) fm [12] and R_{Be}=2.519(12) fm [86, 98] were used for cluster radii. The AC error is estimated by its averaging at the range of 5-15 fm where the asymptotic constant is practically stable. In addition to the allowed BS corresponding to the ground state of the ^{10}B nucleus such P-potential has two FS in complete correspondence with the classification of orbital cluster states which was given above.

For the purposes of comparison, we give the results of work [101] for AC where its value equals C_W=2.37(2) fm$^{-1/2}$. To bring these constants to a unified dimensionless form the results should be divided by $\sqrt{2k_0}$, where k_0=0.536 fm^{-1} for the p^{9}Be system. Then we receive for AC in our definition the value of 2.29 which substantially differs from the result given above. However, if we take AC value received in work [101], the charge radius of the ^{10}B nucleus will be somewhat underestimated because the "tail" of the wave function decreases more sharply.

The following parameters for the potentials of the first three excited but not bound in p^{9}Be channel states with J^PT=1$^+$0, 0$^+$1 and 1$^+$0 at energies 0.71835, 1.74015 and 2.1543 MeV [86] were received:

$V_0(0.718350) = -715.162918$ MeV and $\alpha = 0.4$ fm^{-2},

$V_0(1.740150) = -708.661430$ MeV and $\alpha = 0.4$ fm^{-2},

$V_0(2.154300) = -705.935443$ MeV and $\alpha = 0.4$ fm^{-2}.

They describe precisely the values of energy levels given above and shown in brackets which, relative to the p^{9}Be channel threshold, equal to -5.867550, -4.845700 and -4.431600 MeV. They properly lead to charge radii of 2.59, 2.60 and 2.61 fm, asymptotic constant of 2.74(1), 2.46(1) and 2.35(1) at the range 4-5 to 11-13 fm and have two FS and one AS. It seems that these potentials correspond to the triplet ^3P bound levels in the p^{9}Be channels.

The variational method with the expansion of the cluster wave function of the p^{9}Be system in nonorthogonal Gaussian basis (7) is used for an additional control of the accuracy of bound energy calculations. The GS energy of -6.585896 MeV was obtained for this potential with the order of matrix $N=10$ which differs from finite-difference value [90] given above by 4 eV only. Residuals have 10^{-11} order, asymptotic constant at the range 5-10 fm equals 2.95(3), the charge radius does not differ from the previous results. Expansion parameters of the received variational GS radial wave function of the ^{10}B nucleus in the p^{9}Be cluster channel are listed in Table 19.

Table 19. The variational and expansion parameters of the radial WF of the ^{10}B ground state of the p^{9}Be channel in nonorthogonal Gaussian basis [90]. The normalization coefficient of the wave function in the range 0-25 fm equals N=1.000000000000002

i	β_i	C_i
1	7.715930101739352E-002	-2.802002694398972E-002
2	3.224286905853033E-002	-2.092599791641983E-003
3	1.677117157858407E-001	-1.481060223206524E-001
4	3.388993785610822E-001	-5.049291144131660E-001
5	9.389553670123860E-001	4.713342588832875
6	1.999427899506135	-7.632712971301209
7	2.988529100669578	-5.267741895838846E-001
8	6.878703971128334	6.022748751134505E-002
9	23.149662023260950	2.252725100117285E-002
10	100.917699526293000	1.285655220977827E-002

Table 20. The variational parameters and expansion coefficients of the radial WF of the 0$^+$1 state of the ^{10}B nucleus at the energy of 1.74015 MeV in nonorthogonal Gaussian basis [90]. The normalization coefficient of the wave function in the range 0-25 fm equals N=9.999999999999970E-001

i	β_i	C_i
1	6.669876139241313E-002	-2.347210794847986E-002
2	2.667656102033708E-002	-1.775040363036249E-003
3	1.517176918481825E-001	-1.283117981223353E-001
4	3.212149403864399E-001	-4.601647158129205E-001
5	9.260148198737874E-001	4.396116518097601
6	1.968143319382518	-7.091171845894630
7	2.891825315028276	-6.237051439471658E-001
8	6.205147839342107	6.217503950196968E-002
9	20.141061492467640	2.305215376077275E-002
10	86.640072856521640	1.321899076244325E-002

As we repeatedly noted the variational energy decreases as the dimension of the basis increases and gives the upper limit of the true bound energy, but the finite-difference energy increases as the size of steps decreases and the number of steps increases [90]. Therefore, it is possible to use the average value of -6.585898(2) MeV for the real bound energy in this potential. Meanwhile, the calculation error of finding the bound energy of the ^{10}B nucleus in the p^{9}Be cluster channel using two different methods is about ±2 eV.

In the frames of VM we obtained the value of energy equal to -4.845692 MeV, charge radius - 2.61 fm and AC equal to 2.48(2) in the range 5-12 fm for the potential of the second excited state. The expansion parameters of the WF in nonorthogonal Gaussian basis are listed in Table 20. The average energy value -4.845696(4) MeV is obtained for this level using two different methods and two different computer programs and residuals are of the order of 10^{-13}.

8.3. Astrophysical *S*-factor

While considering electromagnetic transitions we will take into account the *E*1 process from the resonance $^3S_1$-wave of scattering to three bound states of the cluster p^{9}Be channel of the ^{10}B nucleus with $J^PT=1^+0$, 0^+1 and 1^+0 [86] denoting it as *E*1(BS), as well as the *E*1 transition from the $^5S_2$-wave of scattering with zero phase shift to the ground bound $^5P_3$-state of this nucleus denoting it as *E*1(GS).

The calculation results and experimental data for the *S*-factor at the energy range 50-600 keV (l.s.) from work [95] are shown in Figure 23 by the solid line. Obviously, the value of the total calculated *S*-factor at the energy range 50-100 keV remains almost constant and equals 1.15(2) keV b, which is in quite a good agreement with the data of work [95], where the average of the first three experimental points at the energy 70-100 keV equals 1.27(4) keV b.

The transition to the ground $^5P_3$-state of the ^{10}B nucleus from the $^5S_2$-wave of scattering leads to the value of 0.81 keV b for the calculated *S*-factor at the energy 50 keV (dashed line in Figure 23). Line extrapolation of the received result to zero energy gives 0.90-0.95 keV b. The sum of transitions from the $^3S_1$-wave of scattering to three bound ^3P levels is shown in Figure 23 by the dotted line.

For comparison we will give some extrapolation results to zero energy for different experimental data. For instance, for the *S*-factor with transition to the GS the value 0.92 keV b was obtained in work [92], which is in complete agreement with the value received here. However, the values 1.4 keV b, 1.4 keV b and 0.47 keV b are given for the transitions to the three levels considered above with $J^PT=1^+0$, 0^+1 and 1^+0 respectively [92] and their sum evidently exceeds our result and data from work [95].

Further, the value 0.96(2) keV b for the total *S*-factor was received in the later work [9] and the value from 0.96(6) to 1.00(6) keV b was found in one of the latest works [100] devoted to this reaction. Both of these values are in good agreement with the values obtained above.

Thus, the *E*1 transitions from $^5S_2$ and $^3S_1$-waves of scattering to the ground bound $^5P_3$ state in the p^{9}Be channel of the ^{10}B nucleus and its three excited states 1^+0, 0^+1 and 1^+0 also bound in this channel are considered in the potential cluster model.

Having made certain assumptions of common character about interactions in the p^{9}Be channel of the ^{10}B nucleus it is possible to describe well the existing experimental data of the astrophysical *S*-factor at the energy range up to 600 keV and to receive its value for zero (50 keV) energy which is in complete agreement with the latest experimental data [95]. The *S*(0)-factor for the transition to GS of the ^{10}B nucleus is obtained quite correctly, too.

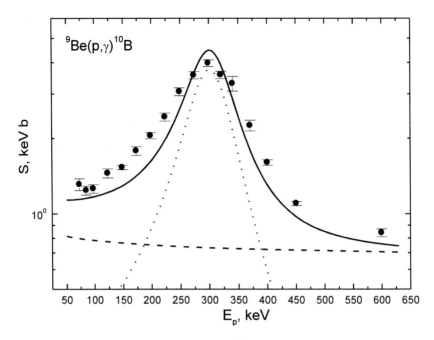

Figure 23. Astrophysical S-factor of the reaction of p^{9}Be radiative capture. Dots: experimental data from work [95]. Lines: calculation results for different electromagnetic transitions with the potentials mentioned in the text.

However, the scattering potentials are built on the basis of some qualitative conceptions because of the absence of the data for the phase shift analysis of p^{9}Be elastic scattering and three BS potentials are obtained only approximately because of the absence of data on the radii and AC of the ^{10}B nucleus in these excited states. Therefore these results should be considered only as a preliminary estimate of the possibility to describe the astrophysical S-factor of the reaction under consideration on the basis of the PCM with FS. But it is possible to obtain quite acceptable results for the astrophysical S-factor in spite of approximate consideration of the p^{9}Be → ^{10}Bγ radiative capture process.

In conclusion, it should be mentioned that the available S-factor experimental data for this reaction differ significantly from each other and it seems that more accurate study of the p^{9}Be → ^{10}Bγ radiative capture at astrophysical energy is required.

CONCLUSION

The description of behavior of the S-factors in all considered systems at low energies may be viewed as a certain evidence in favor of the potential approach in cluster model. The inter-cluster interactions including FS are constructed on the basis of the phase shifts of the cluster elastic scattering, and each partial wave is described by its potential, for example of the Gaussian form, with certain parameters.

The splitting of the general interaction into the partial waves allows detailing its structure and the classification of the orbital states according to Young's schemes allows identifying the presence and the number of the forbidden states. It gives the possibility to find the number of nodes of the WF of cluster relative motion and leads to a definite depth of the interaction allowing to avoid the discrete ambiguity of the potential depth as it is the case in the optical model.

The form of each partial phase shift of scattering can be correctly described only at a certain width of such a potential which saves us from the continuous ambiguity also characteristic of the well-known optical model. As a result, all the parameters of such a potential are fixed quite uniquely, and the "pure" according to Young's schemes interaction component allows describing the basic characteristics of the bound state of the lightest clusters correctly, which is realized in the light atomic nuclei with a high probability.

However, all the above-mentioned is correct provided that the phase shifts of scattering are obtained correctly from the experimental data of the elastic scattering. Up to present, for the majority of the lightest nuclear systems the phase shifts of scattering have been received with rather big errors, sometimes reaching 20-30%. This makes the construction of the exact potentials of the inter-cluster interaction very difficult and, finally, leads to significant ambiguities in the final results obtained in the potential cluster model.

In this connection it is very urgent to raise the accuracy of experimental measurements of elastic scattering of light nuclei at astrophysical energies and to perform a more accurate phase shift analysis. The increase in the accuracy will allow making more definite conclusions regarding the mechanisms and conditions of thermonuclear reactions, as well as understanding better their nature in general [102].

REFERENCES

[1] Barnes, C. A.; Clayton, D. D.; Schramm, D. N.; *Essays in Nuclear Astrophysics Presented to William A. Fowler*; Publisher: Cambridge University Press, Cambridge, UK, 1982; pp 562.

[2] Kukulin, V. I.; Neudatchin, V. G.; Smirnov, Yu. F. *Fiz. Elem. Chastits At. Yadra* 1979, vol. 10, 1236-1255; *Sov. J. Part. Nucl.* 1979, vol. 10, 1006.

[3] Baktybayev, M. K. et al. *The Fourth Eurasian Conference "Nuclear Science and its Application"*, October 31-November 3, Baku, Azerbaijan; 2006, p. 62; Burtebayev, N. et al. The new experimental data on the elastic scattering of protons from ^{6}Li, ^{7}Li, ^{16}O and ^{27}Al nuclei; *Book of Abstracts the Fifth Eurasian Conference on "Nuclear Science and its Application"*, October 14-17, Ankara, Turkey, 2008; p 40; Dubovichenko, S. B.; Burtebayev, N.; Zazulin, D. M.; Kerimkulov, Zh. K.; Amar, A. C. A. *Phys. Atom. Nucl.* 2011, vol. 74, 984-1000.

[4] Zazulin, D. M. et al. Scattering of protons from ^{12}C; *The Sixth International conference "Modern Problems of Nuclear Physics"* September 19-22, Tashkent, Uzbekistan, 2006; p 127; Baktybaev, M. K. et al. Elastic scattering of protons from ^{12}C, ^{16}O and ^{27}Al; *The 4th Eurasia Conf., "Nuclear Science and its Application"*, Baku, Azerbaijan, 2006; p 56.

[5] Neudatchin, V. G. et al. *Phys. Rev.* 1992, vol. C 45, 1512-1527.

[6] Dubovichenko, S. B. *Phys. Atom. Nucl.* 1995, vol. 58, 1174-1180.

[7] Fowler, W. A.; Caughlan, G. R.; Zimmerman, B. A. *Annu. Rev. Astron. Astrophys.* 1975, vol. 13, 69-112.

[8] Mohr, P. J.; Taylor, B. N. *Rev. Mod. Phys.* 2005, vol. 77(1), 1-107.
[9] Angulo, C. et al. *Nucl. Phys.* 1999, vol. A 656, 3-183.
[10] Dubovichenko, S. B.; Dzhazairov-Kakhramanov, A. V. *Phys. Part. Nucl.* 1997, vol. 28, 615-641.
[11] Varshalovich, D. A.; Moskalev, A. N.; Khersonskii, V. K. *Quantum Theory of Angular Momentum*; Publisher: World Scientific, Singapore, 1989.
[12] http://physics.nist.gov/cgi-bin/cuu/Value?mud|search_for=atomnuc!
[13] Plattner, G. R.; Viollier, R. D. *Nucl. Phys.* 1981, vol. A 365, 8-12.
[14] Dubovichenko, S. B. *Phys. Atom. Nucl.* 2008, vol. 71, 66-75.
[15] Fowler, W. A. *Experimental and Theoretical Nuclear Astrophysics: the Quest for the Original of the Elements*; Nobel Lecture, Stockholm, 8 Dec. 1983.
[16] Snover, K. A. *Solar p-p chain and the $^7Be(p,\gamma)^8B$ S-factor*; Publisher: University of Washington, CEPRA, NDM03, 1/6/2008.
[17] Dubovichenko, S. B.; Dzhazairov-Kakhramanov, A. V. *Euro. Phys. Jour.* 2009, vol. A 39, 139-143.
[18] Schmelzbach, P. et al. *Nucl. Phys.* 1972, vol. A 197, 273-289; Arvieux, J. *Nucl. Phys.* 1967, vol. A 102, 513-528; Chauvin, J.; Arvieux, J. *Nucl. Phys.* 1975, vol. A 247, 347-358; Huttel, E. et al. *Nucl. Phys.* 1983, vol. A 406, 443-455.
[19] Dubovichenko, S. B.; Dzhazairov-Kakhramanov, A. V. *Yad. Fiz.* 1990, vol. 51, 1541-1550; *Sov. J. Nucl. Phys.* 1990, vol. 51, 971.
[20] Griffiths, G. M.; Larson, E. A.; Robertson, L. P. *Can. J. Phys.* 1962, vol. 40, 402-411.
[21] Ma, L. et al. *Phys. Rev.* 1997, vol. C 55, 588-596.
[22] Schimd, G. J. et al. *Phys. Rev.* 1997, vol. C 56, 2565-2681.
[23] LUNA Collaboration; Casella, C.; Costantini, H. et al. *Nucl. Phys.* 2002, vol. A 706, 203-216.
[24] Tilley, D. R.; Weller, H. R.; Hasan, H. H. *Nucl. Phys.* 1987, vol. A 474, 1-60.
[25] Tilley, D. R.; Weller, H. R.; Hale, G. M. *Nucl. Phys.* 1992, vol. A 541, 1-157.
[26] Blokhintsev, L. D.; Borbely, I.; Dolinskii, E. I. *Fiz. Elementar. Chastits Atom. Yadra* 1977, vol. 8, 1189-1245; *Sov. J. Part. Nucl.* 1977, vol. 8, 485.
[27] Bornard, M. et al. *Nucl. Phys.* 1978, vol. A 294, 492-512.
[28] Platner, G. R.; Bornard, M.; Viollier, R. D. *Phys. Rev. Lett.* 1977, vol. 39, 127-130.
[29] Kievsky, A. et al. *Phys. Lett.* 1997, vol. B 406, 292-296.
[30] Ayer, Z. et al. *Phys. Rev.* 1995, vol. C 52, 2851-2858.
[31] Schimd, G. J. et al., *Phys. Rev. Lett.* 1996, vol. 76, 3088-3091.
[32] Schimd, G. J. et al. *Phys. Rev.* 1995, vol. 52, R1732-R1733.
[33] Viviani, M.; Schiavilla, R.; Kievsky, A. *Phys. Rev.* 1996, vol. C 54, 534-553.
[34] Warren, J. B. et al. *Phys. Rev.* 1963, vol. 132, 1691-1692.
[35] Berman, B. L.; Koester, L. J.; Smith, J. H. *Phys. Rev.* 1964, vol. 133, B117-B129.
[36] Fetisov, V. N.; Gorbunov, A. N.; Varfolomeev, A. T. *Nucl. Phys.* 1965, vol. 71, 305-342.
[37] Ticcioni, G. et al. *Phys. Lett.* 1973, vol. B 46, 369-371.
[38] Geller, K. N.; Muirhead, E. G.; Cohen, L. D. *Nucl. Phys.* 1967, vol. A 96, 397-400.
[39] Dubovichenko, S. B. *Russian Physics Journal* 2008, vol. 51, 1136-1143.
[40] Berg H., et al. *Nucl. Phys.* 1980, vol. A 334, 21-34; Kavanagh, R. W.; Parker, P. D. *Phys. Rev.* 1966, vol. 143, 779-782; Morrow, L.; Haeberli, W. *Nucl. Phys.* 1969, vol. A 126, 225-232.

[41] Dubovichenko, S. B.; Neudatchin, V. G.; Sakharuk, A. A.; Smirnov, Yu. F. *Izv. Akad. Nauk SSSR Ser. Fiz.* 1990, vol. 54, 911-916; Neudatchin, V. G.; Sakharuk, A. A.; Dubovichenko, S. B. *Few Body Systems* 1995, vol. 18, 159-172.
[42] Dubovichenko, S. B. *Phys. Atom. Nucl.* 1995, vol. 58, 1295-1302.
[43] Tombrello, T. A. *Phys. Rev.* 1965, vol. 138, B40-B47.
[44] Yoshino, Y. et al. *Prog. Theor. Phys.* 2000, vol. 103, 107-125.
[45] McSherry, D. H.; Baker, S. D. *Phys. Rev.* 1970, vol. C 1, 888-892.
[46] Drigo, L.; Pisent, G. *Nuovo Cimento* 1967, vol. BLI, 419-436.
[47] Szaloky, G.; Seiler, F. *Nucl. Phys.* 1978, vol. A 303, 57-66.
[48] Tombrello, T. A. et al. *Nucl. Phys.* 1962, vol. 39, 541-550.
[49] McIntosh, J. S.; Gluckstern, R. L.; Sack, S. *Phys. Rev.* 1952, vol. 88, 752-759.
[50] Frank, R. M.; Gammel, J. L. *Phys. Rev.* 1955, vol. 99, 1406-1410.
[51] Kankowsky, R. et al. *Nucl. Phys.* 1976, vol. A 263, 29-46.
[52] Arkatov, Yu. M. et al. *Yad. Fiz.* 1970, vol. 12, 227-233; *Sov. J. Nucl. Phys.* 1971, vol. 12, 123.
[53] Hahn, K. et al. *Phys. Rev.* 1995, vol. C 51, 1624-1632.
[54] Canon, R. et al. *Phys. Rev.* 2002, vol. C 65, 044008-044014.
[55] Lim, T. K. *Phys. Lett.* 1975, vol. B 55, 252-254; Lim, T. K. *Phys. Lett.* 1973, vol. B 44, 341-342.
[56] Gibson, B. F. *Nucl. Phys.* 1981, vol. A 353, 85-98.
[57] Perry, J. E.; Bame, S. J. *Phys. Rev.* 1955, vol. 99, 1368-1375.
[58] Balestra, F. et al. *Nuovo Cimento* 1977, vol. 38A, 145-166.
[59] Meyerhof, W. et al. *Nucl. Phys.* 1970, vol. A 148, 211-224.
[60] Feldman, G. et al. *Phys. Rev.* 1990, vol. C 42, R1167-R1170.
[61] Skill, M. et al. *Nucl. Phys.* 1995, vol. A 581, 93-106.
[62] Ishkhanov, B. S.; Kapitonov, I. M.; Tutyn', I. A. *Nucleosynthesis in the Universe*; Publisher: MSU, Moscow, RU, 1998, (in Russian); Peacock, J. A. Cosmological Physics; Publisher: Cambridge University Press, Cambridge, UK, 1999.
[63] Hodgson, P. E. *The Optical Model of Elastic Scattering*; Publisher: Clarendon Press, Oxford, UK, 1963.
[64] Petitjean, C.; Brown, L.; Seyler, R. *Nucl. Phys.* 1969, vol. A 129, 209-219.
[65] Dubovichenko, S. B.; Dzhazairov-Kakhramanov, A. V.; Sakharuk, A. A. *Yad. Fiz.* 1993, vol. 56, 90-106; *Phys. At. Nucl.* 1993, vol. 56, 1044.
[66] Neudatchin, V. G.; Sakharuk, A. A.; Smirnov, Yu. F. *Fiz. Elem. Chastits At. Yadra* 1992, vol. 23, 479-541; *Sov. J. Part. Nucl.* 1992, vol. 23, 210; Neudatchin, V. G.; Struzhko, B. G.; Lebedev, V. M. *Physics of Particles and Nuclei* 2005, vol. 36, 468-497.
[67] Tilley, D. R. et al. *Nucl. Phys.* 2002, vol. A 708, 3-163.
[68] Switkowski, Z. E. et al. *Nucl. Phys.* 1979, vol. A 331, 50-60; Bruss, R. et al. Astrophysical S-Factors for the Radiative Capture Reaction ^{6}Li(p,γ)^{7}Be at Low Energies; *Proc. 2-nd Intern. Symposium on Nuclear Astrophysics, Nuclei in the Cosmos*, Karlsruhe, Germany, 6-10 July, 1992, in: F. Kappeler, K. Wisshak (Eds.); Publisher: IOP Publishing Ltd., Bristol, UK, 1992; vol. 1093, p 169.
[69] Arai, K.; Baye, D. *Nucl. Phys.* 2002, vol. A 699, 963-975.
[70] Prior, R. M. et al. *Phys. Rev.* 2004, vol. C 70, 055801-055808.
[71] Burker, F. C. *Austr. J. Phys.* 1980, vol. 33, 159-176.
[72] Cecil, F. E. et al. *Nucl. Phys.* 1992, vol. A 539, 75-96.

[73] Dubovichenko, S. B. *Russian Physics Journal* 2009, vol. 52, 294-300.
[74] Jackson H. L. et al. *Phys. Rev.* 1953, vol. 89, 365-369.
[75] Jackson H. L. et al. *Phys. Rev.* 1953, vol. 89, 370-374.
[76] Moss, S. J.; Haeberli, W. *Nucl. Phys.* 1965, vol. 72, 417-435.
[77] Barnard, A. C. L. et al. *Nucl. Phys.* 1966, vol. 86, 130-144.
[78] Neudatchin, V. G.; Smirnov, Yu. F. *Nucleon associations in light nuclei*; Publisher: Nauka, Moscow, RU, 1969; pp 1-414 (in Russian); Kukulin, V. I.; Neudatchin, V. G; Obukhovsky, I. T.; Smirnov, Yu. F. *Clusters as subsystems in light nuclei*; in: K. Wildermuth, P. Kramer (Eds.); *Clustering Phenomena in Nuclei*; Publisher: Vieweg, Braunschweig, 1983; vol. 3, p. 1.
[79] Ajzenberg-Selove, F. *Nucl. Phys.* 1991, vol. A 523, 1-101.
[80] Ajzenberg-Selove, F. *Nucl. Phys.* 1990, vol. A 506, 1-186.
[81] Burtebaev, N.; Igamov, S. B.; Peterson, R. J.; Yarmukhamedov, R.; Zazulin, D. M. *Phys. Rev.* 2008, vol. C 78, 035802-1-035802-11.
[82] Dubovichenko, S. B.; Dzhazairov-Kakhramanov, A. V. *Russian Physics Journal* 2009, vol. 52, 833-840.
[83] Dubovichenko, S. B.; Zazulin, D. M. *Russian Physics Journal*, 2010, vol. 53, 458-464.
[84] Dubovichenko, S. B. *Phys. Atom. Nucl.* 2010, vol. 73, 1573-1584.
[85] Dubovichenko, S. B. *Russian Physics Journal* 2010, vol.53, 1254-1263.
[86] Tilley, D. R. et al. *Nucl. Phys.* 2004, vol. A 745, 155-363.
[87] Warters, W. D.; Fowler, W. A.; Lauritsen C. C. *Phys. Rev.* 1953, vol. 91, 917-921.
[88] Brown, L. et al. *Nucl. Phys.* 1973, vol. A 206, 353-373.
[89] Tilley, D. R. et al. *Nucl. Phys.* 2002, vol. A 708, 3-163.
[90] Dubovichenko, S. B. *Calculation methods of the nuclear characteristics*; Publisher: Complex, Almaty, Kazakhstan, 2006; pp 311; http://xxx.lanl.gov/abs/1006.4947 (Russian).
[91] Zahnow, D. et al. *Z. Phys.* 1995, vol. A 351, 229-236.
[92] Cecil, F. E. et al. *Nucl. Phys.* 1992, vol. A 539, 75-96.
[93] Godwin, M. A. et al. *Phys. Rev.* 1997, vol. C 56, 1605-1612.
[94] Spraker, M. et al. *Phys. Rev.* 1999, vol. C 61, 015802-015808.
[95] Zahnow, D. et al. *Nucl. Phys.* 1996, vol. A 589, 95.
[96] Itzykson, C.; Nauenberg, M. *Rev. Mod. Phys.* 1966, vol. 38, 95-101.
[97] Dubovichenko, S. B. et al. *Russian Physics Journal* 2010, vol.53, 743-749.
[98] Ajzenberg-Selove, F. *Nucl. Phys.* 1988, vol. A 490, 1-225.
[99] Wulf, E. A. et al. *Phys. Rev.* 1998, vol. C 58, 517-523.
[100] Sattarov, A. et al. *Phys. Rev.* 1999, vol. C 60, 035801-035808.
[101] Mukhamedzhanov, A. M. et al. *Phys. Rev.* 1999, vol. C 56, 1302-1312.
[102] Dubovichenko, S. B. *Property of light atomic nuclei in potential cluster models*; Publisher: Daneker, Almaty, Kazakhstan, 2004; pp 247; http://xxx.lanl.gov/abs/1006.4944 (Russian).

In: The Big Bang: Theory, Assumptions and Problems
Editors: J. R. O'Connell and A. L. Hale

ISBN: 978-1-61324-577-4
© 2012 Nova Science Publishers, Inc.

Chapter 2

THE BIG BANG AND WHAT IT WAS

E. E. Escultura[1]

GVP – Professor V. Lakshmikantham Institute for Advanced Studies
and Departments of Mathematics and Physics,
GVP College of Engineering, JNT University
Madurawada, Visakhapatnam 540041, AP, India

ABSTRACT

This chapter takes as its premise the occurrence of the Big Bang, a colossal explosion that gave rise to the birth of our universe. Then through sequences of mathematical and scientific reasoning called rational thought using the methodology of qualitative modelling the grand unified theory (GUT) is developed on laws of nature which establishes what it was that exploded. This is a new approach. In earlier development both the occurrence of the Big Bang and what caused it were taken as premises.

In its present development GUT builds its pillars – quantum and macro gravity and thermodynamics. Through the discovery of appropriate laws of nature, GUT well defines its fundamental physical concepts such as the superstring, basic constituent of matter; matter and energy; dark matter; and charge and gravity. It also well-defines and explains other physical concepts: primum, photon, matter-anti-matter interaction, atom, cosmic waves, brittle and malleable materials, metal fatigue, turbulence, galaxy, black hole and supernova along with the conversion of dark to visible matter.

The cosmology of our universe is traced from its birth at Cosmic Burst 1.5 billion years from the start of the Big Bang through its evolution to the present as a super…super galaxy 1010 billion light years across and a local bubble in the timeless boundless Universe and predicts its destiny as black holes back in dark matter. The existence of universes other than ours is established from available evidences. Then it is deduced from this cosmology what the Big Bang was and what caused it.

---

[1] http://users.tpg.com.au/pidro; http://edgareescultura.wordpress.com.

1. Introduction

Scientists agree that the Big Bang was a colossal explosion, release of staggering amount of energy 8 billion years ago [19,24]. Moreover, all accounts about our universe (nature) trace its origin to the Big Bang as if its emergence was a spectacular violation of the first law of thermodynamics with neither rhyme nor reason. We take a new approach by assuming our universe as part of a grand order and the Big Bang a natural occurrence in it. Our task is to discover that order by working through sequences of mathematical and scientific reasoning called rational thought and analysis subject to the highest standards of precision that the language of science – mathematics – can offer until we determine beyond reasonable doubt what it was that exploded. To give direction to our search for that grand order we need to put at its core the discovery of the laws of nature to shed light on how it works. This is a new approach; in previous development of that grand order by the author, the Big Bang and what it was were assumed and the birth of our universe was traced to the Big Bang. Here, its occurrence is still taken as a premise or axiom of macro gravity [8] but we proceed further: after having discovered or built the order of our universe based on or defined by its laws, we establish what it was that exploded as the Big Bang. We call the grand order grand unified theory (GUT) anchored on its three pillars – quantum gravity, thermodynamics and macro gravity.

2. Mathematical Requirements

Current mathematics is mainly quantitative or computational at the base of which are the real number system and its foundations whose extension includes the broad and rapidly expanding field of nonlinear analysis. The complement of quantitative mathematics is qualitative mathematics that includes abstract mathematical spaces and reasoning, axiomatic systems and the search for natural laws. We bring this complementary mathematics to bear on our task of developing GUT as a comprehensive physical theory anchored on the laws of nature that unify the forces and interactions of nature and the natural sciences as well.

Qualitative mathematics, the representation of rational thought, includes the following daily activity of the mathematician and scientist:

Making conclusions, visualizing, abstracting, thought experimenting, learning, doing creative activity, intuition, imagination, trial and error to sift out what is appropriate, negating what is known to gain insights into the unknown, altering premises to draw out new conclusions, thinking backwards, finding premises for a mathematical space and devising techniques that yield results.

Since qualitative mathematics broadens the sources of information and liberalizes admissibility of concepts and validity of conclusions, it also raises the chance of error creeping into mathematics and its applications. Therefore, we tighten the filter of admissibility by way of rectifying the ambiguity and inadequacy of the real number system and its foundations and raising the precision of mathematical-scientific reasoning. The inadequacy of both mathematics and physics is revealed by the failure to solve or resolve long standing problems of mathematics and physics such as the 360-year-old Fermat's conjecture (popularly known as Fermat's last theorem) [5, 6, 63] and the 200-year-old gravitational n-body and turbulence problems [17, 23] and resolve fundamental questions of physics such as what the basic constituent of matter and the structure of an elementary particle are [24]. The rectification of mathematics undertaken in [9] is built on the first major rectification of the foundations of mathematics by David Hilbert a century ago (item (1) below to which we add more:

(1) Hilbert was the first to recognize that since individual thought is inaccessible to others and, therefore, cannot be studied and analysed collectively with precision, the subject matter of mathematics can only be representation of thought by objects in the real world called concepts subject to consistent premises or axioms. Such representation is called a mathematical space.
(2) Every concept of a mathematical space must be well-defined (distinguished from dictionary or encyclopaedic definition), i.e., its existence, behaviour or properties and relation to other concepts must be specified by the axioms. Undefined concepts or symbols are inadmissible as they introduce ambiguity in a mathematical space; therefore, the choice of the axioms is not complete until every concept is well defined.
(3) The rules of inference must follow from the axioms and external or universal rules of inference like formal logic are not valid because they have nothing to do with the axioms. In other words, the axioms are the foundations of mathematical-scientific reasoning or rational thought.
(4) Among the sources of ambiguity and errors are: large and small numbers, vacuous and ill-defined concepts and infinity [9,38]. This has some bearing in physics that deals with the very large and the very small such as the radius of our universe and the Planck's constant h [24].

These are the mathematical requirements by which we build scientific knowledge. A physical theory is nothing more than a mathematical space built on laws of nature as its axioms; it is the form by which we express, store and apply scientific knowledge. This alters the task of the scientist from computation and measurement such as solving mathematical equations to the discovery of laws of nature upon which to build physical theory.

3. THE NEW METHODOLOGY

Methodology is part of rational thought and, therefore, mathematics. The present methodology of science, particularly, physics, is mathematical modelling (now called quantitative modelling) that describes the appearances of nature mathematically. For instance, mathematical physics today is a collection of mathematical principles, equations and inequality, which are rather disparate, that describe the appearances and motion of physical systems or natural phenomena and where reasoning is mainly by analogy based on such description. This is the reason mathematical physics failed to solve and resolve its long standing problems and fundamental questions [8].

The remedy is qualitative modelling (formerly called dynamic modelling) that explains not only the appearances of nature but also its forces, interactions and behaviour in terms of natural laws, introduced and the main contribution in [28] and applied to physics for the first time to solve the gravitational n-body problem [23]. Naturally, its main tool is qualitative mathematics. However, qualitative and quantitative mathematics and modelling are complementary and indispensable to each other. The quantitative mathematics that play a major role in GUT are found in [5,6,7,9,10,11,12,13,14,23].

4. OUR STRATEGY

The present methodology of quantitative modelling does not well define fundamental physical concepts such as matter and energy, the atom, galaxy, black hole and Big Bang. It only describes their appearances. For example, black hole is defined as physical singularity which is ambiguous and non-operative in the sense that the concept is not amenable to computation and even mathematical reasoning. Therefore, our strategy is to discover appropriate laws of nature upon which to build a physical theory that well defines physical concepts and systems and explains their nature, especially, their forces, interactions and appearances. In particular, the theory must explain the Big Bang, where it came from, why it occurred and what caused it. That will be the ultimate goal of this article.

However, there is no short cut to it and we need to develop quantum and macro gravity, thermodynamics and even cosmology in the chain of scientific-mathematical reasoning and theoretical development leading to knowledge of what the Big Bang was in the course of which we find that order of our universe in GUT, the same approach taken in the solution of the gravitational n-body problem [23] where a scientific problem serves as catalyst for the development of physical theory that provides its solution. In the present case, the problem is to find out what the Big Bang was.

5. QUANTUM GRAVITY

We first make a tentative definition of energy as motion of matter where amount of matter is measured by its mass in suitable unit. Obviously, this definition is partial until we know what matter consists of which requires the discovery of its basic constituent. However, this definition insures that matter and energy are never separate and there is no such thing as pure mass or pure energy. Then it follows that any form of energy has mass including the photon and neutrino. Thus, the well defining of a physical concept which resolves its ambiguity yields new information. In current physics the photon is massless and the mass of the neutrino is unknown; the assumption that the photon has no mass certainly contradicts energy conservation [24].

5.1. The Superstring

First in our agenda is the search for the basic constituent of matter, the superstring, which, for now is only a name. However, we give it *substance* and embellish it with structure, properties and behaviour by first establishing its existence and well defining them in terms of the laws of nature.

We start with the first law of thermodynamics that says, energy cannot be created or destroyed. At once we detect a flaw in this formulation as it does not take latent or dark (non-observable) energy into account. For example, when gasoline is ignited, where does the burst of energy come from? The first law is valid here only if we assume that gasoline has latent energy that converts to kinetic energy by ignition.

To fix this law, consider this thought experiment:

Shoot a beam of light into a vacuum and turn it off. The beam vanishes without trace since there are no gas molecules that absorb the energy of the beam and turn it into kinetic energy, e.g., higher temperature, along the beam. We interpret the disappearance of light in this case as conversion to latent or dark energy. Therefore, we enrich the first law of thermodynamics into what we take as the most fundamental law of nature [24]:

Energy Conservation. *In any physical system and its interaction, the sum of kinetic and latent energy is constant, gain of energy is maximal and loss of energy is minimal.*

This is the starting point in our search for the superstring and in the event that we encounter a physical system or natural phenomenon that seems to violate this law we find another natural law that reconciles them. The existence of such natural law follows from the grand order. For example, observation by the Hubble reveals that matter forms steadily in the Cosmos, first as cosmic dust that gets entangled into cosmological vortices and collect as stars at their cores at the rate of one star per minute [1,49,54]. The only remedy for this apparent contradiction with energy conservation is the following natural law [24]:

Existence of two fundamental states of matter. *There exist two fundamental states of matter, dark and visible or ordinary; the former is indirectly observable but known only by its impact on visible matter.*

With this law we now ask this non-vacuous question: what does dark matter consist of? The answer: superstrings. For now the superstring is just a name but we embellish it with properties and behaviour and how it converts to visible matter through the laws of nature. From the perspective of materialist science and rational thought this law is quite crucial since searching for something that does not exist is a contradiction known as Perron Paradox [62, 63].

We next state a natural law that is quite important in deriving details of the structure, behaviour and properties of a physical system as well as its interaction with other physical systems [12, 24, 25]. A physical system is any motion or configuration of matter. Thus, a wave or vibration is a physical system [8, 26, 27].

Energy Conservation Equivalence. *Energy conservation has many expressions or forms: order, symmetry, periodicity, economy, least action, optimality, efficiency, stability, self-similarity (nested fractal), coherence, resonance, quantization, synchronization, smoothness, uniformity, motion-symmetry balance, non-redundancy, non-extravagance, evolution to infinitesimal configuration, helical and related configuration, circular, helical, spiral and sinusoidal and, in biology, genetic encoding of characteristics, reproduction, and specialization and order in diversity and complexity of functions, configuration and capability.*

Each component of this law is called physical principle. The principles of non-redundancy and non-extravagance established the *gluon* in 2004 which turned out to be the −quark discovered in Fermilab near Chicago several decades earlier.

We introduce more physical concepts. *Flux* is energy, i.e., motion of matter. Flux with identifiable direction of motion at each point is called *turbulence*; *chaos* is mixture of order none of which is identifiable [17].

We next state a natural law central to quantum and macro gravity and earthly turbulence, e.g., hurricane and tornado [17,22].

Flux-Low-Pressure Complementarity). *Low pressure sucks matter and the initial chaotic rush of dark matter towards a region of low pressure stabilizes into local or global coherent flux; conversely, coherent flux induces low pressure around it.*

A typical chaos is the transitional phase of the standard dynamics [11, 17, 29] such as typhoon. It starts as a typical calm summer day in the Pacific (phase 1 of turbulence). Then suitable broad contiguous packets of warm ocean surface called *el niño* emerges due to under-ocean volcanic activity warming up the lower atmosphere and causing low pressure or depression. By flux-low-pressure complementarity, the depression sucks trillions of gas molecules towards it their collisions pushing the system into phase 2 or the transitional phase of chaos due to the uncertainty of large number [9, 11, 17]. Concretely, the position and direction of motion of every molecule cannot be identified or monitored due to the immensity of the molecules involved (due uncertainty or ambiguity of large numbers) and yet it is subject to natural laws. However, this phase is energy dissipating due to collisions. Therefore, by energy conservation, it evolves to Phase 3, the phase of turbulence, a vortex flux called tropical cyclone or typhoon, the final phase of this standard dynamics.

Now we recount the discovery of the superstring and determination of its structure, properties, behaviour and conversion to visible matter [12, 24]. First, we limit our choices to pin it down precisely and well direct our search. Since the superstring is matter it must be indestructible; otherwise, our universe would have collapsed a long time ago. Moreover, as basic constituent, it must be unique since every piece of matter must be reducible to it. Like the electron it takes different forms convertible to each other and replicated everywhere in the Cosmos.

The next pivotal phase in our search is finding an indestructible structure or configuration in the real world. Consider an egg shell that contains an egg shell that contains an egg shell,..., etc., ad infinitum. This configuration is called nested generalized fractal sequence [8, 31]. If we hit it with a hammer, can we destroy its fractal sequence structure? The answer is *no*; since the hammer must first destroy the first term then the second, etc., but at most only a finite number of them in the sequence. Therefore, the tail end of the sequence remains intact and its nested fractal structure survives. This is the structure the superstring has that makes it indestructible.

We introduce an important physical concept: *wave*. Like "motion" of neon light due to synchronized switching, a wave is propagated by suitably synchronized vibration of the medium and its energy is imparted by the generator augmented by the vibration of the medium. Wave propagates through the medium via vibration resonance [8]. Thus, wave does not travel and its appearance of propagation is an illusion. This is true of water and air waves where the mediums are water and gas molecules, respectively, vibrating in fix places. In the Cosmos, there are two kinds of waves across dark matter: (a) basic cosmic or electromagnetic waves generated by the normal vibration of atomic nuclei and (b) seismic waves generated by the micro component of turbulence at its interface [17]. The medium for both is dark matter. Basic cosmic waves are propagated from and travel in all directions across the Cosmos while seismic waves come from interface of turbulence such as at conservative tectonic plate boundaries and compressed volcanic lava flow [17] and at the core of cosmological vortices. [17,39]. The impact of basic cosmic waves account for the normal vibration of physical systems their characteristics determined in accordance with the following natural law [24].

Internal-External Factor Dichotomy. *The interactions and dynamics of a physical system are shaped by the internal and external factors; in general the internal is principal over the external and the latter works through the former.*

(Incidentally, this law applies to social science where, in a country, the internal factor is the socio-political-economic-cultural milieu and the external factor the complex of international relations; this is the reason it is quite difficult for one country to impose its rule on another)

We now derive the nested fractal configuration of the superstring from the laws of nature. The only force that interacts with or has impact on the superstring is cosmic wave.

Consider a *non-agitated* superstring (defined later). When hit by suitable basic cosmic wave this scenario may occur: (a) it is thrust into collision with other non-agitated superstrings bouncing with them until the energy imparted on it is exhausted and the superstring grinds to a halt as non-agitated superstring or (b) if it gets near its previous path, it gets sucked by the latter, by flux-low-pressure complementarity, forming a loop with the former travelling through it at 7×10^{22} cm/sec [2] called its toroidal *flux*. Energy conservation and energy conservation equivalence converts its path into a uniform circular helical loop, a semi-agitated superstring, like a lady's spring bracelet [16] (Figure 5.1.1). Another scenario is: (b) the first term of the nested fractal non-agitated superstring expands into a semi-agitated superstring (a superstring is equivalently referred to as the first term of its nested fractal sequence). Thus, when hit by suitable basic cosmic wave a non-agitated superstring may become semi-agitated or remain non-agitated as toroidal flux of a new semi-agitated superstring.

We now have the full structure of the semi-or non-agitated superstring derived from the laws of nature which we also state as a law of nature [8, 24].

Existence of Basic Constituent of Dark Matter and its Generalized Nested Fractal Structure. *The basic constituent of dark matter is the superstring. It is a helical loop and nested fractal sequence of superstrings or toroidal fluxes, with itself as first term; each toroidal flux in the sequence is a superstring having toroidal flux, a superstring, travelling at speed beyond that of light along its cycles, etc.; each superstring except the first, is contained in and self-similar to the preceding term in structure, behaviour and properties.*

Clearly, the superstring is a generalized nested fractal sequence of superstrings [8, 24].

When hit by suitable basic cosmic wave, the first term of a semi-agitated superstring either (a) breaks in which case its toroidal flux remains a non-agitated superstring or its segment expands or bulges into an agitated superstring called primum, unit of visible matter. By energy conservation equivalence, its profile is sinusoidal of even power [24] and its configuration is its full rotation about the base [Figure 5.1.2]). We state this result as a law of nature which governs dark to visible matter conversion.

Figure 5.1.1. An artist's conception of a semi-agitated superstring. A non-agitated superstring, its toroidal flux, has a similar nested fractal structure as itself and travels through its helical cycles at the speed of 7×10^{22} cm/sec. The helix winds around the torus rapidly its cycles infinitesimally close and at dark distance from its other [8]).

Dark-to-Visible-Matter Conversion. *When suitable shock wave hits a semi-agitated superstring one of these occurs: (a) the outer superstring breaks, its toroidal flux remaining non-agitated; (b) a segment bulges into a primum, unit of visible matter.*

We well define the phases of the cycle of a superstring. It is dark if its cycle length is less than 10^{-14} meters, non-agitated if CL $< 10^{-16}$ meters, semi-agitated if $10^{-16} <$ CL $< 10^{-14}$ meters and agitated or visible if CL $> 10^{-14}$ meters. De-agitated, i.e., left alone without agitation, a superstring shrinks steadily for it shortens the helical paths of the toroidal fluxes, by energy conservation, its tail end becoming infinitesimal physical continuum [12], while maintaining the toroidal flux speed of 7×10^{22} cm/sec as a constant of nature [2]. The boundaries between agitated, semi-agitated and non-agitated superstrings will be refined by experimental and mathematical physicists as constants of nature but will affect neither GUT's validity nor its explanation of how nature works. Since only the first term of the nested fractal superstring sequence has impact on visible matter we identify the superstring with it.

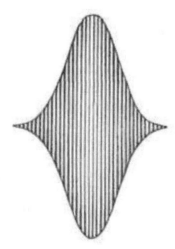

Figure 5.1.2. A simple primum, bulged segment of semi-agitated superstring and a magnet. Its polarity conforms to the right-hand-rule of electromagnetism: when the index finger points to the direction of its toroidal flux, the thumb points to the N-pole.

The next law complements flux-low-pressure complementarity, both central to quantum and macro gravity.

Flux Compatibility. *Two prima of opposite toroidal flux spins attract at their equators but repel at their poles; otherwise, they repel at their equators but attract at their poles. Two prima of the same toroidal flux spin are joined equatorially only through a primum of opposite toroidal flux spin called connector.*

5.2. The Primum

Computations on the primum and its interactions including the generalized fractal and integral as well as qualitative-quantitative modelling of the superstring belong to quantum algebra.

A qualitative-quantitative model of a simple primum in cylindrical coordinates is given by the equation $x = t$, $y(t) = \beta(\sin n\pi t)(\cos^m k\pi t)$, $\theta = n\pi t$, $t \in [-1/k, 1/k]$, n, m, k, integers, $n \gg k$, m even (n much larger than k) [12]. The first factor of $y(t)$ consists of rapid oscillations and since this is in cylindrical coordinates, $y(t)$ consists of rapid spirals about the x-axis. The cycle energy of the spiral is Planck's constant $h = 6.64 \times 10^{-34}$ Joules [4,24]. Scooped up and carried by cosmic wave, its cycles flatten to rapid oscillations, $z = 0$, $x = t$, $y(t) = \beta(\sin n\pi t)(\cos^m k\pi t)$ due to dark viscosity. It becomes a photon, $z = 0$, $y(t) = \beta(\sin n\pi t)(\cos^m k\pi t)$, when it breaks off from its loop. The energy of one full cycle of the primum or one full arc of the photon it converts to is h, its toroidal flux speed 7×10^{22} cm/sec [2]. The configuration of the primum and photon is based on the universality of oscillation and related motion of matter and the uniformity principle of the energy conservation equivalence. (Different prima convert to light (photons) of different colours)

It would seem that since the energy of the photon is known and the energy of its full arc as rapid oscillation or one cycle of the primum it comes from is the Planck's constant h [8], we can divide the former by h to find the number of oscillations the photon or the number of cycles the primum it comes from has. Unfortunately, it is not possible in view of the uncertainty of large and small numbers [9]. At best, we can only know its energy density using the generalized integral [7,32].

The extreme cycle of the bulge of a primum is called the *equator*. The primum and its vortex flux is an eddy in dark matter. While dark matter is inert, i.e., dark superstring does not interact with anything its induced flux being infinitesimal, this is not the case with the primum. Hit from all directions by cosmic waves its toroidal flux is thrust into erratic motion called *spike* in the neighbourhood of its helical cycles as it speeds through them at 7×10^{22} cm/sec pulling the superstrings around the primum into a vortex flux with its eye along the axis and turning it into a magnet with polarity in accordance with the right hand rule of electromagnetism, i.e., when the index finger points in the direction of the toroidal flux, the thumb points to the N- or north pole; otherwise, it points to the opposite pole, the S- or south pole. The vortex flux is its magnetic flux, its energy measured as charge. Thus, charge is the energy of the primal induced vortex flux. A primum is positive if its vortex flux spins counterclockwise viewed from its N-pole, negative otherwise. The electron is a basic negative primum, its charge -1 (1.6×10^{-19} coulombs [4]) the unit of charge of electromagnetism by convention. The +quark, another basic primum, has charge +2/3 while the −quark, the third basic primum, has charge –1/3 [37]. These prima are basic because they comprise every light isotope of an atom; they are produced at enormous quantity in the Cosmos and cellular membranes of living organisms [8,30,32,33,34,35,36]. A heavy isotope has neutrino.

The speed of the toroidal flux of a primum was deduced from the linear speed of the toroidal and induced vortex flux of the proton measured at 7×10^{22} cm/sec [2]. By the principle of synchronization of the energy conservation equivalence law, this linear speed must be true of all toroidal fluxes and their induced fluxes, i.e., it is a constant of nature, and applies to all toroidal and induced fluxes of superstrings including split off from them like electric current. Otherwise, there will be much dissipation of energy and violation of energy conservation. In electric current, the speed of the electron that rides on the toroidal flux through the conductor is diminished by the resistance of the conductor due to collision with its atoms and molecules. Electrical power generation works on this principle. When a close circuit electrical conductor cuts across a magnetic line of force its vortex flux splits from the vortex fluxes around the atomic nuclei and goes through the conductor (since it offers much less resistance than air) and becomes electric current.

A simple primum is charged so that a neutral primum must be coupled. Thus, by the principles of optimality, non-redundancy and non-extravagance the neutral neutrino is coupled pair of simple prima of opposite but equal charges, say, +q and –q so that its charge is +q + –q = 0, i.e., neutral.

Primal interaction is governed by flux-low-pressure complementarity and flux compatibility.

The proton consists of two positive quarks joined by a negative quark equatorially, by flux compatibility (Figure 5.2.1). By energy conservation, their axis are coplanar; its charge: 2/3 – 1/3 + 1/3 = +1. Thus, there is net coherent counterclockwise vortex flux around the proton viewed from its N-pole.

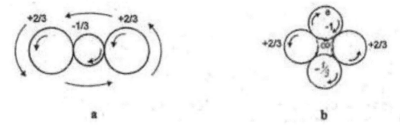

Figure 5.2.1. The proton: two +quarks joined by a –quark at their vortex fluxes' rims by flux compatibility (left). The neutron consisting of a proton, electron and neutrino (right); the neutrino is represented by a figure 8 since it is a coupled primum of opposite but numerically equal charge joined likewise at their fluxes' rims.

By flux compatibility the electron can attach itself to a positive quark of the proton at any point but energy conservation and optimality of the energy conservation equivalence attaches it to both +quarks beside the negative quark as the most stable position but pushes the negative quark a bit by flux compatibility so that their centres viewed from the N-pole form the vertices of a quadrilateral. In its interior are coherent vortex fluxes of the +quarks, –quark and electron that make it a region of low pressure or depression. By flux-low-pressure complementarity its interior sucks neutral primum around it since charged primum is repelled by primum of the same charge already in the coupling. Therefore, only suitably light neutral primum fits in and that is the neutrino. We have just composed the neutron consisting of a proton, electron and neutrino (Figure 5.2.1). Its charge: +2/3 − 1/3 + 2/3 − 1 + 0 = 0, i.e., neutral, and there is no net coherent vortex flux around it. The vortex flux of a coupled primum is also discular for the same reason as the simple primum's is due to greater centrifugal force at the equator.

Since the masses of the neutron, proton and electron are known [24] we compute the mass of the neutrino.

$$\begin{aligned} &\text{Neutron:} & &1.674 \times 10^{-27} \text{ kg} \\ &\text{Proton:} & &1.672 \times 10^{-27} \text{ kg} \\ &\text{Electron:} & &9.611 \times 10^{-31} \text{ kg.} \end{aligned} \qquad (1)$$

Converting to atomic mass unit (amu) we obtain their masses:

$$\begin{aligned} &\text{Neutron:} & &1.0087 \text{ amu} \\ &\text{Proton:} & &1.0073 \text{ amu} \\ &\text{Electron:} & &5.486 \times 10^{-8} \text{ amu} \end{aligned} \qquad (2)$$

and the mass of the neutrino:

$$\eta = 8.5 \times 10^{-8} \text{ amu or } 1.55 \text{ times electron mass.} \qquad (3)$$

The neutrino has been assumed to have no mass which contradicts energy conservation; still, its mass is a subject of hot pursuit [52].

5.3. The Atom

All the forces and interactions of quantum gravity originate in the atom. Therefore, we consider it the core of quantum gravity.

The nucleus consisting of protons alone is the first to form in the atom. Their induced toroidal fluxes add up to a coherent vortex flux around it which is discular, thick at the nucleus and thin at the rim like the gravitational flux of a galaxy, its dual traced by the latter's visible halo of stars, dust cloud and other visible matter. The eye (of any vortex) is a region of calm but since the atom, like the primum, has spin some forces come into play like centrifugal force that affects the arrangement of the nucleons. Since the vortex flux of the proton is counterclockwise, the combined vortex flux spin of the nucleus is counterclockwise. By Newton's action-reaction law, the nucleus and, naturally, the atom rotate or spin clockwise as a unit. It is this spin that provides centrifugal force on the nucleons of a free atom and determines their arrangement.

When there is only one proton in the nucleus it coincides with the eye of its vortex flux. If there are two they are joined equatorially by a negative quark in the eye of their combined vortex flux; if three they form an equilateral triangle, by centrifugal force and energy conservation, and are joined pair-wise equatorially by negative quarks (see Figure 5.3.1 for light nucleus). For nucleus with more protons they form rings parallel to the equatorial plane joined pair-wise equatorially and polarly between rings as much as possible. By energy conservation, there is a threshold of number of protons in a ring so that another ring forms beyond it since a single large ring is unstable and violates energy conservation and energy conservation equivalence. The threshold for the number of protons in a ring can be determined experimentally. Large atom like uranium has several layers of such rings so that their profile viewed from the equatorial plane is similar to that of a primum, sinusoidal of even power. This arrangement of the nucleons conforms to the combination of energy conservation, energy conservation equivalence (e.g., optimal symmetry, universality of oscillation and related arrangements) and centrifugal force imparted by the spin of the nucleus where the rings lie at the inner boundary of the eye. They form several strings of prima joined polarly depending on the geometry of the equally spaced strings of protons joined N to S and S to N around the axis stretching from pole to pole and forming a bulge of sinusoidal profile of even power like that of a primum.

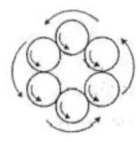

Figure 5.3.1. The nucleus of a light isotope viewed from the north-pole. Not shown are the –quarks that join the protons at the rims of their vortex fluxes shown as circular rings, by energy conservation.

The charges of the protons add up to the charge of the atom so that their combined nucleonic charge is equal to the number of protons in the nucleus. However, the nucleus itself is neutral since the protons' induced toroidal fluxes cancel each other's charges there. As positive coupled primum, the nucleus is a magnet of positive polarity with the vortex flux around it providing the magnetic field. With the right-hand rule and index finger pointing in the direction of the vortex flux the thumb points to the N-pole. Viewed from its N-pole the vortex flux of a free atom spins counterclockwise so that, as above, by Newton's action-reaction law the nucleus rotates in the opposite direction.

By flux compatibility and flux-low-pressure complementarity, electrons are attracted to the nucleus away from the eye but being light it is swept into orbit by its vortex flux. A stable atom has orbital electrons equal in number to the protons in the nucleus. Otherwise, it is an ion, positive (when there is a deficiency in orbital electrons) or negative (when there is an excess of orbital electrons). The most stable elements are inert. Moreover, by Flux-low-pressure complementarity, non-agitated superstrings steadily accumulate in the centre of the nucleus; they are the principal source of kinetic energy in nuclear fission.

5.3.1. Formation of Heavy Isotope

By flux-low-pressure complementarity, the nucleus can suck neutron only to form heavy isotope. The chance that a neutrino is sucked by it is quite remote since it whizzes by at the speed of light being carried by basic cosmic wave. The number of neutrons sucked depends on the holding energy of the atomic vortex flux provided by the protons. The periodic table shows that the nucleus can hold slightly more neutrons than protons except hydrogen which has only one proton in the nucleus. Hydrogen has three known isotopes with 0, 1, 2 neutrons;

deuterium has one neutron and tritium two. The most common in the atmosphere has one proton alone. Two atoms of deuterium combine with one atom of oxygen to form a molecule of heavy water.

How are the neutrons arranged in heavy nucleus? By centrifugal force, energy conservation and energy conservation equivalence, the neutrons of heavy isotope of an atom pile up along and on opposite sides of the nuclear equatorial plane arranged symmetrically with respect to it as much as possible. Then the complete profile of the nucleus viewed from the equatorial plane consists of a string of protons along a sinusoidal envelope of even power with the equatorial plane. By centrifugal force the neutrons are just outside the proton rings on both sides of the nuclear equatorial plane.

In the immediate vicinity of the eye dark superstrings are also sucked, by flux-low-pressure complementarity, and accumulate inside the proton rings, the inner eye, since, being dark, hence, weightless, they are unaffected by centrifugal force. They form a mini black hole. In fissionable nuclei they convert to prima and photons released in nuclear explosion. Thus, the energy of nuclear explosion comes not from the splitting of the atom *per se* but from the agitation and conversion of superstrings in the nucleus to visible matter and energy – prima, photons and radiation.

Being negatively charged, electrons are attracted to the nuclear vortex flux away from the eye, by flux compatibility and flux-low-pressure complementarity. But, being light, they are swept into orbit by the vortex flux and become orbital electrons. By centrifugal force, the most energetic orbital electrons are those closest to the equatorial plane. They form the outermost orbital shells. The least energetic ones cluster near the poles and form the lowest orbital shell. A subshell may consist of one or two electrons; when there is one electron in a subshell physicists give it $+1/2$ spin when there are two the second is given spin $-1/2$ so that those belonging to the same subshell have opposite spins, by flux compatibility, since they both oscillate within the subshell. Their orbits (probability expectations), however, are on opposite boundaries of the vortex flux due to centrifugal force and symmetry. Thus, the spin of quantum physics is not just a filler to distinguish the quantum numbers between two electrons in one subshell but has physical significance. The spin of the electron is opposite that of its toroidal flux, by Newton's action-reaction principle. This completes the qualitative model of the atom.

When orbital electrons are expelled by turbulence or cosmic waves the atom becomes a positive ion. It can also become negative ion when more electrons are drawn into orbit by flux-low-pressure complementarity in addition to flux compatibility.

In a ferromagnetic material, e.g., iron, the atoms can be aligned and joined at their poles north-to-south and south-to-north and form a string. Then a string can be joined equatorially with other strings by –quarks to form a bundle, the maximum number of strings of a bundle depending on the material, by quantization. In ordinary magnet each bundle determines a line of force whose induced vortex fluxes follow the right hand rule. Both the N- and S- poles of a bar magnet have little poles corresponding to the lines of force. This can be verified by ocular inspection using iron filings or a paper placed over the magnet. The coherent vortex flux of the bundles provides its magnetic field. Iron is ferromagnetic. To make, say, a bar magnet, one wraps a coil around it connected to a direct current. Then another coil is wrapped over the first connected to an alternating current. The second coil shakes the atoms, the first aligns them polarly to form strings and bundles and the iron bar becomes a magnet. A compass needle is made by simply rubbing it on parallel to a bar magnet of opposite polarity. The needle becomes a magnet of opposite polarity. (Since magnets are familiar we need no figure here)

Prima that form atoms are converted in great numbers from superstrings in the Cosmos [1,49,54] and in the cellular membranes of living organisms that form their cells, tissues and chemicals through the genes' brain waves [30,32,33,34,35,36]. They also form carbon that replenishes the supply in oil fields and they are continually produced in the atmosphere as well (this has bearing on the issue of carbon dioxide concentration there). In the Cosmos superstrings are converted to prima that get entangled in cosmological vortices and collect as cosmological bodies. The micro component of turbulence at the inner core of a cosmological vortex generates seismic waves that similarly convert superstrings to prima and photons [8,17,22,49,54].

In molecular bonding the valence electron serves as connector between atoms at their outer subshells, by flux compatibility. Some atoms have weak ionization energy that the outer or valence electrons are easily knocked off by cosmic waves or even atmospheric turbulence in the case of gases and become free. This is true of malleable materials like metal that is endowed with free electrons making it good electrical conductor.

Among gases the most energy conserving arrangement is diatomic, i.e., coupled pair of atoms of gas. Ions are unstable because they interact with other atoms. It follows that neutral clusters including neutral atoms are the most stable clustering and this is the basis of chemical replenishment including that of oil.

6. THERMODYNAMICS

Thermodynamics in the broad sense is concerned with the generation, conversion, transfer or conduction and utilization of energy. Here, *conversion* refers to latent (dark) to kinetic (visible) energy conversion. Thermodynamics lies at the intersection of quantum and macro gravity and includes electromagnetism. It brings us to the threshold of a new technological epoch of GUT technology based on utilization of the kinetic energy of vortex fluxes of superstrings and cosmic waves. Even today there are technologies belonging to this epoch like the magnetic train and the electric power plant. Just like the steam engine that was invented before the development of the science of thermodynamics the magnetic train and electric power plant were invented before the development of GUT. Appropriate GUT technology can now be devised for the treatment of genetic diseases such as cancer, systemic lupos erythematosus, diabetes, muscular dystrophy and mental disorder, e.g., depression, without injury to normal cells, by way of genetic alteration, modification and sterilization [33].

While high technology is based on quantum physics, a subtheory of quantum gravity, GUT technology is based on two of its pillars, thermodynamics and quantum. The new element in GUT technology is the utilization of dark matter, direct or indirect.

6.1. Electromagnetism

Much of large scale electrical power generation today falls under electromagnetism. One exception is generation by photo-voltaic cells based on quantum gravity. We recall that quantum gravity is the dynamics of primal induced vortex fluxes of superstrings [25]. As in previous section much imagination is required here since we are dealing with dark matter and energy.

6.2. Direct Current Generation

This subsection is quite familiar to the general reader; therefore, we shall simply provide a general summary for completeness. Direct current generation is mainly through the battery. How does the battery work? It consists of two electrodes, a metal which has low ionization energy, hence, free electrons like copper, and a semi-metal like lead that can absorb electrons beyond orbital electrons in the vortex fluxes of its atoms, by flux-low-pressure complementarity. Both are soaked in liquid electrolyte that ionizes the metallic electrode creating free electrons. By introducing direct current through the electrolyte from the metallic to the semi-metallic electrode the free electrons are pushed to the semi-metallic electrode creating voltage between the electrodes. The battery is fully charged when the free electrons in the metallic electrode have been transferred to the semi-metallic electrode. Since the electrolyte is a poor conductor, when a conductor connects the terminals outside the electrolyte a flux of superstrings in which the electrons ride and flow with the flux of superstrings as split off from the vortex fluxes of the atoms in the conductor like a fan belt all the way back to the metallic electrode replenishing the lost electrons there. When the metallic electrode has been replenished the battery is discharged. Then it can be recharged as before. Such a battery commonly used in cars is called storage battery. Another kind of storage battery is lithium used in lap top computers but it does not need liquid electrolyte.

There is theoretically no limit on the power or capacity of a storage battery. One can connect several storage batteries in series as a unit to attain desired voltage. Then several units of this voltage can be connected as a single storage battery in parallel circuits to desired capacity and power. However, this kind of electrical generation is impractical for transmission over long distances since much energy is dissipated by the resistance of the conductor.

6.3. Magnet and Electromagnet

In ferromagnetic material, e.g., iron, the atoms can be joined polarly to form strings and strings can be joined equatorially to form a bundle, i.e., a magnet with the coherent vortex fluxes of the bundles of atoms comprising its magnetic flux or field, its polarity in accordance with the right hand rule of electromagnetism. By the quantization principle the number of strings in a bundle is fixed depending on the material. In a magnet the lines of force are the bundles. In a bar magnet the bundles extend beyond the north pole, curve outward away from its longitudinal centre and bend over toward and into the south pole and join their respective south poles to form loops of bundles. It has its mirror image of loops of bundles with respect to the longitudinal centre. Their profiles are seen when a piece of paper is placed over a bar magnet and iron filings poured over it that trace the lines of force. Taking a single bundle and applying the right hand rule on it and moving in the direction of its north pole the index finger points to the direction of the vortex flux around the bundle as the thumb and index finger slide through it, bends left away from the longitudinal centre (right on the other side of this centre) and bends towards the south pole, across it and along the magnet. By energy conservation, the magnetic fluxes of the bundles form a coherent magnetic flux around them. Again, applying the right hand rule with the thumb and index finger sliding through the bundles across the north pole and curving down (left on one side of the longitudinal centre and right on the other side) towards the south pole and across it through the magnet, we trace half of the almost annular profile of the magnetic flux. We note that the bundles along the magnet and those outside and parallel to it have opposite polarity so that, by energy conservation, there is a cylindrical eye around the magnet between the inner and outer flux. The magnet itself lies in the eye of its main or inner flux.

A permanent magnet is made by winding a double helix around it and passing through direct current in one and alternating current in the other. The alternating current shakes the atoms of the material and the direct current aligns them to form the strings and bundles and the whole magnetic flux. When a coil of conductor with direct current through it is wound around a piece of metal it aligns the atoms of the metal, by flux compatibility, to form strings and bundles and, therefore, magnetic lines of force that form the magnetic flux of the metal turning it into a magnet. Here lies a kind of duality between electricity and magnetism – electromagnetic duality – that has many technological applications.

When a conductor is wound as a coil around an iron rod or bar and direct current is passed through it, the rod or bar becomes an electromagnet its polarity in accordance with the right hand rule.

6.4. Electric Power Generation

We apply the electromagnetic duality to the construction of the electric generator. First we note that when a conductor cuts across a line of force direct current through it is produced. How do we explain it? The magnetic flux of the bundles flows through and resonates with the magnetic fluxes of the atoms of the conductor like a fan belt through the circuit when open. This is the basic mechanism for generating direct current through electromagnetism. We can amplify the generation of direct current by letting several conductors cut through the entire magnetic lines of force of a magnet. However, since direct current is impractical for high power transmission, we generate alternating current in a generator.

The generator consists of a C-shaped magnet with magnetic lines of force going from the north pole at its end, say, the upper end, to its south pole on the lower end. The axis of the magnet extends through the gap between the poles forming a loop. When a conductor cuts across the lines of force between the poles the direct current reverses direction as it cuts across the magnetic axis. If we repeatedly move the conductor back and forth across the lines of force, we generate an alternating current. We can amplify the alternating current by winding the conductor many times around an armature and rotating it at high peed so that the rate of cutting across the lines of force is raised considerably. The magnet can be an electromagnet its magnetic flux provided by the generator. The armature can be spun by an electric motor, a waterfall or turbine in coal fired or nuclear generation. The electric power plant is an up-scaled generator.

The electric power plant has several advantages aside from higher power capacity. The rotating armature results in slicing of fluxes which agitates and converts the superstrings to electrons raising the intensity of generated power. The voltage can be raised by a transformer and electric power transmitted at great distances with minimal dissipation of energy, the voltage reduced at desired level for the users. Clearly, electric power generation is GUT technology that preceded the development of GUT just as the steam engine preceded the development of thermodynamics. So is the magnetic train that preceded GUT. Both utilize the natural engine of vortex flux of superstrings, the magnetic flux. In the case of the magnetic train a fixed magnet on the track has the same polarity as the electromagnet on the train so that their magnetic fluxes push each other.

7. MACRO GRAVITY

The Universe of dark matter is boundless. For, if there were a boundary that separates dark matter in our side from another side that is empty, then, by flux-low-pressure complementarity, dark matter will cross and demolish it in due course. Moreover, since the superstring is generalized nested fractal sequence it has no initial element and, therefore, no beginning. Therefore, the Universe of dark matter has no beginning. Furthermore, since the superstring is indestructible, it has no end; it goes through its cycle endlessly – from its non-agitated phase through semi-agitated, agitated and back to non-agitated. Therefore the Universe is timeless and boundless.

7.1. Ordinary Universe

An ordinary or usual universe arises from the steady shrinking of superstrings due to energy conservation that forms nested fractal sequences of depression, by the law of uneven development [12]. By flux-low-pressure complementarity; each depression sucks dark matter around it and enters the phase of chaos of the standard dynamics. Since chaos is energy dissipating, in this case involving random collisions, energy conservation and energy conservation equivalence induce its evolution to the third phase – turbulence – a cosmological vortex like planet, star or galaxy. Since this happens to every term of the fractal sequences, the nested fractal sequences of depression (e.g., regions of depression each region containing regions of depression, each region containing regions of depression, etc., ad infinitum) evolve to nested fractal sequences of cosmological vortices, by flux-low-pressure complementarity [29].

For example, a galaxy as cosmological vortex flux of superstrings has collected mass around the eye as its core and minor cosmological vortices in its vortex flux under its influence such as stars; each minor cosmological vortex has collected mass around the eye as its core and minor cosmological vortex fluxes such as planets and moons; ..., etc.

How does an ordinary universe form? By the quantization principle there is an optimal spread of nested fractal sequences of depression that forms nested fractal sequences of cosmological vortices as an ordinary universe. For our purposes the basic unit of a universe is the galaxy which is an ordinary universe by itself formed the usual way in terms of structure. A galaxy consists of a main vortex and its eye at its core, the collected mass around it. It has minor cosmological vortices revolving around the core along orbits at the balance between suction by the main eye, by flux-low-pressure complementarity, and centrifugal force of spin on it as a cosmological vortex of superstrings, i.e., minor cosmological vortex. Its minor cosmological vortices include stars, planets and moons (minor cosmological vortices of a planet).

The solar system is an average minor cosmological vortex of the Milky Way; its minor cosmological vortices are the planets and their moons. Some galaxies have galaxies among its minor cosmological vortices. For example, the Milky Way has 11 one of which a dead galaxy now the Sagittarius cloud of stars [40]. Another is Andromeda with 22, all young galaxies. In a young galaxy called spiral nebula the minor vortices fall towards the spinning core along spiral trajectories that wind around the eye due to its suction (gravity) and the effect of core spin and dark viscosity. The spiral trajectory of falling minor cosmological vortices rides on and revolves around it with the gravitational flux (of superstrings) of the main vortex around the main eye.

7.2. Our Universe

Only a universe launched by a big bang evolves to a super...super galaxy due to infusion of great energy by the explosion and a second but more powerful explosion called cosmic burst [19,21,39,48]. Our universe is special and originated from the Big Bang, an enormous burst of energy in the Cosmos that occurred 8 billion years ago [24]. It is a local bubble, a super...super galaxy 10^{10} billion light years across [19,21,38] in the timeless boundless Universe. Its core is a cocoon shaped galaxy cluster 650 million light years across discovered by French astronomers in 1994 [38].

There is evidence of existence of some universe other than ours [55,57]. The Big Bang did not create galaxies other than the super...super galaxy, our universe. All the rest formed the usual way. There is evidence that the Milky Way formed before the Big Bang and was far from it when it occurred [50]. It was drawn into our universe as the latter expanded to a super...super galaxy.

The Milky Way is the oldest galaxy in its neighbourhood. A young galaxy, e.g., spiral nebula, is bright with prominent spiral streamlines of minor vortices falling into its core due to gravity. The Milky Way is dim, a tell-tale sign of old age, having faint spirals of falling minor vortices most of which already sucked by the main eye and have joined the spinning core. Another part of the evidence is our being able to see our young universe (through Hubble) when it was only 3% of its present age. Still another is the discovery of a star in it older than the Big Bang [50].

7.3. Its Birth

We trace the origin of our universe from Big Bang. In traditional cosmology it presumably occurred spontaneously with neither rhyme nor reason as if our universe emerged from spectacular violation of energy conservation. As a matter of fact, the Big Bang was a natural phenomenon subject to natural laws.

The Big Bang created two physical systems: a super...super depression in dark matter and an expanding spherical wave front at accelerated rate called Cosmic Sphere pushed by the explosion. During its initial phase, 0 < t < 1.5 years, the Cosmic Sphere was compressed layer of dark matter trapped and pressed between the force of explosion and suction by the super...super depression, by flux-low-pressure complementarity, and pounded and agitated by less energetic shock waves (concentrated cosmic wave with enhanced latent energy) bouncing between its inner and outer boundaries [12,19,21]. This agitation endowed the Cosmic Sphere and the superstrings in it with enormous latent energy. However, compression kept them from conversion to prima, only semi-agitated superstrings. The more energetic shock waves pierced the Cosmic Sphere and converted dark to visible matter in the immediate exterior of the once Cosmic Sphere which got entangled in cosmological vortices there. The expanding Cosmic Sphere weakened and, combined with outward pressure from the compressed semi-agitated superstrings, burst at t = 1.5 billion years from the start of the Big Bang called the Cosmic Burst or second big bang [19,21,48], much more powerful than the Big Bang due to enormous infusion of latent energy by the semi-agitation of the trapped superstrings.

The Cosmic Burst released the semi-agitated superstrings that converted to simple prima at very high temperature, the first visible matter of our young universe that formed the bright radioactive clusters called quasars which peaked at t = 2.5 billion years from the start of the Big Bang [21,55]. Dark viscosity and energy conservation reduced their kinetic energy and temperature and allowed formation of coupled prima such as proton, neutron and neutrino and light elements that got entangled into usual cosmological vortices in the vicinity of the once Cosmic Sphere. This marks the birth of the early galaxies of our universe. To use a biological analogy, the Big Bang was only the mitosis of the fertilized egg while the Cosmic Burst gave birth to our universe. Thus, the quasars evolved to the galaxies of our universe.

7.4. Its Evolution

We take a qualitative model of the evolution of a galaxy not just the super...super galaxy but we focus on the latter.

The Cosmic Burst added to the breadth and depth of the super...super depression that sucked cosmological vortices around it, formed the transitory phase of chaos of the standard dynamics [8,17] and, by energy conservation and flux-low-pressure complementarity, evolved into a super...super cosmological vortex that started the evolution of our universe into a super...super galaxy. It pulled cosmological vortices along the way by gravity (suction by the eye of the developing super...super galaxy). The super...super depression gave rise initially to a local vortex but as visible matter formed by the agitation of the spinning core, conversion of dark to visible matter by the micro component of turbulence of the spinning inner core augmented by falling visible matter around it that plunged into the core imparted momentum on and raised the power of its spin. However, greater momentum and spin were added mainly and instantly by the conversion of dark to visible matter due to the micro component of turbulence at the inner-outer core of our universe. As our universe increased its spin it imparted greater centrifugal force on the galaxies but suction by the eye balanced it and induced them to form elliptical orbits around the core. As its power rose further, centrifugal force surpassed gravitational suction and catapulted the galaxies outward. This explains its present accelerated radial expansion [50]. Some special universe apart from ours must have catapulted the galaxy clusters traversing our universe [59,60]. An ordinary universe could not have done it for it lacks the power imparted by a big bang both at its initial and cosmic burst.

As visible matter falls into the core of a galaxy its dark component's resonance with dark matter pulls and thins out the gravitational flux and reduces dark viscosity and suction by the eye on the minor vortices. Suction reaches its peak, declines and isolates the core once again from its minor vortices as it treks to its destiny. This dynamics is replicated by the minor vortices.

7.5. The Present State of Our Universe

Based on extensive direct measurement of the separation of galaxies from Earth, Edwin Hubble formulated his law that expresses the rate of separation of a galaxy from us at distance s from Earth and, in effect, the radial expansion of our universe:

$$ds/dt = \rho s, \qquad (4)$$

where s is the distance of the receding galaxy and $\rho = 1.7 \times 10^{-2}$/km is the rate of recession. For convenience, we measure distance S along a great circle in the spherical dark halo of our universe. Then,

$$dS/dt = \rho S. \qquad (5)$$

Since this discovery, estimates of the age of our universe increased from the original 8 billion to the present 14.7 billion and raising it to 20 billion is being considered. Each time an old star is discovered the estimate is adjusted to accommodate it. This star-chasing game is based on the wrong premise that only our universe exists. There are others. One evidence of it is the presence of galaxy clusters traversing our universe [55] and another is the collision of galaxies coming from different directions [57]. Galaxies in our universe travel along outward radial trajectories and cannot collide among themselves. Still another is the discovery of stars in the Milky Way older than the Big Bang [50].

Therefore, we stick to the original estimate of 8 billion to solve (5) and find the radius r as function of t. Since $dS/dt = 2\pi dr/dt$ and (5) is independent of the distance between us and other galaxies it holds when $S = r$. Then,

$$2\pi dr/dt = \rho r \text{ or } dr/r = (\rho/2\pi)dt. \qquad (6)$$

Solving r, reckoning time from the Big Bang and taking 1 light year and 1 billion years as units, we have,

$$r(t) = 10^{10} e^{(\rho/2\pi)(t-8)} \text{ light years},$$
$$r'(t) = (\rho/2\pi) 10^{10} e^{\rho/2\pi\,(t-8)} \text{ light years/billion year},$$
$$r''(t) = (\rho/2\pi)^2 10^{10} e^{\rho/2\pi\,(t-8)} \text{ light years /(billion year)}^2 \text{ [24]}. \qquad (7)$$

Using standard units we have, at $t = 8$,

$$r(8) = 3.2 \times 10^{22} \text{ km},$$
$$r'(8) = 840 \text{ km/sec},$$
$$r''(8) = 3 \times 10^{-10} \text{ km/secsec}. \tag{8}$$

Since r" > 0, our universe is on the young phase of its cycle, its power still rising. With this radial expansion and acceleration, the rate of radial expansion of our universe will eventually surpass the speed of light unless it reaches its destiny as cluster of black holes sooner [12,53,56]. The value of ρ is based on direct observation and analysis of the Doppler effect of receding light source.

Now, Encarta Premium 2007 has this value: ρ = 260,000 km/hr/3.3 million light years, i.e., the receding galaxy moving away from Earth faster by 260,000 km/hr for every 3.3 million light years distance away from us [4]. This data was calculated or inferred from past records going back or projected least 3.3 million years ago which, obviously, did not exist. In other words, the premise is vacuous. Nevertheless, regardless of where this value came from, does it make sense?

Converting to standard units and simplifying, we get ρ = 3 × 10^{-19} /km; inserting this value in (7) with the value of ρ replaced by 3 × 10^{-19} /km we obtain, r'(t) = 5 × 10^{-14} km/sec, the supposed rate of radial expansion of our universe, and acceleration of 3 × 10^{-32} km/secsec which point to a static universe at odds with present observation and measurements [50]. If it were correct we would have been roasted by intense heat coming from the steady formation of stars in the Cosmos at one per minute [1,8,49,54] and emergence of new galaxies (two baby galaxies were discovered since 2004). On the contrary; the average temperature of the Cosmos remains close to 0 degree Celsius and this is expected to reduce steadily as our universe expands and cools off. Thus, the rise in temperature is offset by the actual rapid expansion of our universe [51].

7.6. Formation and Evolution of a Cosmological Vortex

It is clear that the Big Bang did not create a cosmological vortex other than the super...super galaxy. Some cosmological vortices including stars and galaxies preceded the Big Bang but a lot more have continued to form at rapid rate. In fact there are regions in the Cosmos called star nests that yield stars more rapidly than others [49, 54].

The core of a cosmological vortex is initially dark and isolated but the kinetic energy of its spin and the micro component of turbulence raise its temperature that agitates and converts the superstrings to prima. Thus, its expansion and enhancement of visible mass come mainly from within. As turbulence, the core's micro component generates seismic waves [8,17] that convert dark to visible matter in it and its vicinity, in the former mainly simple prima and in the latter also simple prima that forms atoms and cosmic dust in the cores of micro cosmological vortices and get entangled in minor cosmological vortices. Converted visible matter in the core instantly gains momentum that augments core spin and angular momentum and, combined with momentum imparted by falling visible matter, raises the power of spin, expands its influence outward by dark viscosity and pulls and catapults outlying cosmological vortices into rotating spiral streamlines of falling minor vortices. The same dynamics is replicated in minor vortices. Cosmic dust in a cosmological vortex also emerges from what are called cosmic ripples, energetic cosmic waves some of which gamma-ray bursts [54]. Then it gets entangled with cosmological vortices that collect at their cores as stars, planets, moons, etc. This phenomenon of populating a cosmological vortex is dramatically illustrated by the baby galaxy discovered in 2004 and another one a couple of years later showing the spirals of visible matter just forming and falling into its core. Once minor vortices form they become self-sustaining, i.e., they do what the main vortex and its core do. The spinning matter around their eyes also generates seismic waves [17,22] that convert dark to visible matter in and around their neighbourhoods.

In any cosmological vortex the lucky few minor vortices that lie at the balance between suction by the eye and centrifugal force take their orbits around the eye along rotating spirals and escape suction by the eye. In the solar system they are the planets and planetoids that orbit the Sun. The Sun is a minor vortex of the Milky Way and what we see is its solid core of collected mass around the eye. In an average galaxy the minor vortices are the stars and in a planet its moons if any but they are all minor vortices of the Milky Way. The minor cosmological vortices are pulled by suction by the main eye and as its power of spin rises and gravitational flux expands the peripheral minor vortices are the first to take their orbits while those closer to the main eye that did not attain orbits are sucked by gravity. This explains why in the solar system there are no planetoids in the orbital corridors of the inner planets up to the orbital corridor of Mars but there is abundance of planetoids in the orbital corridors of Neptune and Jupiter, the outer planets. If the spin of a cosmological vortex continues to rise beyond a threshold, the peripheral minor vortices may be catapulted beyond its influence. This explains the presence of stars travelling along straight lines [4] and the galaxy clusters traversing our universe [55].

Since a cosmological vortex rotates at great speed, greatest at the equator and 0 at the poles, centrifugal force throws visible matter outward at the equator. Then it becomes a thin disc of visible matter consisting of minor vortices and their cores and clouds of cosmic dust riding on the gravitational flux thick and concentrated around the eye. The thin rim of a cosmological vortex is confirmed by the thin rings of Saturn and the other massive planets with powerful vortices that throw debris that forms these rings. The discular shape of a cosmological vortex is also seen in pictures of galaxies. The solar system is also discular with the planetary orbits along the solar equatorial plane. Mercury is the only planet that lies on its thicker portion just off the solar equatorial plane. This explains its perihelion shift of 1.67 seconds of an arc, i.e., the angle between the solar equatorial plane and the equatorial plane of Mercury. Although the dark halo is spherical being unaffected by gravity and centrifugal force, resonance with the dark components of the visible halo along with flux-low-pressure complementarity leads to its greater concentration in the discular visible halo.

7.7. Cosmological Vortex Dynamics and Interactions

Flux compatibility and flux-low-pressure complementarity have direct bearing on vortex interaction. However, energy conservation and energy conservation equivalence are always at work in any interaction. Other natural laws are their consequences. They are highlighted because of the insights they provide in understanding the fractal principle and uneven development and resonance laws.

Spin determines interaction between cosmological vortices mediated by their gravitational fluxes by virtue of flux compatibility: two vortices of opposite spins are attractive through the common coherent induced flux at their rims along their equatorial planes; they are repulsive otherwise. If they have the same spin and their masses have the same order of magnitude, they evolve into binary vortices each revolving around the other and mutually riding on each other's spiral flux; centrifugal force prevents them from falling into each other. If they have the same spin, regardless of their relative masses, they have mutual repulsion unless one is a giant compared to the other in which case the more massive one may gobble up the other by gravity. However, if one is large compared to the other and has opposite spin, the latter rides as minor vortex or an eddy on the gravitational flux towards and merges smoothly with the core of the former unless the centrifugal force on the smaller vortex balances the main gravitational flux pressure in which case it takes elliptical orbit around the main core. Otherwise, if centrifugal force exceeds gravitational pull on a body, it may get catapulted off the vortex's influence. The galaxy clusters traversing our universe [55] reveals the existence of powerful universe elsewhere with the power to catapult them.

Elliptical orbit, being due to radial oscillation is the most probable orbital configuration since perfect balance that yields circular orbit is unstable, by uneven development. A minor vortex along the main spiral streamline that spins opposite that of the main vortex either forms elliptical orbit around it as an eddy or gets sucked into and is crushed by the core and becomes part of it. As an eddy a vortex has relative autonomy. Two contiguous vortices of comparable masses with the same spin do not crash into each other due to mutual repulsion of opposite fluxes. Here, again, we see quantum-macro gravity duality.

As in a game of chance, an even game is unlikely over a period of time. While a pair of vortices may have initially the same mass and vortex power, once one vortex gains advantage, by uneven development, it builds up over time until it is more massive than the other. Then one becomes a minor vortex of the other. Thus, the most likely configuration of nested fractal sequences of vortices is one with a single large core vortex and many minor vortices of diverse masses along its rotating flux spirals. There are, of course, binary stars that form when the balance is attained at the tapering of their increase in mass.

Among the intriguing questions arising from this theory is the possibility of tampering natural object to break global flux coherence and quash its capability to exert gravitational pull on other objects. (Local flux coherence cannot be eliminated since every atom has it) Moreover, by flux-low-pressure complementarity, such tampering cannot shield objects from the gravitational pull of another. However, like the stealth bomber that breaks coherence of reflected radar beams to evade detection, a sufficiently tampered body, e.g., debris like asteroids, may lose global coherent fluxes that, while acted upon by gravity, may no longer exert gravitational pull or push on other bodies. They are bodies that have lost cosmological history. To verify, we use some natural laboratory: the asteroid belt along the orbital corridors of Jupiter, Neptune and Uranus [43,46,58] (there should be asteroid belts also along the orbital corridors of the other powerful planets). The irregular shape of asteroids and the objects that form the planetary rings reveals lack of cosmological history, meaning, lack of coherent gravitational vortex flux; they are debris rather than matter collected at vortex cores. They do not form gravitational clusters either, that is, they do not exert gravitational pull or push among themselves and yet they have masses. They resolve the above question and at the same time serve as counterexamples to Newton's law of gravitation.

Recent study reveals that cosmic dust particles are oblong, confirming they have cosmological history, i.e., like a planet, a piece of cosmic dust is accumulated mass at the core of a micro vortex. Its axis of rotation wobbles like the summer and winter solstices. Like Earth it has crust and mantle. It is estimated that interstellar dust constitutes one thousandth of the Milky Way's mass and hundreds of times more than the mass of the galaxy's planets [1]. This means that cosmic dust is a significant factor in the mass enhancement of a planet. Cosmic dust continues to form and collect into stars at the cores of stellar vortices. While our universe is the first term of its nested fractal sequences as a super…super galaxy, the last terms of its cosmological vortex sequences are cosmic dust. Appending the molecules, atoms and superstrings yields the full stretch of our fractal universe all the way from the super…super galaxy through the atoms, prima and superstrings of dark matter.

The fractal-reverse-fractal algorithm [18] locates any vortex in our fractal universe starting from any cosmological body including cosmic dust particle where one can trace a fractal sequence up into the macro scale (reverse-fractal) and end up in the super…super galaxy; or go down the sequence at the micro scale and end up at cosmic dust. Conventional science takes the view that these dust clouds formed during the last 1.5 billion years. GUT provides physical explanation of their existence and origin. Conversion to prima that form cosmic dust occurs all the time due to superstring agitation by cosmic waves, cosmic ripples and γ-ray bursts [53]. However, agitation by high temperature at cosmological vortex core and micro component of turbulence are the principal generators of prima in and around its immediate neighbourhood.

7.8. The Earth as a Cosmological Vortex

The increase in mass of a cosmological body is necessarily proportional to its mass (which depends on the power of its vortex spin). The Earth's gravity was 67% of its present gravity 65 million years ago [42] which is roughly the same percentage of mass then relative to its present mass. The increase in mass comes from agitation by the hot spinning inner core by the micro component of turbulence, the principal factor in the formation and build up of visible matter within and around it so that the Earth becomes more massive over time to which falling matter is added, e.g., meteors and debris of light asteroids that explode in the atmosphere. Visible matter formation in and around the Earth's core exerts outward pressure on its mantel and forces magma to ooze out of the surface, fuel volcanic eruptions and pile up mountains of lava along constructive tectonic plate boundaries under the oceans that congeal into and join the Earth's crust [8,16,20,58].

Galileo was amazed by his discovery that the rate of acceleration of a free-falling body above Earth is constant regardless of mass. What appears to be a free falling object on Earth between the poles actually follows a counterclockwise spiral when viewed from the North Pole. The constant acceleration is enlightened by a simple experiment: In a water vortex, say, a sink full of water with objects of different weights floating on it; release the water through an orifice at the centre-bottom of the sink. A vortex will form and the floats will be accelerated at the same rate along spirals towards the orifice proportional to its cross-sectional area. In Galileo's experiments the bodies were falling into the Earth's core along gravitational flux spirals. The rate of acceleration is specific to the cosmological vortex, specifically, gravity or suction by the eye; thus, the Earth and Moon have different gravitational acceleration.

7.9. "Cannibalistic" Activity of Giant Galaxies

In our neighbourhood there are two giant galaxies belonging to the Constellation Virgo. One is Andromeda, the brightest and farthest object that can be seen by the naked eye 2.2 million light years from us in the Milky Way, the other giant. This is an interesting combination because Andromeda is special in the sense that its initial visible matter comes from the Cosmic Burst, but Milky Way is ordinary. Both are average giants and have similar features except that Andromeda is young.

Andromeda's visible discular halo is 200 million light years across its mass equivalent to 3,500 billion Suns [40]. It has a double nucleus or core at the centre. The discular halo of a galaxy is spherical since it is unaffected by gravity and centrifugal force. However, the visible halo within it is discular in shape due to centrifugal force, thick at the centre where visible matter collects due to suction by the eye but thin at the rim along the equatorial plane due to stretching by centrifugal force. Its profile seen from its equatorial plane away from the rim is sinusoidal of large even power comprised of two full sinusoidal arcs joined and tangent to each other at the ends and round but narrow at their crests. This profile is similar to the primum's [12,15,24,25] another feature of quantum-macro gravity duality. Two of Andromeda's 22 minor galaxies are at opposite sides of and near its visible discular halo and appear headed for gravitational gobbling [40].

Milky Way contains 400 billion stars including our Sun [4,40]. Its visible discular halo along its galactic equatorial plane is 100 million light years across, its visible core, crater or metropolis 10 million light years thick [41]. Like Andromeda its dark halo has greater concentration in the visible discular halo due to resonance and flux-low-pressure complementarity. Sagittarius, now a cloud of stars has been cannibalized by Milky Way that has gobbled some of its stars through the "saw-tooth" action by the rim of its visible halo that slices the Sagittarius' cloud of stars and throws them into a sector between the tangent and normal to flux rim [40].

7.10. What Is Our Universe's Destiny?

We answer this question with respect to any cosmological vortex including our universe. They differ only in scale. As soon as a cosmological vortex reaches its peak of power and leaves its minor vortices free, each one treks home to its destiny back in dark matter along the same cosmological path. With the thinning of its dark and visible halo the contribution of visible matter falling into the core in augmenting its kinetic energy (that includes all visible energy – heat, light, motion of mass, etc.) declines but mass, spin and angular momentum continue to rise because of dark-to-visible matter conversion that introduces instant momentum to the spinning core. However, the increase in mass absorbs and puts a break on kinetic energy and agitation and reduces the rate of dark-to-visible matter conversion and energy of spin inducing steady deterioration of the kinetic energy of the primal toroidal and vortex fluxes. This results in weakening of primal bonding leading to their separation as simple prima. By energy conservation, the prima collapse to semi-agitated superstrings over a long period of time. Both the prima and semi-agitated superstrings remain around the eye due to the latter's suction.

When significant level of prima in a star has collapsed to semi-agitated superstrings the core becomes a neutron star, a misnomer since there is no such thing. Rather, the core has lost energy, specifically, the primal charge, that it has become neutral. Consequently, their bonding vanishes and they remain only around the eye due to suction by the eye. Further de-agitation by the eye at its boundary pushes the semi-agitated superstrings to the non-agitated phase, layer by layer, and the non-agitated superstrings join the black hole in the eye. Then the core of the once cosmological vortex has reached its grave and destiny, a black hole back in dark matter. The black hole becomes naked and there is no longer suction but absence of visible matter that was sucked by its graveyard, the eye that nurtured it. Many such "voids" in the sky have been mapped and catalogued. This dynamics is replicated in the minor cosmological vortices of a cosmological vortex. Clearly, a black hole being dark never sucks matter. It is the eye of the vortex that nurtures it that does.

In a galaxy the core transitions as huge star. This was verified in 1997 with the discovery of a giant star, observed through the Hubble, 10 million times the mass of our Sun; more massive ones as much as 200 million times the Sun's mass have been discovered since then. Each star has destiny: black hole in its eye. In a massive galaxy cluster or super…super galaxy the core evolves into galaxy clusters, each galaxy evolving to its destiny as black holes. This is the scenario of evolution of the core of our universe [8].

8. EXPLANATION OF NATURAL PHENOMENA AND CLARIFICATION OF ISSUES

We explain presently unexplained or misunderstood phenomena, identify long standing issues and explain and resolve them from GUT's perspective [8]. We also make explicit predictions some of which where already implicit in this author's earlier works.

8.1. Primal Angular and Linear Momentum and Energy

A primum at rest has angular momentum provided by the toroidal flux that travels through its helical cycles. However, when scooped up by suitable basic cosmic wave, breaks off from its loop and becomes a rapid oscillation (planar, polarized light) due to dark viscosity its angular momentum converts to linear momentum which is augmented by the linear momentum of the carrier; so does the photon. Viscosity encountered by a moving physical system is due to collision with the physical constituents of the medium. For example, electrical resistance in a conductor is due to collision of the flux of electrons with its atoms and molecule. When electric current simply consists of flux of dark superstrings the conductor offers no resistance, a phenomenon called superconductivity.

Experiments show that both the primum in flight and the photon have angular momentum. How do we explain it? Vortex fluxes are everywhere in their neighbourhood and they impart angular momentum on them. So does the Earth's gravitational field which is a vortex flux of superstrings [8]. Even outside the Earth's gravitational field, the Sun's does and beyond it, the Milky Way's, etc. In fact, there is no place in the Cosmos free from the impact of some gravitational flux. In regions between galaxies, for example, gravitational flux is provided by our universe [8].

As pointed out in [8], in any energy exchange, linear momentum does not necessarily convert to linear momentum nor does angular momentum to angular momentum. For example, the wind's linear momentum converts to angular momentum that rotates the sensor of the anemometer or the weather vane. Conversely, the angular momentum of the vortex flux of superstrings around a magnetic line of force of a magnet converts to the linear momentum of the electric current when the conductor cuts across it in an electric generator.

It has long been established experimentally that the smallest unit of energy is the Planck's constant $h = 66.61 \times 10^{-34}$ J [12]. Where does this energy come from? In a primum, h is the energy of its toroidal flux as it travels though one cycle and by the principle of uniformity it is uniform for all its cycles, i.e., independent of the cycles. In a photon h is the energy of the toroidal flux as it travels through one cycle of the primum it comes or through one full arc of its rapid oscillation. In conventional physics, the energy of the photon in flight is given by $E = h\nu$, where ν is misinterpreted as the frequency of the carrier basic cosmic wave or the reciprocal of the arc length of its envelope. This value underestimates the energy of the photon as it does not take into account the energy imparted by its carrier basic cosmic wave, its envelope.

It is tempting to compute the number of cycles of a primum or photon by dividing its known energy by h. This is impossible, however, due the uncertainly of small and large number, e.g., the small value of h and the large quotient. At best only its energy density can be computed using the generalized integral [7,32].

It is known that for long wavelengths and very low frequencies less than 1 Hertz, quantized energies less than h in magnitude are obtained (information provided by Prof. C. G. Jesudason, University of Malaya, in private correspondence). This contradicts the fact that h is the irreducible unit of energy. How do we account for it? The equation is valid only when ν is the frequency of the photon as rapid oscillation. Sufficiently low-frequency waves do not carry photon by the resonance law discovered in the course of analysing the disastrous final flight of the Columbia Space Shuttle [22]:

Resonance. *Maximum resonance between waves, oscillation and vibration occurs when they have exactly the same characteristics but arc length or its reciprocal, frequency, is the principal factor for resonance. The degree of resonance declines drastically with the difference in orders of magnitude of frequency. However, negligible resonance between waves that differ by orders of magnitude add up to significant level at critically high order of magnitude of frequencies.*

Basic cosmic wave of low frequency, i.e., longer wavelength than the envelope (carrier) of any photon by at least an order of magnitude does not carry a photon (non-photonic), by resonance law. Its energy comes from the vibration of the generating source and the synchronized vibration of the medium, dark matter, which reinforces it. Therefore, the energy equation of such wave must have terms that reflect the energy of the generating source and vibration of dark matter that resonates with and reinforces it. It must be a function of wavelength since long wavelength means less vigorous vibration and low energy of both source and medium. That equation which is unknown at this time should apply to non-photonic waves, e.g., ordinary water wave whose energy comes from wind motion and gravity, their combined force causing the synchronized vibration of the water molecules [8]. At any rate, the energy of water wave does not depend on the Planck's constant h; so is the energy of the guitar string because they are not photonic.

8.2. Wave

Wave in general is due to suitably synchronized motion of the medium. Where does its energy come from? It comes from the motion of the generator, e.g., wind motion, tectonic plate subduction at suitable distance from the ocean surface creating tsunami. It is reinforced by the induced synchronized vibration of the medium via resonance. In a sense, a wave is an illusion for no mass travels with it. A wave is like a "travelling" neon light; it does not really travel its apparent motion being due to synchronized switching on and off. Consider a spherical stone thrown into a pool. It pushes a cylindrical column of water downward. Water pressure pushes it back upward beyond the surface due to momentum while being pulled back by gravity until it grinds to a halt at its crest (highest point) and reverses downwards due to gravity beyond the water surface due to momentum. Then water pressure pushes it back upward again and the cycle is repeated. How do we account for its sinusoidal profile? It is due to water viscosity at the cylindrical boundary of the column that exerts a retarding pull on the rising column and gives it that profile. The vibration or oscillation of the water column resonates with the surrounding molecules of water and induces similar sinusoidal motion of concentric cylindrical columns of water propagating outward. Ocean waves are generated mainly by wind thrust oblique to the ocean surface that creates micro vibration of the water molecules (called micro-component of turbulence [17]) sustained and reinforced by continued wind flow and gravity until large waves form and travel in the direction of forward resultant of wind flow. They break when they become too large and heavy to sustain.

8.3. Primal Polarity

The core of the Earth is the collected mass around its eye including the atmosphere. Technically, the Earth includes its gravitational flux. The Earth's gravitational flux shields Earth from massive asteroids and separates positive prima from their negative anti-matter that otherwise would create instability. It also shields Earth from charged particles rushing towards it from outer space.

The Earth's gravitational flux goes from West to East, is fastest at the Equator and slows down to 0 at the extremities of the Earth's eye at either Pole. Naturally, its pull on the Earth's material also slows down from the Equator to either Pole. This polar lag and its pull on a cosmological body's material was discovered in the Sun where there is a similar lag; halfway from the Equator to either Pole of the Sun the lag is 30%, a constant of nature in macro gravity, by the synchronization principle and energy conservation. It explains why typhoons and tornadoes spin counterclockwise in the Northern Hemisphere and clockwise in the Southern Hemisphere [17].

Consider a simple free primum as it pops out of dark matter when agitated by suitable basic cosmic wave. If its equatorial plane is oblique to the direction of the gravitational flux it rotates counterclockwise in the Northern Hemisphere (clockwise in the Southern Hemisphere) due to the polar lag and aligns its equatorial plane in the direction of the gravitational flux making its vortex flux and eddy in the gravitational flux, its optimal energy-conserving alignment. By flux compatibility, a positive primum is pushed up so that there is abundance of free positive prima in the upper atmosphere (confirmed by shower of fragments of protons smashed by ultra-energetic cosmic waves that fall on Earth [43]). Free neutral prima are oriented randomly.

Free positive ions are counterclockwise eddies in the Earth's gravitational flux; they are also pushed upwards, by flux compatibility. However, being heavy, they remain in the lower atmosphere. The electron as clockwise eddy in the Earth's gravitational flux is pushed downwards, by flux compatibility. Thus, there is abundance of free electrons on the ground. Other free negative prima including the negative quarks should be abundant on the ground also but we do not know where they are; this needs investigation. When the voltage between the positive ions in the lower atmosphere and the electrons on the ground reaches critical level they rush towards each other, collide and explode as lightning. This separation between positive and negative prima contributes to the stability of cosmological vortices like Earth. Outside the Earth's gravitational flux primal orientation is determined by the dominant gravitational flux there; between planetary gravitational fluxes it is the Sun's gravitational flux that prevails. Outside the solar system it is Milky Way's, etc; outside any galaxy it is our universe's gravitational flux that prevails.

The basic prima – the quarks, electrons and neutrino – emerge steadily, especially, in living things and produce their tissues. Since they are trapped in the cells of living organisms they are not polarized and separated by the Earth's gravitation flux. They form atoms and molecules of tissues of living things. They are converted from superstrings in the cellular membranes by brain waves radiated by the genes [30,32,3,34,35,36].

8.4. Matter-anti-matter Interaction

Two simple prima are anti-matter to each other if the toroidal flux of one is the mirror image of the toroidal flux of the other with respect to a normal to their common equatorial plane between their equators. With opposite toroidal flux spins a primum and its anti-matter attract each other at their equators. When they get close, the momentum of their attraction forces their cycles to overlap and their fluxes to collide leading to explosion that throws them apart as two photons heading into opposite directions parallel to their respective axis, by flux compatibility. The positron is the anti-matter of the electron. The logic of GUT says that every simple primum has anti-matter. However primal polarity and the Earth's gravitational flux separate them from each other. In a controlled environment positive anti-matter can be produced on the ground and interact with its anti-matter to mutual destruction [8] or the positive anti-matter may be pushed up by the Earth's gravitational flux, flux compatibility.

Does a coupled primum have anti-matter? When it has suitable symmetry it does. For example, since the proton is linear, i.e., the two positive quarks and the negative quark that join them have coplanar axes, it can have a coupled primum as anti-matter consisting of two negative quarks joined by a positive quark since they are mirror image of each other with respect to a plane between their common equatorial plane. Then they are attractive; when they get close the momentum of their approach forces their cycles to overlap and mutually destroy each. However, the Earth's gravitational flux separates them and minimizes such occurrences.

Neutral prima have no anti-matter. However, individual component of a coupled primum has its respective anti-matter that can mutually destroy each other. The speculation that our universe has its anti-universe somewhere has no scientific basis nor does a cosmological vortex or atom. It is, however, possible for some symmetric coupled primum to have anti-primum, e.g., the proton.

8.5. Wave-particle Duality

The wave-particle duality of quantum physics applies to both the electron and the photon. By energy conservation the electron in flight which is a rapid oscillation due to dark velocity that flattens it behaves as a particle that rides embedded between a suitable pair of parallel basic cosmic waves as its envelope (Figure 8.4.1). It is this wave envelope that gives it wave characteristics which is confirmed by shooting a beam of electrons through slits in a thin plate where the beam becomes waves emanating from the slits on the other side of the plate [40]. This is the explanation of wave-particle duality.

The notion that a particle is both a wave and particle is a contradiction. A particle is an autonomous physical system independent of any medium while a wave is suitably synchronized vibration of the medium that projects the appearance of linear motion analogous to the linear "motion" of neon lights due to synchronized switching of their switches. That medium is unknown in conventional physics; therefore, this contradiction cannot be resolved there.

We look at this resolution closely. Recall that the electron is a bulged segment of a semi-agitated (hence, visible) superstring formed by a sinusoidal helix (rapid spiral with sinusoidal profile; see profile of the primum in Figure 5.2.1), where the helical cycles are infinitely close and its toroidal flux, a non-agitated superstring, travels through the cycles uniformly at the speed of 7×10^{22} cm/sec [2,8]. However, when the electron is in flight and rides embedded between the full arcs of a pair of parallel basic cosmic waves (Figure 8.5.1), its carrier and envelope, the sinusoidal helix flattens into rapid oscillation due to dark viscosity with arcs infinitesimally close together, a thin solid figure of sinusoidal shape that gives it particle properties and the toroidal flux retaining the speed of 7×10^{22} cm/sec, by energy conservation. At the same time, the embedding pair of sinusoidal arcs of the embedding waves that serves as its envelope gives it wave characteristics.

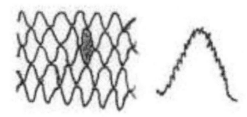

Figure 8.5.1. A segment of beam of basic cosmic waves of same order of magnitude, a primum or photon in flight (shaded) lodged between two basic cosmic waves of opposite crests with wave length of the same order of magnitude as its envelope (left). An arc of basic cosmic wave showing the first two terms of its nested fractal sequence (right).

The solidity of suitably close toroidal fluxes is analogous to the solidity of an ordinary object, say, a piece of iron. All that we have in the latter are toroidal and vortex fluxes of superstrings. In fact, the atomic nuclei which consist of toroidal fluxes are so far apart that at the micro scale iron is practically a vacuum.

Recall also that a photon is a primum in flight that breaks off from its loop, the helical cycles converting similarly to rapid oscillation inside the carrier sinusoidal envelope, its arcs infinitesimally close. The toroidal flux travels through the arcs of the rapid oscillation at the same speed of 7×10^{22} cm/sec, its forward flux speed equal to the speed of the carrier basic cosmic wave which is the speed of light $c = 10^{10}$ cm/sec. Like the primum in flight the rapid oscillation inside the embedding carrier basic cosmic wave gives the photon particle characteristics while its embedding basic cosmic wave envelope gives it wave characteristics. The difference between a primum in flight and a photon is that the former is always stable being a loop while the photon is only stable when the forward toroidal flux speed (forward component) is equal to the carrier basic cosmic wave speed $c = 10^{10}$ cm/sec, the speed of light in dark matter. Otherwise, the photon disintegrates, its toroidal flux remaining in dark matter. This explains why the photon has no rest mass. At the same time, the disintegration of the photon conforms to energy conservation since its toroidal flux, the repository of its latent energy, survives and remains dark matter.

The photon rides only on basic cosmic wave that fits it, by resonance and radiation of long wave length does not carry a photon. Its kinetic energy is imparted by its generator and reinforced by the synchronized vibration of dark matter.

8.6. Nuclear Explosion

It is known that the splitting of a heavy nucleus, e.g., the nucleus of uranium, results in formation of lighter elements (derivatives) where the combined masses of the resulting elements is less than the mass of the original nucleus. Present knowledge attributes this to the release of the binding force that supposedly holds the nucleus together, specifically, the supposed force that keeps the protons together in the narrow confines of the nucleus despite the enormous repulsion between them. Our qualitative model of the atom says that such a force does not exist. The stability of the nucleus simply follows from flux compatibility specifically a −quark joins every pair of protons in the nucleus.

Now we have two issues to explain: (a) where do the energy and (b) light elements released by nuclear fission come from? Recall that every nucleus sucks and accumulate superstrings, by flux-low-pressure complementarity. When a neutron of suitable kinetic energy (0.25 calories) penetrates the nucleus it agitates the superstrings in the nucleus and converts them to prima and photons creating enough instability to split it. (The energy of 0.25 calories provides just enough interaction to generate enough kinetic energy for instability to split the nucleus. Fission releases neutrons of 0.25 calories kinetic energy each that trigger fission of other nuclei. When sufficiently enriched ball of uranium fuel of critical mass (slightly more than 2 pounds) is split into two hemispheres at suitable distance apart a stable chain reaction is maintained. When the two hemispheres are pushed towards each other to form a solid ball of enriched uranium fuel geometric sequences of fission and geometrically increasing amount of energy is generated capped by explosion or release of enormous quantities of prima and radiations, some photonic and others non-photonic. The explosion recreates the high temperature and kinetic energy of our early universe that allows only formation of light elements from the converted prima.

(**Historical note.** Controlled sustained stable nuclear reaction using uranium fuel was achieved for the first time in an improvised reactor under the football stadium at the University of Chicago early in the Second World War. It was the predecessor of the present nuclear reactor).

8.7. Thermonuclear Reaction

This phenomenon is currently misread the reason there is no breakthrough in fusion research that leaves nil possibility for utilization of thermonuclear energy. It is believed that when hydrogen atoms are suitably compressed (e.g., by forcing them through narrow tunnel mechanically) their nuclei merge, shed their energy in the form of heat and form heavier nuclei like helium. In the hydrogen bomb compression is supposed to be effected by the explosion of the trigger atom bomb that presses the hydrogen atoms against the bomb shell and among themselves. This understanding is at odds with energy conservation. There is no natural law that supports it. Even in the formation of heavy isotope once the protons form a nucleus no proton can join in because charged prima are repelled by the positive or negative prima already in it. To merge two nuclei would mean introducing a proton into each other's nucleus. This amounts to reverse alchemy unsupported by the laws of nature. The repulsion between two protons alone in the narrow confines of the nucleus has been estimated at 27 tons from the perspective of conventional physics. Of course, such merging can be done at great infusion of energy which would hardly generate significant energy, the aim of fusion research. This scenario is more likely:

The explosion of the trigger atom bomb agitates the superstrings in the hydrogen nuclei and converts them to simple prima (initially), heat, radiation (photonic or non-photonic) and shock waves released in thermonuclear explosion. The converted simple prima form neutrino and light nucleons like those of helium. The trigger atom bomb in the hydrogen bomb cannot be duplicated and controlled in a reactor and, therefore, the chance of a breakthrough in utilization of thermonuclear energy is remote.

8.8. Unsatable Elementary Particles

There are now about 200 unstable elementary particles (prima) produced in the laboratory. Except for the basic prima and neutrino, they vanish in split second. How do we explain it? The natural principles that provide the answer are non-redundancy and non-extravagance [8] that, in effect, say nature does not need them. They are man-made prima produced by agitation of superstrings in the nucleus due to the impact of the energized proton in the collider. The free simple positive prima like the positron and the +quark and anti −quark are separated from the negative prima by the Earth's gravitational flux and remain in the mesosphere and possibly beyond [22, 39].

8.9. Earthlights and Balls of Fire

The interface of turbulence where the two parts press against each other, e.g., at tectonic plate boundary, vibrating atoms and molecules of interfacing materials generate seismic waves known to soften metal and crack or pulverize concrete and produce balls of fire along geological faults and around volcanoes [16,21]. Lightning in the lower atmosphere also generates seismic waves that convert dark matter to earthlights –sprites, elves, blue jets – and gamma rays, in the mesosphere [44].

8.10. Brittle and Malleable Material

We summarize briefly the nature of brittle and malleable materials. Brittle material has no free electrons; when vibrated vigorously say, by whizzing it through the atmosphere at 12,500 mph, e.g., Columbia Space Shuttle's final return flight, the smoothness of the insulation panel of the Space Shuttle becomes rough with vibrating molecules so that there is collision between them and the molecules of the atmosphere (friction) [22]. Consequently, their valence electrons are expelled; the material cracks or even pulverizes and the heat of friction burns it. This explains one of the puzzles of the Columbia disaster [8,22]: the burning of the insulation panels that had been going on during the previous decades of space exploration. Moreover, brittle material is electrical insulator since there are no free electrons to serve as carrier of charge. It is also heat insulator because heat is due to vibration of material and brittle material does not vibrate, it cracks or pulverizes, e.g., concrete and ceramics.

The interface of turbulence between the atmosphere and insulation panels of the Columbia Space Shuttle generated seismic waves that softened or melted the malleable materials like metal due to the greater rate of expulsion of valence electrons than their rate of replacement. This was the effect on the metal chassis and attachments of the Shuttle that explains its separation and eventual crash, the second puzzle of that disaster.

Malleable material like metal has free electrons. When distorted, e.g., bent, valence electrons are expelled. Distortion is stored when free electrons replace them. This is elasticity. Steel has almost perfect elasticity and is commonly used as spring. Vibration is rapid distortion and "restorsion" of malleable materials. When metal is vibrated and the rate of expulsion of valence electrons is greater than the rate of replacement it softens or even melts. Valence electrons may be expelled also by seismic waves and soften metal attachments of building foundations and concrete reinforcement during earthquake that knock out buildings and crack or even pulverize concrete. It is not really the generally gentle rocking action during earthquake that causes much destruction but the impact of seismic waves. There is technological possibility here: making suitable alloy resistant to the softening impact of seismic waves and composite resistant to cracking and pulverization by seismic waves. It offers the possibility for constructing earthquake-proof structures.

8.11. Metal Fatigue

This phenomenon is not presently understood. Bridges suddenly collapse, a plane's pylon that connects the engine to the wing falls and metallic casing of a train's wheel breaks off, gets stuck dangling under the train and gets caught at rail junction causing terrible derailment accident at rail junction. The last tragedy happened in Germany five years ago. Investigators attributed the accident to metal fatigue but they did not know what this phenomenon was. An earlier tragedy happened to an American Airlines plane just after take-off from O'Hare International Airport in Chicago in the 70s when a pylon broke and the engine fell off causing the plane to crash and killing all 392 passengers.

Metal is not perfectly elastic; when subjected to repetitive distortion like vibration, the net loss of valence elections accumulates over time, reaches a critical point and the metal snaps. This explains the cause of the above accidents involving repetitive motion, mainly vibration. One can experiment on it using metallic paper clip by bending it back and forth; it will eventually snap and break. This infinitesimal but progressive deterioration of vibrating metal is not detectable. Therefore, test of material should be made at appropriate conditions for its use to determine when metal fatigue sets in to avoid accident.

8.12. Brownian Motion and Pressure

Brownian motion is due to the impact of cosmic waves on the atoms and molecules of gas and liquids that throws them in all directions and, combined with their collisions, they are sent into erratic motion. When confined their motion creates pressure on the container, by momentum conservation. Since the atoms and molecules of gas are light and widely dispersed the effect of gravity on pressure is minimal. However, for liquids at least as dense as water pressure from gravity is considerably higher that a metallic pipe at suitable depth under water can be flattened. Brownian motion can be seen through a beam of light in a dark dusty room.

8.13. Physical Dimension

It is easy to define dimension in a vector space: the maximum number of linearly independent vectors in the space. Mathematical dimension can be infinite. Thus, the dimension of the Hilbert cube is countably infinite.

In physics this concept is quite vague and sometimes viewed with mysticism. For instance, a black hole is supposed to suck matter from our universe and bring it to another dimension. The space of relativity is generally considered four-dimensional, three space dimensions and time. However, the Lorentz transformation reduces it to three for it establishes dependence between time and a space coordinate.

How do we well define this notion in physics? We first make some clarification. We qualify that a physical concept exists in the real world, e.g., some object including symbol or measurable entity like temperature, pressure and humidity, which are not man-made concepts of thought. How about *time*? Is it a physical concept? Obviously, not since we do not find it in the real world. It is a man-made concept of thought that expresses relationship between events. So is the concept *distance* that expresses a relationship between physical objects or events. However, as mathematical concepts they describe natural phenomena. In particular, they can be used for devising a coordinate system. Thus, dimension depends on the subject of investigation. In meteorology, for instance, we can have the three space dimensions plus time, temperature, pressure, visibility and humidity for a total of 8 dimensions.

Since every physical system is bombarded by cosmic waves from all directions which resonate with the superstring, the latter's vibration has large dimension in the dark region of matter. At the micro (atomic scale) and macro scales, however, the dimension of vibration is limited by the internal-external dichotomy law [24]. For example, the nucleus of the atom vibrates normal to its equatorial plane. Therefore, its motion has four dimensions, i.e., one more than the three dimensions of ordinary space. Cosmic waves do not resonate with physical systems beyond the micro scale and, therefore, has only indirect measurable impact on natural phenomena. In other words, the dimension of physical space in the visible region is limited to four but quite large in the dark region. It should be noted that the dimension of Einstein's relativistic space is reduced to only three by the Lorentz transformation.

8.14. Celestial Spectacle

Jet outflow. Many interesting dynamics are displayed by galaxies and stars. Among the spectacular dynamics in the Cosmos is jet outflow of hot gas or pure prima from the cores of nascent stars and galaxies [8]. Jet outflow was the first known case of matter speeding as fast as 75 times faster than light [3]. How do we explain it? The eye of a vortex is cylindrical and normal to the plane of its discular halo along the equatorial plane, much like that of the tornado or primum. The rapid spin of the core (pure prima) around the boundary of the eye (event horizon) builds up tremendous kinetic energy and accumulation of hot gas or prima that must find a soft spot to escape through and that is the eye itself. Thus, jet outflow pops out of the eye of a young galaxy or star in opposite directions at great speed as much as 75 times the speed of light [3]. As the accumulated mass at the core cools down, the eye extremities suck the mass inward leaving spherical mass slightly flattened at the poles. This is quite evident in the Earth's flattened polar region (its curved rim seen at Lookout Restaurant near Sydney).

Supernova. It is commonly understood that a supernova is explosion of a star as the terminal phase of its evolution. There is no scientific basis for it since any cosmological vortex is stable. Moreover, if this where true, supernova would have been a rather common occurrence with trillions upon trillions of stars in the Cosmos. On the contrary, it is a very rare occurrence. The only plausible explanation of a supernova is collision between two stars of comparable masses in the same galactic equatorial plane, by flux compatibility. When the vortex rims of two stars of comparable masses and opposite spins are sufficiently close the momentum of their equatorial attraction forces their vortex fluxes to overlap smoothly at first since their fluxes have the same direction but when the vortex rim of each goes beyond the eye of the other their fluxes being at opposite directions explode, by flux compatibility. This is confirmed by the double ring of wave fronts traced by visible matter (see picture of expanding double ring of wave fronts caused by the double explosion in [59]). Simultaneously, with the double explosion is the crash of flux barrier between their eyes resulting in a third and even more powerful explosion whose wave front is traced by a bigger ring of wave front [59]. However, when they have the same spin they repel each other, by flux compatibility. Supernova is the dual of matter-anti-matter interaction of quantum gravity. Supernova may also occur between two stars coming from different galaxy.

Ultra-energetic cosmic waves. Cosmic rays of energy level as much as 10^{21} eV have been reported recently [50]. Traditional theories require them to be heavy elementary particles, possibly protons, coming from outside a 100-million-light-year radius from Earth. The estimate is based on supposed absence of possible source within that radius from the perspective of traditional theories. Acceleration of material object to great speed e.g., proton, is possible through centrifugal force imparted by the powerful spin of some galaxy. If some stars are catapulted so are protons. However, charged particle encounters resistance in flight and the neutron is too heavy to ride on cosmic wave and cannot be sustained at great speed by the natural vibration of dark matter. These energetic cosmic rays are not necessarily particles but cosmic waves that pack huge latent energy through their fractal configuration. They are known to smash protons in the mesosphere [46] that, incidentally, confirms our prediction that positive prima are pushed high up by the Earth's gravitational flux [50]. Such cosmic waves could have come from the cores of powerful galaxies. Then there are energetic gamma-ray bursts coming from distant regions of the Cosmos and some scientists theorize that they are due to black hole explosion.

9. PREDICTIONS

Since GUT has complete explanation of natural phenomena, particularly, the forces and interactions of nature from micro/quantum to macro/galactic scales including the strong and weak forces, electromagnetism and gravity we now make the following predictions:

(1) It was predicted by the author in 1999 [15] that since the stability of the nucleus was fully explained by GUT the gluon did not exist. Then in 2004 physicists were excited by the discovery of the third quark in the nucleus outside the proton in 2004 which they thought was the gluon. The author was then the Science Editor of the Manila Times and there were calls asking what he thought about the discovery. He explained that since it is the negative quark that joins two positive quarks of the proton and the third quark also joins two positive quarks, one from each of two protons then, by nonredundancy and non-extravagance, the third quark must be the negative quark, discovered much earlier in Fermilab near Chicago. The excitement fizzled out quickly. Thus, the original prediction of 1999 that the gluon does not exist holds. A pair of protons in the nucleus is coupled and joined together by a −quark through two +quarks, one from each proton, by flux compatibility. This coupling accounts for the strong force.

(2) Since the strong force in the nucleus of the atom is already accounted for by flux compatibility with the −quark as the connector between atoms the Higgs boson does not exist, by non-redundancy and non-extravagance.

(3) The graviton does not exist. Charge is explained by flux compatibility which, together with flux-low-pressure complementarity accounts for attraction and repulsion between two prima. The motion of the orbital electrons is governed mainly by flux compatibility and flux-low-pressure complementarity and centrifugal

force. Gravity is suction by the eye of a cosmological vortex. Interaction between cosmological vortices is governed by flux compatibility and flux-low-pressure complementarity and, together with centrifugal force, also governs the orbital motion of minor cosmological vortices of a cosmological vortex, another instance of duality between quantum and macro gravity.

(4) Just as thermonuclear reaction which cannot be replicated in man-made laboratory (without its destruction) because the impact of the trigger atom bomb cannot be replicated there, the Big Bang cannot be replicated in man-made laboratory because dark depression and cosmic sphere, the impact of the Big Bang cannot occur there. Moreover, non-basic prima created in man-made laboratory are unstable and short-lived, by the principles of nonredundancy and non-extravagance.

(5) Given that the Earth's gravitational flux has significant shielding effect on asteroids whizzing towards the Earth and its vicinity [20], the chance of a disastrous asteroid collision here on Earth during the present millennium is quite remote.

10. THE COMPLETE THEORY

At this point we have solved the long standing unsolved problems of physics, explained natural phenomena and resolved fundamental questions except one: what was the Big Bang? This is the only remaining question for GUT to resolve and become a complete theory.

We have developed GUT based on natural laws including the main premise of macro gravity: the occurrence of the Big Bang. Using GUT we have established that the destiny of every cosmological vortex is a black hole in its eye. Therefore, the destiny of the core of our universe as a cosmological vortex – a super…super galaxy – is a super…super massive black hole. We know that a black hole is massive concentration of non-agitated superstrings. Can a black hole explode? Some physicists think that the gamma ray bursts come from distant massive explosion of black holes. However, without valid analysis based on the laws of nature it remains pure speculation.

Let us return to our question. We know that a suitable pair of cosmic waves converts a superstring to primum or photon. The first cosmic wave converts the non-agitated superstring to semi-agitated superstring and the second coverts the semi-agitated superstring to a primum or photon. Denote by K_c and K_s the kinetic energy of the second cosmic wave and the latent energy of the semi-agitated superstring converted to kinetic energy, respectively, and K_p the kinetic energy of the primum or photon. By energy conservation, we have $K_p = K_c + K_s$ so that there is a net gain δ of kinetic energy over the energy imparted by the second cosmic wave. Let us call this process trans-conversion. As in nuclear explosion where there is a critical mass that triggers geometric sequences of fission that end up in nuclear explosion, there is a critical level of accumulated energy that triggers geometric chain reaction, i.e., geometric sequences of trans-conversions that culminate in explosion. Reaching the critical level depends on the number of pairs of cosmic waves that generates trans-conversion.

When the accumulated energy does not reach critical level it dissipates in dark matter over time. Clearly black hole explosion is rare at least in the vicinity of our galaxy.

Now, the final question is: what sort of black hole exploded and created the super…super depression that triggered the birth of our universe and sustained its evolution into the present super...super galaxy? By energy conservation, it must have been a super...super massive black hole, the destiny of the core of a previous universe that was at least as powerful and massive as our super...super galaxy. That was what exploded as the Big Bang.

Therefore, we finally complete axiom about the Big Bang that makes GUT a complete theory:

The Big Bang. *The Big Bang was explosion of a super...super massive black hole, the destiny of the core of a previous universe that occurred 8 billion years ago.*

REFERENCES

[1] Astronomy (a) August 1995, (b) January 2001, (c) June 2002.
[2] Atsukovsky, V.A. *General ether-dynamics; simulation of the matter structures and fields on the basis of the ideas about the gas-like ether*; Energoatomizdat: Moscow, 1990.
[3] *Discover*, (a) Feb. 1996, (b) July 1999, (c) Nov. 1998; (d) Dec. 1999.
[4] *Encarta Premium*, 2008.
[5] Escultura, E. E. *Diophantus: Introduction to Mathematical Philosophy (With probabilistic solution of Fermat's last theorem)*, Kalikasan Press: Manila, 1993.
[6] Escultura, E. E. Exact solutions of Fermat's equation (A definitive resolution of Fermat's last theorem, *J. Nonlinear Studies*, 1998, 5, 2, pp. 227 - 254.
[7] Escultura, E. E. The generalized integral as dual of Schwarz distribution, invited paper, *J. Nonlinear Studies*.
[8] Escultura, E. E. Scientific Natural Philosophy, in press, Bentham Science Publishers.
[9] Escultura, E. E. The new real number system and discrete computation and calculus, *J. Neural, Parallel and Scientific Computations*, 2009, 17, pp. 59 – 84.
[10] Escultura, E. E. The new mathematics and physics, *J. Applied Mathematics and Computation*, 2003, 138, 1, 145 – 169.
[11] Escultura, E. E. Chaos, turbulence and fractal, *Indian J. Pure and Applied Mathematics*, 2001, 32, 10, pp. 1539 – 1551.
[12] Escultura, E. E. The mathematics of the grand unified theory, Proc. 5th World Congress of Nonlinear Analysts, *J. Nonlinear Analysis, A-Series: Theory: Method and Applications*, 2009, 71, pp. e420 – e431.
[13] Escultura, E. E. The mathematics of the new physics, *J. Applied Mathematics and Computations*, 2002, 130, 1, pp. 149 - 169.
[14] Escultura, E. E. Dynamic Modelling and the new mathematics and physics, *J. Neural, Parallel and Scientific Computations* (NPSC), 2007, 15, 4, PP. 527 – 538.
[15] Escultura, E. E. Superstring loop dynamics and applications to astronomy and biology, *J. Nonlinear Analysis, A-Series: Theory: Methods and Applications*, 1999, 35, 8, pp. 259 – 285.
[16] Escultura, E. E. From macro to quantum gravity, *J. Problems of Nonlinear Analysis in Eng'g Systems*, 2001, 7, 1, pp. 56 – 78.
[17] Escultura, E. E. Turbulence: theory, verification and applications, *J. Nonlinear Analysis, A-Series: Theory, Methods and Applications*, 2001, 47, 8, pp. 5955 – 5966.

[18] Escultura, E. E. Vortex Interactions, *J. Problems of Nonlinear Analysis in Engineering Systems,* 2001, 7, 2, pp. 30 – 44.
[19] Escultura, E. E. Dynamic Modeling of Chaos and Turbulence, Proc. 3rd World Congress of Nonlinear Analysts, *J. Nonlinear Analysis, A-Series: Theory: Method and Applications,* 2005, 63, 5-7, pp. e519-e532.
[20] Escultura, E. E. Global geology and oceanography, invited, *International J. Earth Sciences and Engineering.*
[21] Escultura, E. E. *Dynamic Modelling and Applications*, Proc. 3rd International Conference on Tools for Mathematical Modeling, 7, State Technical University of St. Petersburg, St. Petersburg, 2003, pp. 103 - 114.
[22] Escultura, E. E. The Pillars of the new physics and some updates, *J. Nonlinear Studies,* 2007, 14, 3, pp. 241 – 260.
[23] Escultura, E. E. The solution of the gravitational n-body problem, *J. Nonlinear Analysis, A-Series: Theory, Methods and Applications,* 1997, 30, 8, pp. 5021 - 5032.
[24] Escultura, E. E. The grand unified theory, contribution to the Felicitation Volume on the occasion of the 85th birth anniversary of Prof. V. Lakshmikantham, *J. Nonlinear Analysis, A-Series: Theory: Method and Applications,* 2008, 69, 3, pp. 823 – 831.
[25] Escultura, E. E. Qualitative model of the atom, its components and origin in the early universe, *Proc. 5th World Congress of Nonlinear Analysts, J. Nonlinear Analysis, B-Series: Real World Applications,* 2009, 11, pp. 29 – 38.
[26] Escultura, E. E. *Dynamic and mathematical models in physics*, Proc. 5th International Conference on Dynamic Systems and Applications, June 30 – July 5, 2007, Atlanta.
[27] Escultura, E. E. The basic concepts and dynamics of quantum gravity with applications, invited, *J. Nonlinear Studies.*
[28] Escultura, E. E. The trajectories, reachable set, minimal levels and chain of trajectories of a control system, *Ph. D. thesis*, University of Wisconsin, 1970.
[29] Escultura, E. E. Basic Physical Concepts and Dynamics of the Hybrid Grand Unified Theory, in press, *Proc. International Conference on Recent Advances in the Mathematical Sciences*, GVP College of Engineering, JNT University, Visakhapatnam, AP, India, December, 2009
[30] Escultura, E. E. The Unified Theory of Evolution, invited, *International Journal of Biological Sciences and Engineering.*
[31] Escultura, E. E. Extended geometrical generalized fractals, Chaos and Applications, a chapter in, *Fractal Classification and Application*, ebook in press, Nova Science Publishers.
[32] Escultura, E. E. Superstring interactions and computation, in press, *Neural, Parallel and Scientific Computations.*
[33] Escultura, E. E. Genetic Alteration, Modification, Sterilization With Applications to the Treatment of Genetic Diseases, accepted , *J. Science of Healing Outcomes.*
[34] Escultura, E. E. The origin and evolution of biological species, *J. Science of Healing Outcomes,* 2010, 6-7, pp. 17 - 27.
[35] Escultura, E. E. The physics of the mind, accepted, *J. Science of Healing Outcomes.*
[36] Theory of Intelligence and Evolution, *Indian Journal of Pure and Applied Mathematics,* 2002, 33(1), pp. 1 – 18.
[37] Gerlovin, I. L. *The Foundations of United Theory of Interactions in a Substance,* Energoattomizdat: Leningrad, 1990.

[38] *Guinness Book of Records*, 2002.
[39] Lakshmikantham, V.; Escultura, E. E.; Leela, S. *The Hybrid Grand Unified Theory, Atlantis (World Scientific)*: Paris, March 2009.
[40] Merchbacher, E. *Quantum Mechanics* (2nd ed): John Wiley & Sons, Inc. New York, 1970.
[41] New Scientist, (a) *Cannibalism by giant galaxies*, July 1997; (b) *Fractal universe*, November 1999.
[42] Nieper, H. A. *Revolution in Technology, Medicine and Society*, Management Interessengemeinschaft für Tachyon-Feld-Energie: Odenburg, FRG, 1, 1984.
[43] *Our Solar System*, A Reader's Digest Young Families Book, *Joshua Morris Publishing*, Inc., Westport, CT, 1998.
[44] Pendick D. Fires in the Sky, *Earth*, June 1996, 20, pp. 62 – 64.
[45] Pontrjagin, L. S.; Boltyanskii; V. G., Gamkrelidze; R. V.; Mischenko, E. F. *The Mathematical Theory of Optimal Processes*, (K. N. Trirogoff, Tran.; L. W. Neustadt, Ed.), Interscience Publishers: New York, 1962.
[46] Ridpath, I., *Atlas of Stars and Planets*, George Philip Ltd.: London, 1999.
[47] Science, *Ultraenergetic particles*, Aug. 1998, 281, 5379, pp. 241 – 242.
[48] Science, *Watching the universe's second biggest bang*, March 1999, 283, 5410, pp. 2013 – 2014.
[49] Science, *Glow reveals early star nurseries*, July 1998, pp. 332 – 333.
[50] Science, *Galaxy's oldest stars shed light on Big Bang* (hydrogen, helium lithium and others no heavier than boron), Nov. 1998, 282, 5382, pp. 1428 – 1429.
[51] Science, (a) *Cosmic motion revealed*, December 1998, 282, 1998,*5397*; (b) *No Backing off from accelerating universe*, Nov. 1998, 282, 5392, pp. 1248 – 1249.
[52] Science, (a) *Weighing in on neutrino mass*, 1997, 280 pp. 1689 – 1691; (b) *Neutrinos weigh in*, Dec. 1998, 282, *5397*, pp. 2158 – 2159; (c) *Search for neutrino mass is a big stretch for three labs*, February 1999, 283, 5405, pp. 928 – 929.
[53] Science, *New clues to the habits of heavy weights* (black holes at craters of galaxies), 1999, 283, 5401, pp. 480 – 481.
[54] Science, (a) *Starbirth, gamma blast hint at active early Universe*, Dec., 1998 282, 5395, p. 1806; (b) *Gamma burst promises celestial reprise*, Jan. 1999, 283, 5402, p. 616; (c) Powerful cosmic rays tied to far off galaxies (100 million times reached in largest particle accelerators), Nov. 1998, 282, 5391, p. 1023.
[55] Science, The mystery of the migrating galaxy clusters, Jan. 1999, 283, 5402, pp. 625 – 626.
[56] Scientific American, *The quasars and early galaxies*, April 1983, pp. 708 – 745.
[57] Scientific American, *Galactic collision*, April 1995, pp. 11 – 14.
[58] *The Earth Atlas*; (b) *The Oceans Atlas*, Dorling Kindersley: London, 1994.
[59] *The Great Discoveries*, Time Books, 2001.
[60] Young, L. C. Generalized curves and the existence of an attained absolute minimum in the calculus of variations, *Camp. Rend. Soc. Sci. Letter, Cl III*, 30, Varsovie, 1937, pp. 211 – 234.
[61] Young, L. C. On Generalized surfaces of finite topological types, *Memoirs of the AMS*: Philadelphia, 1955.
[62] Young, L. C. *Lectures on the Calculus of Variations and Optimal Control Theory*, W. B. Saunders: Philadelphia, 1969.

[63] Young, L. C. *Mathematicians and Their Times*, North-Holland: Amsterdam, 1980.
[64] Young, L. C.; Nowosad, P. Generalized curve approach to elementary articles, *JOTA*, 1983, 41, 1, pp. 261 – 273.

In: The Big Bang: Theory, Assumptions and Problems
Editors: J. R. O'Connell and A. L. Hale

ISBN: 978-1-61324-577-4
© 2012 Nova Science Publishers, Inc.

Chapter 3

COSMIC STRUCTURE FORMATION AFTER THE BIG BANG

L. M. Chechin[1]

V.G.Fessenkov Astrophysical Institute,
National Center for Cosmic Researches and Technologies,
National Space Agency, Almaty, Kazakhstan

ABSTRACT

In the new version of the Universe large-scale structures forming based on the refuse of analyses, only the gravitational instability of cosmological substrate is proposed. It is shown that vacuum is the dominant of non-baryonic matter in the Universe which creates the anti-gravitational instability of baryonic cosmic substrate itself and causes the galaxies' formation.

The growth of a baryonic substance's perturbations that is caused by the non-stationary equation of state of a non-baryonic matter in the very early Universe is searched. It was shown that initially their amplitudes are drastically larger than perturbations of the baryonic matter in the framework of the standard scenario of gravitational instability.

For a deeper understanding of the process of baryonic matter evolution in the expanding Universe, it is necessary to know the physical property of the concrete field that represents the background of the substrate type of dark energy. Besides, it is necessary to explore in detail the influence of such a field on the continuous medium of baryonic matter. These statements were realized for the quintessence field. As the result, we describe the quintessence field by two gravitating scalar fields. They give contributions at the total pressure and at the total mass density of baryonic matter. It allowed to show that the evolution of baryonic matter's density perturbations obeys the equation of forced oscillations and admits the resonance case, when amplitude of baryonic matter's density perturbations gets the strong short-time splash. This splash is interpreted as a new macroscopic mechanism of the initial matter density perturbations appearance.

The stochastic equation of a thread's motion in the background of the massive cosmic string whose linear mass density is subjected to stochastic perturbations was investigated. It was shown that stochastic movement of the thread generates, in a cosmological substrate, the long perturbing waves that lead to the galaxies formation, in its turn.

---

[1] E-mail address: chechin@aphi.kz.

The interval of a weak-oscillating cosmic string with standing flat disturbance waves on it was deduced and the brief discussion of some physical processes occurring in the vicinity of oscillating massive cosmic strings followed. Among them, the dynamics of open "probe" cosmic thread performing the constraint oscillations. The nature of these oscillations, as it was pointed out, is drastically different from the previous well-known mechanisms.

It is shown that domain wall pumps a cosmic string by gravitational energy if it is located in its vicinity. This leads to the tension energy increasing and cosmic string stretching. The discussing process of straighten is the new physical mechanisms of cosmic string stretching.

The rotational galaxy movement produced by vacuum anti-gravitation force was searched. The expression for suitable angular velocity is found and its estimation for the real elliptical galaxies is done. One cosmological consequence of the vacuum angular velocity effect for the Universe description – the Birch effect – is discussed also.

Keywords: baryonic matter's density perturbations, quintessence field, non-stationary equation of state of the Universe, cosmic string stretching, dynamics of oscillating cosmic strings, vacuum-induced galaxy rotation, the Birch effect.

PACS number: 98.80 – k

1. INTRODUCTION

For discovering the matter of this chapter, we would briefly remain on the basic epochs of the Universe evolution after the Big Bang following (see, for example, some well-known monographs [1 - 5]) and focused our attention on thosewhere the below-mentioned structures were formed.

So, the time-instant in which the Universe began rapidly expanding from an extremely high energy density is known as the Big Bang. It was about 13.7 billion years ago and named the Planck epoch - 10^{-43} seconds after the Big Bang.

The inflationary epoch occurred further - between 10^{-36} seconds and 10^{-32} seconds after the Big Bang. During inflation, the universe is flattened and rapidly became homogeneous and isotropic, in which later the seeds of structure formation are laid down in the form of a primordial spectrum of nearly-scale-invariant fluctuations of baryonic substrate.

The early Universe consists of quark epoch - between 10^{-12} seconds and 10^{-6} seconds after the Big Bang; hadrons' epoch - between 10^{-6} seconds and 1 second after the Big Bang; lepton epoch - between 1 second and 10 seconds after the Big Bang; nucleosynthesis epoch - between 3 and 20 minutes after the Big Bang.

At the matter domination epoch – 70,000 years – the densities of non-relativistic matter (atomic nuclei) and relativistic radiation (photons) are equal. The Jeans length which determines the smallest fluctuations, that can form (due to competition between gravitational attraction and pressure effects) large structures, began to grow in amplitudes.

Later on (till 150 million to 1 billion years), the large cosmic structures began to create. In fact, the Hubble images show an infant galaxy forming nearby, which means this happened very recently on the cosmological timescale. This shows that new galaxy formations in the Universe are still occurring.

Structure formation in the Big Bang model proceeds hierarchically, with smaller structures forming before larger ones. The first structures that form are quasars, which are thought to be bright, early active galaxies, and population III stars. Before this epoch, the evolution of the Universe could be understood on the basis of linear cosmological perturbation theory. In fact, all structures could be understood as small deviations from a perfect homogeneous Universe.

The goal of this chapter is the concretization of the above-cited general picture of large-scale structures formation. We set forth some new physical processes that can be the real reasons for large-scales structures formation and their subsequent evolution. This chapter is organized as follows.

In the new version of the Universe, large-scale structures forming based on the refuse of analyses, only the gravitational instability of cosmological substrate is proposed (Section 2). It was shown that vacuum is the dominant of non-baryonic matter in the Universe created by the anti-gravitational instability of baryonic cosmic substrate itself and caused the galaxies' formation.

The growth of a baryonic substance's perturbations that is caused by the non-stationary equation of state of a non-baryonic matter in the very early Universe is searched in Section 3. It was shown that initially their amplitudes are drastically larger than perturbations of the baryonic matter in the framework of the standard scenario of the gravitational instability.

For a deeper understanding of the process of baryonic matter evolution in the expanding Universe, it is necessary to know the physical property of the concrete field that represents the background of non-baryonic substrate type of dark energy. Besides, it is necessary to explore in detail the influence of such a field on the continuous medium of baryonic matter.

These statements were realized for the quintessence field (Section 4). As the , we describe a quintessence field by two gravitating scalar fields. They give contributions at the total pressure and at the total mass density of baryonic matter. It allowed to show that evolution of baryonic matter's density perturbations obeys the equation of forced oscillations and admits the resonance case, when amplitude of baryonic matter's density perturbations gets the strong short-time splash. This splash interprets as a new macroscopic mechanism of the initial matter density perturbations appearance.

The stochastic equation of a thread's motion in the background of a massive cosmic string whose linear mass density is subjected to stochastic perturbations was investigated in Section 5. It was shown that stochastic movement of the thread generates, in a cosmological substrate, the long perturbing waves which lead to the galaxies formation, in its turn.

In Section 6, we deduced the interval of a weak-oscillating cosmic string with standing flat disturbance waves on it and give the brief discussing of some physical processes occurring in the vicinity of oscillating massive cosmic strings. The dynamics of an open "probe" cosmic thread performing the constraint oscillations over its influence among them. The nature of these oscillations is drastically different from the previous well-known mechanisms.

It is shown (Section 7) that domain wall pumps the cosmic string by gravitational energy if it locates it in its vicinity. This leads to the tension energy increasing and the cosmic string stretching. The discussing process of straighten is the new physical mechanisms of cosmic string stretching.

Lastly, the rotational galaxy movement produced by vacuum anti-gravitation force is researched in Section 8. The expression for suitable angular velocity is found and its estimation for real elliptical galaxies is done. One cosmological consequence of the vacuum angular velocity effect for the Universe description – the Birch effect – is also discussed.

2. ANTI-GRAVITATIONAL INSTABILITY OF COSMIC SUBSTRATE IN THE NEWTONIAN COSMOLOGY

2.1. Introduction

The newest achievements of modern cosmology – the discovery of vacuum [1], dark matter and dark energy [2], fossil and invisible galaxies [3] – radically change the problem's organization about the observable structures' forming in the Universe. If the traditional approach in the galaxies' creation was the gravitational instability of baryonic substrate (see, for example, [4]), then the fact of essential dominance of the non-barionic matter in the Universe over the baryonic matter (their relation is about 95% and 5%, respectively) leads to the necessity of searching an anti-gravitational instability of cosmic substrate.

Being more concrete, this means the following – can vacuum (or other types of dark energy) itself be the reason of large-scale structures forming in the Universe? However, until now, the role of vacuum at the large-scale structures' forming was reduced to searching: the vacuum influence on the dynamics of galaxies' clusters [5], the dynamics of perturbations under action of vacuum in the standard cosmological model [6], the growth of substrate's fluctuations in the presence of vacuum [7] and dark matter [8], etc.

Instead of these works, vacuum is considered as the basic reason for growing perturbations in cosmological substrate in this article. For better clearness, we neglect the gravitational self-interaction of substrate and consider the vacuum's influence on it, only. Besides, to make the essence of this process, we search the cosmic substrate's anti-gravitational instability in the framework of Newtonian cosmology.

The matter of this chapter has been partially published in [9].

2.2. The Newtonian Equations of Hydrodynamics in the Vacuum Background

Write down the equations of hydrodynamics for baryonic matter (cosmic substrate) in Newtonian approximation in the vacuum background. For doing this, it is necessary to bear in mind that except for surface force, the additional vacuum force F_v acts at each elementary volume of baryonic substrate. So, in accordance with [4], we have

$$\frac{\partial \rho_m}{\partial t} + div(\rho_m \vec{v}) = 0, \qquad (2.1)$$

$$\frac{\partial \vec{v}}{\partial t} + (\vec{v} grad)\vec{v} + \frac{1}{\rho_m} grad P_m - \vec{F}_v = 0. \qquad (2.2)$$

The relation between these forces may be different – it is determined by representative scales of baryonic matter distribution. Really, the analyses of dynamics of Local Group showed [7] that at distances smaller than one megaparsec ($r_0 \leq 1.0 Ìpñ$), the attractive force dominates in it; at distances up to several megaparsecs ($1.0 Ìpñ \leq r_0 \leq 10.0 Ìpñ$), the attractive force, is about repulsive force; but at distances larger than ten megaparsecs ($r_0 \geq 10.0 Ìpñ$,) the repulsive force, determined by vacuum, dominates in it. Later on, we will be interested in the case when vacuum is the basic perturbed factor of the structural Universe evolution. For that reason, the dynamical processes are searched at the minimal permissible space scales, namely at distances $r_0 \approx 1.0 Mpc$.

For search equations (2.1) – (2.2), descry that vacuum satisfies the following equation of state $p_v + \rho_v = 0$. Thus, it produces the anti-gravitation with negative gravitational energy is [1], i.e.

$$\rho_G = \rho_v + 3p_v = -2\rho_v. \tag{2.3}$$

Pick out now in the baryonic matter, that is moving in vacuum background, a sphere of radius r_0. Then at any fraction of matter inside of the sphere, located at distance r from its center $(r < r_0)$, the gravitational force

$$F_v = -\frac{4\pi G \rho_G}{r^2} \int_0^r \xi^2 d\xi = \frac{8\pi G \rho_v}{r^2} \int_0^r \xi^2 d\xi. \tag{2.4}$$

will active. The integration (2.4) gives the gravitational vacuum force - $F_v = \frac{8\pi G \rho_v}{3} r$. Putting it into (2.1) – (2.2), we find the equations of hydrodynamics of baryonic substrate in vacuum background

$$\frac{\partial \rho_m}{\partial t} + div(\rho_m \vec{v}) = 0, \tag{2.5}$$

$$\frac{\partial \vec{v}}{\partial t} + (\vec{v} grad)\vec{v} + \frac{1}{\rho_m} grad P_m - \frac{8\pi G \rho_v}{3} \vec{r} = 0. \tag{2.6}$$

Solution of equations (2.5) – (2.6), we'll look for by the perturbation method, putting that no perturbed state of motionless baryonic substrate yields the conditions - $\rho_{m_0} = const$, $P_{m_0} = const$ and $\vec{v}_0 = 0$, as in the Jeans theory. This enable us to write down the perturbed solution in the form of plane waves put on non-perturbed solution. Thus, we have

$$\rho_m(\vec{r},t) = \rho_{m_0} + \delta \rho_m = \rho_{m_0}\left[1 + \delta(t) \cdot \sin(\vec{k}\vec{r})\right], \tag{2.7}$$

$$\vec{v}(\vec{r},t) = 0 + \vec{w}(t) \cdot \cos(\vec{k}\vec{r}). \tag{2.8}$$

The pressure of baryonic substrate due to its general expression $P_m = P_m(\rho_m)$ may be decomposed into Taylor series

$$P_m = P_{m_0} + \left(\frac{\partial P_m}{\partial \rho_m}\right)_0 \delta\rho_m + \cdots = P_{m_0} + b^2 \delta\rho_m + \cdots, \tag{2.9}$$

where b - speed of sound in baryonic substrate. At last, in accordance with the problem's organization, we set the following limitation $\frac{\rho_v}{\rho_m} \ll 1$ and should regard that all additions are of the same order, i.e. $\delta(t) \sim \frac{w(t)}{v(r,t)} \sim \frac{\rho_v}{\rho_m}$. Under these conditions, we get the system of equations for finding additions of the first order

$$\left. \begin{array}{l} \dfrac{d\delta(t)}{dt} - \vec{k}\vec{w}(t) = 0 \\ \cos(\vec{k}\vec{r}) \dfrac{d\vec{w}(t)}{dt} + \vec{k} b^2 \cos(\vec{k}\vec{r}) \delta(t) - \dfrac{8\pi G \rho_v}{3} \vec{r} = 0 \end{array} \right\}. \tag{2.10}$$

This system is equal to differential equation

$$\frac{d^2\delta(t)}{dt^2} + A^2 \delta(t) = \Phi, \tag{2.11}$$

with coefficients

$$A^2 = k^2 b^2, \quad \Phi = \frac{8\pi G \rho_v}{3} \cdot \frac{(\vec{k}\vec{r})}{\cos(\vec{k}\vec{r})}. \tag{2.12}$$

For integrating equation (2.11), it is necessary to take into account the fact of baryonic substrate's expansion, created by vacuum. In another words, it is necessary to set the concrete functional $f(\vec{r}(t), \dot{\vec{r}}(t), t) = 0$, basing on Hubble law of expansion. Then equations (2.5), (2.6), together with such functional, will be the closed system for three variables - $\delta(t), \vec{w}(t), \vec{r}(t)$.

2.1. For an arbitrary cosmological model, the distance between any pare of points is described by functional $\dot{\vec{r}} - H\vec{r} = 0$. Thus, it evolves as

$$\vec{r}(t) = \vec{r}_0 \exp Ht, \tag{2.13}$$

where $\vec{r}_0$ is the given initial scale of baryonic substrate distribution. Hence, coefficients (2.12) take on the form

$$A = kb, \quad \Phi = \Phi(t) = \frac{8\pi G \rho_v}{3} \cdot \frac{(\vec{kr}_0) \exp Ht}{\cos(\vec{kr}_0 \cdot \exp Ht)}. \quad (2.14)$$

Physically, the most interesting is non-periodical solution of equation (2.11), because it enables to examine the evolution of baryonic substrate's perturbations during the cosmologically significant time.

The general form of such solution is

$$\delta(t) = \frac{8\pi G \rho_v}{3} \cdot \frac{(\vec{kr}_0)}{kb} \int_0^t \frac{\exp H\tau}{\cos(\vec{kr}_0 \cdot \exp H\tau)} \cdot \sin kb(t-\tau)d\tau \quad (2.15)$$

but its integration is rather difficult. Thus mark, that for arbitrary instant of time, the condition $\cos(\vec{kr}_0 \cdot \exp H\tau) \leq 1$ takes place. Hence, the perturbation of baryonic substrate's density will be larger than magnitude

$$\delta_{min}(t) = \frac{8\pi G \rho_v}{3} \cdot \frac{\vec{kr}_0}{kb} \cdot \int_0^t \exp H\tau \cdot \sin kb(t-\tau)d\tau. \quad (2.16)$$

Calculation of (2.16) gives

$$\delta_{min}(t) = \frac{8\pi G \rho_v}{3} \cdot \frac{\vec{kr}_0}{(H^2 + k^2 b^2)} \cdot \left(\exp Ht - \frac{H}{kb} \sin kbt - \cos kbt \right) \quad (2.17)$$

So, after every period of oscillations T, the perturbation of baryonic substrate's density increases at

$$\delta_{min}(T) = \frac{8\pi G \rho_v}{3} \cdot \vec{kr}_0 \cdot \frac{\exp HT}{H^2 + k^2 b^2} \quad (2.18)$$

times. Concerning the perturbation of baryonic substrate's velocity, we find that its magnitude after one period will be equal

$$\vec{w}(T) = \frac{8\pi G \rho_v}{3} \cdot \frac{H}{H^2 + k^2 b^2} \cdot \exp HT \cdot \vec{r}_0. \tag{2.19}$$

It is interesting to consider equation (2.11) in the concrete cosmological model. Considering the expansion of the uniform and isotropic cosmological model for the case $\rho_m < \rho_c$, where ρ_c - critical density of the Universe, it is possible to show [4] that the suitable functional has the form $\frac{dr}{dt} - \sqrt{\frac{8\pi G}{3} r_0^2 (\rho_c - \rho_m)} = 0$. Hence, the searching dependency is written down as

$$r(t) = r_0 \sqrt{\frac{8\pi G}{3}(\rho_c - \rho_m)} \cdot t. \tag{2.20}$$

Here, r_0 - the given above-mentioned magnitude of the sizes of baryonic substrate distribution. Introducing expression (2.20) into coefficients (2.12), for the given cosmological model we get,

$$A^2 = k^2 b^2, \quad \Phi(t) = \left(\frac{8\pi G}{3}\right)^{3/2} \cdot \frac{\vec{k}\vec{r}_0 \cdot \rho_v \sqrt{(\rho_c - \rho_m)} \cdot t}{\cos\left(\vec{k}\vec{r}_0 \sqrt{\frac{8\pi G}{3}(\rho_c - \rho_m)} \cdot t\right)}. \tag{2.21}$$

The solution of equation (2.11) with an account of non-periodical terms only, is

$$\delta(t) = \left(\frac{8\pi G}{3}\right)^{3/2} \cdot \frac{\vec{k}\vec{r}_0}{kb} \cdot \rho_v \sqrt{(\rho_c - \rho_m)}$$
$$\cdot \int_0^t \frac{\tau \cdot \sin kb(t - \tau) d\tau}{\cos\left(\vec{k}\vec{r}_0 \sqrt{\frac{8\pi G}{3}(\rho_c - \rho_m)} \cdot \tau\right)} \tag{2.22}$$

The calculation of this integral may be given approximately, thus we get

$$\delta(t) \approx \left(\frac{8\pi G}{3}\right)^{3/2} \cdot \frac{\vec{k}\vec{r}_0}{kb} \cdot \rho_v \sqrt{(\rho_c - \rho_m)} \cdot \left(\frac{t^2}{2} \sin kbt - kb \frac{t^3}{3} \cos kbt\right) \tag{2.23}$$

under the condition

$$b = r_0 \sqrt{\frac{8\pi G}{3}(\rho_c - \rho_m)}. \qquad (2.24)$$

From (2.23), it is possible to see that perturbation of baryonic matter's density, generally speaking, is growing in time infinitely and guarantying the fragmentation of cosmic substrate. But this leads to violation of the condition - $\delta(t) << 1$, at which the expression (2.23) was achieved. Nevertheless, the solution (2.23) describes the baryonic substance's evolution – its permanent growth – correctly in the limits of admitted quantities. Therefore, the anti-gravitational instability of baryonic cosmic matter leads to the effect of amplitude growing and must result in galaxies and its clusters forming.

As for the approximate expression of perturbation of baryonic substrate's velocity, it has the form

$$\vec{w}(t) \approx \left(\frac{8\pi G}{3}\right)^{3/2} \cdot \rho_v \sqrt{(\rho_c - \rho_m)} \cdot \left(t \frac{\sin kbt}{kb} - \frac{t^2}{2}\cos kbt\right) \cdot \vec{r}_0 \qquad (2.25)$$

testifying its permanent growth.

2.3. Some Cosmological Consequences

Consider some cosmological consequences of the obtained results with its relation to the standard results [4]. The Newtonian theory of galaxies' creation, as it is well-known, is applicable at matter-dominance epoch. At that time, the equation of state of baryonic substance (one-atomic ideal gas) characterizes by conditions: $P_m << \rho_m$, and speed of sound - $b^2 = \frac{5k_B T_R}{3m_H}$. Here, T_R - the temperature of hydrogen recombination; m_H - mass of atomic hydrogen, k_B - Boltzmann constant. Thus, speed of sound is $b \approx 7.5 \cdot 10^5 \, cm/\sec$. So far, the epoch of recombination begins from instant of time $t_R \sim 10^{13}$ sec (and elapses up to the present time), then maximum numerical value of the wave vector is - $k = k_R = \frac{2\pi}{\lambda_R} = \frac{2\pi}{ct_R} \propto 2.1 \cdot 10^{-23} \, cm^{-1}$. Hence, $kb \approx 1.5 \cdot 10^{-17} \sec^{-1}$ and thus $H >> kb$, if Hubble constant $H \sim 10^{-13} \sec^{-1}$. (At small time intervals, as it is known [4], $H = \frac{2}{3}t^{-1}$). So, instead of (2.18) and (2.19), we get approximately

$$\delta_{\min}(T) \approx \frac{8\pi G \rho_v}{3} \cdot \frac{k\vec{r}_0}{H^2} \cdot \exp HT, \quad \vec{w}(T) \approx \frac{8\pi G \rho_v}{3} \cdot \frac{\vec{r}_0}{H} \cdot \exp HT. \qquad (2.26)$$

As $\rho_v \approx 0.4 \cdot 10^{-29} g/ñm^3$, then for $r_0 \sim \lambda \sim ct_R$ and $T \to 0$, we get the following estimation of the relative initial addition to substance's density – $\delta_{min}(0) \propto 10^{-9}$. As for the numerical value of the initial velocity, it is equal to $w(0) \sim 10^1 cm/sec$ and obviously has the non-relativistic character. Hence, the minimum initial mass of the spherical fluctuation (the Jeans mass) may be classically estimated as –

$$M_f = \frac{4}{3}\pi\lambda_R^3 \cdot \delta\rho_m = \frac{4}{3}\pi\lambda_R^3 \cdot \delta_{min}(0) \cdot \rho_m \approx M_\otimes \approx 10^{33} g \tag{2.27}$$

Here, $\rho_m \sim 10^{-30} g/cm^3$ is the density of the pre-galactic baryonic substance, $M_\odot$ - mass of Sun. But rather quickly – to the instant of time $t \sim (10^{19} - 10^{20})сек$ – mass of fluctuation reaches the value $10^8 M_\odot$, related to the mass of dwarf galaxy.

Return now to the expression (2.23). Bearing in mind the above-mentioned estimations, we get that $kbt_R \approx 1.5 \cdot 10^{-4} \ll 1$. Therefore, it simplifies and takes on the form

$$\delta(t_R) \approx \left(\frac{8\pi G}{3}\right)^{3/2} \cdot \frac{\vec{k}\vec{r}_0}{2kb} \cdot \rho_v \sqrt{(\rho_c - \rho_m)} \cdot t^2. \tag{2.28}$$

To find its numerical magnitude, calculate r_0, basing on expression (2.24). After substituting needed values into (2.24), we get $r_0 = \sqrt{\frac{3}{8\pi G(\rho_c - \rho_m)}} \cdot b \approx 10^{24} ñm \propto 1 Ípñ$. We note in passing that this estimate coincides practically with the above-mentioned magnitude of the baryonic matter's radius - $r_0 = ct_R \approx 3.0 \cdot 10^{23} см$. Hence, the relative fluctuation of baryonic substance's density calculated in accordance with (2.28), is $\delta(t_R) \propto 10^{-11}$. The mass corresponding to this fluctuation equals - $M_f \propto 10^{33} g \approx M_\odot$, if we put that density as - $\rho_c \sim 10^{-29} g/cm^3$. Upon quadratic growth of fluctuations, as shown in the previous sample, it will increase in time in accordance to the law (2.28) and may form objects mass of galaxy.

On the other side, the fluctuations with the appearance of objects mass of Sun, in its turn, may stimulate the process of stars' creation. (It is necessary to mention here that vacuum not only creates the stars, but make their properties highly unusual. For instance, according to general relativity, the repulsive force appeared around the stars [9].)

The given numerical estimations show, that in spite of explicit simplification of the problem organization – neglecting by gravitational matter self-interaction, omitting its thermodynamic properties, restricted by non-relativistic model - it describes the new physical mechanism of large-scales structures formation in the Universe satisfactorily. Namely, perhaps the vacuum itself creates the anti-gravitational instability of cosmic substance and play the role of factor, to initiate the galaxies and their clusters' creation.

3. THE BEHAVIOR OF BARYONIC MATTER'S PERTURBATIONS IN THE EARLY UNIVERSE DESCRIBED BY THE NON-STATIONARY EQUATION OF STATE

3.1. Introduction

Our Universe, in accordance to the standard cosmological scenario [1], was born from the space-time foam. After creation, the classical space-time inflation stage was $\Delta t_i = (10^{-43} - 10^{-36})$ sec. At this stage, the baryonic matter appeared from the quantum fluctuations of vacuum. In this, the matter distribution was initially homogeneous. Later on, the stage of heterogeneities forming have appeared and their evolving lead to the creation of galaxies and other cosmic structures.

The vast number of references was devoted to searching the evolution of baryonic matter fluctuations (see, for example [2, 3],), but the physical reason of their growing is not clear. This is related to the fact that traditional consideration of this problem is based on the postulate of the gravitational instability of baryonic matter in the Universe. But the recent observable data proved with the high level of confidence in the essential dominance of the non-baryonic matter over the baryonic one in the Universe [4].

Thus, the question of the possibility that non-baryonic substance (dark energy, for example) is the reason of the cosmic structures creation in the Universe arises. Different aspects of this problem were considered in the references [5]. An important place in the analysis of anti-gravitational - vacuum, in particular, - instability secures among them. In doing this in [6], it was shown that vacuum itself can create an objects type of dwarf galaxies.

The present article is continuing researches on the baryonic substrate perturbations' evolving under the non-baryonic matter influence. Namely, we are analyzing the baryonic matter perturbations growing in the Universe with a non-stationary equation of state.

The non-stationary equation of state for a non-baryonic substance was probably proposed first in [7]. Its physical essence, as it seems, consists in that the properties of non-baryonic substrate must change together with the Universe evolving. In doing this, the state of non-baryonic substrate becomes different from the vacuum one and closes to the quintessence or phantom energy states at the very early stages of the Universe evolution [9].

The matter of this chapter has been partially published in [10].

3.2. The Scale Factor Evolution

The Einstein equations that describe the scale factor evolving have the form [1]

$$\ddot{a} = -\frac{4\pi}{3} G(\rho_{nb} + 3p_{nb})a, \qquad (3.1)$$

$$H^2 + \frac{k}{a^2} = \left(\frac{\dot{a}}{a}\right)^2 + \frac{k}{a^2} = \frac{8\pi}{3} G\rho_{nb}. \qquad (3.2)$$

Here, ρ_{nb}- is the density of non-baryonic matter, p_{nb} - its pressure; $G = M_p^{-2}$ - gravitational constant, M_p - Planckian mass. Besides, $H = \dfrac{\dot{a}}{a}$ - Hubble constant that depends on time; k - space curvature that takes on, as it well-known, three magnitudes: - 1 for the close, 0 for the flat, -1 for the open Universe's model.

In order to find the Universe evolving in time, it is necessary to set the equation of state for a non-baryonic substance that connected its pressure and energy density between themselves. For the adiabatic processes in the Universe, it is set as

$$p_{nb} = \omega \cdot \rho_{nb}, \qquad (3.3)$$

where ω is the parameter of state (usually constant). Thus, for the known types of non-baryonic substrates - quintessence, vacuum, phantom energy - it has the magnitudes: $-1 < \omega < 1/3, -1, \omega < -1$ (see, for example, the reference [4]).

According to the problem's organization, we use the following parameterization [11]

$$\omega(a) = \omega_0 + \omega_1 (1 - \dfrac{a}{a_0}) \dfrac{a}{a_0}, \qquad (3.4)$$

with coefficients $\omega_0 = \omega\big|_{a=1}$, and $\omega_1 = -\dfrac{d\omega}{da}\bigg|_{a=1}$, that describes the width class of a non-stationary equations of state for dark energy. In this, $\omega_0 = -1.3$; and $\omega_1 = 4$ or $\omega_1 = -2$.

The law of energy conservation, that follows from equations (3.1) – (3.2), can be written down as

$$\dot{\rho}_{nb} a^3 + 3(\rho_{nb} + p_{nb}) a^2 \dot{a} = 0. \qquad (3.5)$$

Basing on the above-cited equations, it is possible to get the following system for the flat cosmological model

$$\ddot{a} = -\dfrac{4\pi}{3} G (\rho_{nb} + 3 p_{nb}) a, \qquad (3.6)$$

$$\dot{\rho}_{nb} a + 3(\rho_{nb} + p_{nb}) \dot{a} = 0, \qquad (3.7)$$

$$p_{nb} = \omega(a) \rho_{nb}. \qquad (3.8)$$

From (3.7) and (3.8), we get the solution

$$\rho_{nb} = \rho_0 x^{-3\kappa} \exp[\frac{3}{2}\omega_1(x-1)^2], \qquad (3.9)$$

where $x = \dfrac{a}{a_0}$, and $\kappa = 1+\omega_0$ is the constant value. Besides, ρ_0 is the initial mass density for a non-baryonic matter. Putting (3.9), (3.8), (3.4) and magnitude $\omega_0 = -1.3$ into the equation (3.5) now, we get the following non-homogeneous differential equation of the second order

$$\frac{d^2 x}{dt^2} = 3Cx^2 \exp[\frac{3}{2}\omega_1(x-1)^2] \cdot (-1 + \omega_1 x - \omega_1 x^2), \qquad (3.10)$$

where $C = -\dfrac{4\pi}{3} G\rho_0$ is the integrating constant.

For solving this equation, consider the case when $\dfrac{a}{a_0} \ll 1$. Such a condition takes place in the very early Universe. Then, equation (3.10) simplifies and is written down as

$$\frac{d^2 x}{dt^2} = -3Cx^2 \exp[\frac{3}{2}\omega_1 - 3\omega_1 x]. \qquad (3.11)$$

Let us perform the following rearrangements: $\dfrac{dx}{dt} = p,\ \dfrac{d^2 x}{dt^2} = p\dfrac{dp}{dx}$. Substituting them into (3.11), we get the differential equation with the separable variables

$$p\, dp = Fx^2 \exp[-3\omega_1 x] dx \qquad (3.12)$$

where the new constant value $F = -3C \exp(3/2\omega_1)$ is introduced. Then, after returning to the previous variables, anyone can find

$$\frac{dx}{\sqrt{2F \exp(-3\omega_1 x)}} = dt. \qquad (3.13)$$

Integration (3.13) leads to the following expression

$$t = \frac{1}{\sqrt{2F}} \frac{2}{3\omega_1} \exp(3/2\omega_1 x) \qquad (3.14)$$

and after the reversible procedure usage in it, we get the needed result

$$x = \frac{2}{3\omega_1} \ln(\chi t). \tag{3.15}$$

Here, one more designation - $\chi = \frac{3\omega_1 \sqrt{F}}{\sqrt{2}}$ - was introduced. Differentiating (3.15), it is easy to calculate the Hubble constant

$$H = \frac{\dot{x}}{x} = \frac{1}{t \ln(\chi t)}. \tag{3.16}$$

And at last, bearing in mind the explicit expression of the variable x, we find the dependency of scale factor from time

$$a = \frac{2a_0}{3\omega_1} \ln(\chi t). \tag{3.17}$$

3.3. The Perturbation Density Growth in the Very Early Universe

Write down the equation describing the evolving of baryonic matter perturbations' in the Universe [2]

$$\ddot{\delta} + 2H\dot{\delta} + v_s^2 k^2 \delta - 4\pi G \rho_b \delta = 0, \tag{3.18}$$

where ρ_b is the baryonic matter density, v_s is the speed of sound in the baryonic matter, $\vec{k}$ – wave vector. Let that for the relativistic substance $v_s^2 = 1$, and assume that wave vector decreases as $k \propto a^{-1}$ [3]. Hence, $k^2 = \frac{9\omega_1^2}{4a_0^2 \ln^2 \chi t}$ as it follows from expression (3.17). Putting these values into equation (3.18), we get the standard differential equation of the second order with the variable coefficients

$$\ddot{\delta} + P(t)\dot{\delta} + Q(t)\delta = 0; \tag{3.19}$$

they are

$$\begin{aligned} P(t) &= \frac{2}{t \ln \chi t}, \\ Q(t) &= \frac{3\omega_1^2}{4a_0^2 \ln^2 \chi t} - 4\pi G \rho_b \end{aligned} \tag{3.20}$$

For this equation, solving introduce the new function z, that is connected with δ by the relation

$$\delta = u(t)z. \qquad (3.21)$$

Let us twice differentiate (3.21) and, in accordance with the standard procedure [12], put these derivatives into equation (3.20)

$$u\ddot{z} + (2\dot{u} + P(t)u)\dot{z} + (\ddot{u} + P(t)\dot{u} + Q(t)u)z = 0 \qquad (3.22)$$

Then, equates to zero the coefficient at $\dot{z}$:

$$2\dot{u} + P(t)u = 0. \qquad (3.23)$$

From this, it is easy to find the expressions for u and its derivatives. Substituting all of them into (3.22) and making the needed transformations, we get the Euler equation

$$\ddot{z} + I(t)z = 0. \qquad (3.24)$$

Expression $I(t) = -\frac{1}{4}P^2(t) - \frac{1}{2}\dot{P}(t) + Q(t)$ is the invariant of this equation. Introducing $P(t)$ and $Q(t)$ in it and having in mind the numerical estimations of the early Universe epoch, we get approximately

$$\ddot{z} - \frac{1}{t^2}\left(\frac{1}{\ln \chi t} - \frac{2}{3}\right)z = 0. \qquad (3.25)$$

After marking the expression in brackets by $T(t)$, equation (3.27) takes on the canonical form

$$t^2\ddot{z} - T(t)z = 0. \qquad (3.26)$$

For its solving, it is necessary to estimate the magnitude of expression $T(t)$, more precisely. So, the numerical magnitude of χ, as it follows from its definition, depends on the density of non-baryonic matter at the instant of Universe creation. In accordance with [1], its magnitude is of $10^{94} g \cdot cm^{-3}$ order; hence $\chi \approx 10^{41} s^{-1}$. Later on, from (3.17), we get that for condition $x = \dfrac{a}{a_0} \ll 1$ yielding, that takes place in the very early Universe, the time interval must be closed to the time of substance born, i.e. to $t \sim 10^{-36} s$. Thus, as the result, anyone can find that T will be no larger than -0.6, i.e. we put that $T(t) = const$.

Then, the solution of (3.25) with the coefficient $T = -0.6$, may be represented as

$$z = \sqrt{t}\left[C_1 t^{0.9} + C_2 t^{-0.9}\right]. \tag{3.27}$$

And at last, putting it into (2.21), we find the needed function – namely, the perturbation of baryonic matter's density as function of time

$$\delta(t) \sim \frac{t^{1.4}}{\ln \chi t}. \tag{3.28}$$

For estimating the rate of expression's (3.28) growth, compare it with the evolving of baryonic matter's perturbations in the Friedmann model for two well-known expanding regimes.

The baryonic matter's density, according to the relativistic theory of perturbations, growth in time linearly, i.e. as $\delta(t) \sim t$, while in the framework of non-relativistic expanding regime as $\delta(t) \sim t^{2/3}$ [2]. Analysis showed that during the period of substance creation $t \propto 10^{-36} \div 10^{-33} s$, the module of expression (3.28) may drastically be larger then canonical ones. This means, that such types of perturbations have the larger probability for surviving in the future, instead of above-cited standard perturbations. Naturally, future evolving (in the framework of other mechanisms) will lead to creation of cosmic objects-type of galaxies.

3.4. Discussion and Resume

So, in this article, the new physical reason of the perturbations of baryonic matter growing has been discussed. In fact, the standard approach to this problem was based on the gravitational instability of baryonic matter. Their future evolving leads to the objects forming with mass of Jeans. In it turn, these cosmic objects evolve then to the different types of galaxies [2].

But recent discovering of vacuum with its anti-gravitation property is radically changed the problem of perturbations growth. Really, in articles [2, 13] in was shown that vacuum itself can be the reason of such perturbations, due to the so called anti-gravitational instability of a cosmic substrate. Meanwhile, the anti-gravitational property possesses not only the vacuum, but other types of dark energy, also. And some forms of the dark energy describe by the different non-stationary equations of state. Hence, the question arises – do these types of dark energy produce the anti-gravitational instability and what is their evolving nature?

Here it was shown that in this case the perturbations of baryonic matter appeared as well as in other models of the Universe evolving. But the nature of perturbations' evolving in the Universe with non-stationary equation of state is essentially differing, than in previous models. Really, for the case of vacuum anti-gravitational instability the matter perturbations is growing proportionally to $t^2 \sin kbt$ [6], while in our case they growth in accordance to (3.28.) Moreover, it is necessary to mark that perturbation (3.28) decreases in time, while the above cited expressions for perturbation evolving increases in time. This means that the rate of baryonic matter's perturbations is changing in accordance with the changing of the equation of state of non-baryonic substrate. But at every stage of the Universe evolving (remember that we are talking about the very early Universe) the anti-gravitational type of baryonic matter's perturbations is dominate in compare with their gravitational one.

Some new results in this cosmological area have been discussed in articles [14].

4. INFLUENCE OF QUINTESSENCE FIELD AT THE GROWTH RATE OF BARYONIC MATTER'S DENSITY PERTURBATIONS

4.1. Introduction

The evolution of baryonic matter from its density fluctuations appearance up to the processes of galaxies origin is one of the most important problems for modern cosmology [1]. This theme was considered as in Newtonian cosmology and as in the framework of relativistic cosmology from different sides (see, for example, [2]). Some current tendencies in this problem, in particular, are lighting in articles [3].

We'll focus this paper on another aspect of this problem. Namely, in many articles the influence of different cosmological substrates on the evolution of baryonic matter's perturbations was reduced to setting their equations of state, i.e. to setting parameter ω. As the result it leads to setting the different expressions of Hubble constant in the "friction term" of the basic equation - $\dfrac{d^2 \delta\rho_m}{dt^2} + 2H(t)\dfrac{d\delta\rho_m}{dt} + (\vec{k}^2 - 4\pi G\rho_0)\delta\rho_m = 0$, that describes the evolution of baryonic matter's density perturbations. However, this approach allows consider such evolution as the process that elapses on the background of non-baryonic cosmic substrate, only (and even stay in shadow the physical properties of this substrate).

But in reality, this substrate interacts with baryonic matter in a definite way. That is why it is essential to consider its influence on the baryonic matter evolution in detail. It will be shown below that a chosen variant of substance (quintessence field with parameter $\omega \geq -1$) is described as small time-increasing field $\psi(t)$ on the background of constant scalar field φ, while the chosen variant of baryonic matter is described as small wave - type fluctuations $\delta(t) = \dfrac{\delta\rho_m(t)}{\rho_{m_0}}$ on the background of uniformly distributed motionless gas with non-variable mass density ρ_{m_0}. Thus, we have the system of two small fields ($\psi(t)$ and $\delta(t)$) that evolves on the stable background of φ and ρ_{m_0}.

Such problem wording is analogous to those in cosmology where the multi-fluids evolution searches [4] and, more generally, in hydrodynamics where the motion of poly-component media examines (see, for example, [5]). Besides, for the closest to our physical system (scalar field and perfect fluid), the properties of cosmological density perturbations were considered in article [6]. Note that opposite physical situation is also permissible. In fact, the cosmological evolution of two coupled scalar fields in the presence of a barotropic fluid was examined in [7].

This chapter is organized as follows. In Section 4.2, we demonstrate that two scalar fields can describe the physical properties of a quintessence field. Section 4.3 devotes to searching the evolution of scalar field ψ that represents small standing waves on the background of a basic scalar field φ. Section 4.4 is devoted to exploring the influence of field ψ on the evolution of baryonic matter's density perturbations. In Section 4.5, we examine the effect of stimulation of the baryonic matter's density perturbations growing by scalar field ψ. Our conclusions are presented in Section 4.6.

4.2. Scalar Fields Representing the Quintessence Field

One of the actual problems for modern cosmology is the theoretical description of the quintessence field - one type of dark energy. Its observable properties are the scale homogeneity and the absence of clustering [8]. Quintessence is described by an ordinary scalar field minimally coupled to gravity, but with particular potentials that lead to late time inflation [9]. The action for quintessence is given by

$$S = -\int \sqrt{-g} \left[\frac{1}{2} \left(\Delta - \frac{\partial^2}{\partial x^{0^2}} \right) \phi + U(\phi) \right] d^4 x \qquad (4.1)$$

where $-\Delta$ is the 3-dimensional Laplace operator, ϕ - potential of any scalar field. The important theoretical aspect of this problem – to establish the phantom field equation of state.

The simplest equation of state for any type of dark energy is usually in the linear form $p = \omega \rho$, where magnitude of ω lays within the interval $-1 < \omega < -\frac{1}{3}$ [10]. Therefore,

$$\omega = \frac{p}{\rho} = \frac{\dot{\phi}^2 - 2U(\phi)}{\dot{\phi}^2 + 2U(\phi)} \qquad (4.2)$$

Later on we will describe the arbitrary quintessential field ϕ by coupled scalar fields - ordinary φ and Higgs-type ψ.

Thus, consider the self-consistent problem for the mutual evolution of fields and Universe. The corresponding system of Einstein's equations and equations of two interacting scalar fields is

$$\left(\frac{\dot{a}}{a} \right)^2 = \frac{4\pi G}{3} \left(\dot{\varphi}^2 + m_\varphi^2 \varphi^2 + \dot{\psi}^2 - m_\psi^2 \psi^2 + \frac{\lambda_\varphi}{2} \varphi^4 + \frac{\lambda_\psi}{2} \psi^4 + \nu \varphi^2 \psi^2 \right), \qquad (4.3)$$

$$\ddot{\varphi} + 3\frac{\dot{a}}{a}\dot{\varphi} + m_\varphi^2 \varphi + \lambda_\varphi \varphi^3 + \nu\psi^2\varphi = 0, \tag{4.4}$$

$$\ddot{\psi} + 3\frac{\dot{a}}{a}\dot{\psi} - m_\psi^2 \psi + \lambda_\psi \psi^3 + \nu\varphi^2\psi = 0. \tag{4.5}$$

Let masses and fields correlate each other as $m_\varphi << m_\psi, \varphi >> \psi$, while the self-action coefficients fulfill inequality - $\lambda_\varphi << \lambda_\psi << 1$. Hence, the period of oscillations for field φ is essentially larger than the period of oscillations for field ψ ($T_\varphi >> T_\psi$) and - $\ddot{\varphi} << \dot{\varphi}$, accordingly. In another words, in time of field ψ changing the basic field φ doesn't change practically, i.e. we may describe it by the conditions

$$\dot{\varphi} \approx 0, \varphi \approx const. \tag{4.6}$$

After neglecting the fields' self-actions, we get the simplified system of equations

$$\left(\frac{\dot{a}}{a}\right)^2 = \frac{4\pi G}{3}\left(m_\varphi^2 \varphi^2 + \dot{\psi}^2 - \hat{m}_\psi^2 \psi^2\right), \tag{4.7}$$

$$\ddot{\psi} + 3\frac{\dot{a}}{a}\dot{\psi} - \hat{m}_\psi^2 \psi = 0, \tag{4.8}$$

which will be under our analyses. Here, $\hat{m}_\psi^2 = m_\psi^2 - \nu\varphi^2$ is the squared field's ψ effective mass that determines by field mass m_ψ and its interaction with field - φ.

Later on, it is necessary to set masses of scalar fields and their initial amplitudes. According to [11], their typical magnitudes are

$$\begin{array}{l} m_\varphi << \lambda_\varphi^{\frac{1}{2}} M_p, m_\psi << \lambda_\psi^{\frac{1}{2}} M_p, \\ \varphi_0 \approx \lambda_\varphi^{-\frac{1}{4}} M_p, \psi_0 \approx \lambda_\psi^{-\frac{1}{4}} M_p \end{array} \tag{4.9}$$

where M_p is the Planckian mass. Having in mind these constrains, consider the case when

$$m_\varphi \varphi >> \hat{m}_\psi \psi, m_\varphi \varphi >> \dot{\psi}. \tag{4.10}$$

Moreover, for our model, the inequalities (4.15) take place if - $\lambda_\psi << \lambda_\varphi << 1$. Conditions (4.15) indicate that energy of basic field φ is essentially larger than energy of additional field - ψ. Under this assumption, the system (4.12) – (4.13) takes on a more simple form

$$\left(\frac{\dot{a}}{a}\right)^2 = \frac{4\pi G}{3} m_\varphi^2 \varphi^2,$$
$$\ddot{\psi} + 3\frac{\dot{a}}{a}\dot{\psi} - \hat{m}_\psi^2 \psi = 0.$$
(4.11)

Equations (4.11) reduce to one linear differential equation of the second order $\ddot{\psi} + \Im\dot{\psi} - \Re\psi = 0$ with coefficients $\Im = \sqrt{12\pi G} m_\varphi \varphi$, $\Re = \hat{m}_\psi^2$. Its solutions we will look for in the standard form - $\psi = \psi_0 \exp(M \cdot t)$. Hence, we get the algebraic equation $M^2 + \Im M - \Re = 0$ that has two roots:

$$M_{1,2} = -\frac{\Im}{2} \pm \sqrt{\frac{\Im^2}{4} + \Re}.$$
(4.12)

From (4.9) and (4.10), it follows that - $m_\varphi \gg \hat{m}_\psi$. This condition allows to decompose the expression under root sign into the Taylor series with respect to small value - $\hat{m}_\psi / m_\varphi$, and to get two solutions

$$M_1 = \frac{\hat{m}_\psi^2}{2\sqrt{3\pi G} m_\varphi \varphi}, \quad M_2 = -2\sqrt{3\pi G} m_\varphi \varphi.$$
(4.13)

Note also, that the second of them is the approximate solution of zero accuracy with respect to the ratio - $\hat{m}_\psi / m_\varphi$. Thus, the sought-for solutions of field ψ we may take in the forms

$$\psi_1 = \psi_0 \exp(M_1 \cdot t) = \psi_0 \exp\left(\frac{\hat{m}_\psi^2 \cdot t}{2\sqrt{3\pi G} m_\varphi \varphi}\right),$$
(4.14)

$$\psi_2 = \psi_0 \exp(M_2 \cdot t) = \psi_0 \exp\left(-2\sqrt{3\pi G} m_\varphi \varphi \cdot t\right).$$
(4.15)

From (4.9) – (4.11), it is unproblematic to find the additives to energy density and to pressure

$$\delta\rho = \frac{1}{2}\dot{\psi}^2 - \frac{m_\psi^2}{2}\psi^2 + \frac{\nu}{2}\varphi^2\psi^2,$$
(4.16)

$$\delta p = \frac{1}{2}\dot{\psi}^2 + \frac{m_\psi^2}{2}\psi^2 + \frac{v}{2}\varphi^2\psi^2, \qquad (4.17)$$

while the main items are

$$\rho_0 = \frac{1}{2}m_\varphi^2\varphi^2, \; p_0 = -\frac{1}{2}m_\varphi^2\varphi^2. \qquad (4.18)$$

From (4.21) – (4.23), it is easy to verify that these two scalar fields describe the quintessence field. In fact, from (4.2), we get

$$\omega = \frac{p}{\rho} = \frac{p_0 + \delta p}{\rho_0 + \delta \rho} \approx \frac{p_0}{\rho_0}\left(1 + \frac{\delta p}{p_0} - \frac{\delta \rho}{\rho_0}\right). \qquad (4.19)$$

Due to condition (4.6) and (4.18), we see that ordinary scalar field φ is in the vacuum state, i.e. $p_0(\varphi) = -\rho_0(\varphi)$. Hence,

$$\omega = -\left(1 - 2\frac{\dot{\psi}^2 + v\varphi^2\psi^2}{m_\varphi^2\varphi^2}\right) \geq -1. \qquad (4.20)$$

(Note that the same result we get and for the case of two ordinary scalar fields φ and ψ). Therefore, two coupled scalar fields describe the quintessential state of dark energy.

4.3. The Scalar Field ψ Evolution

Let the fields ψ and φ possess any space non-homogeneity. Under the conditions - $\nabla\psi \sim m_\psi\psi$, $\nabla\varphi \ll m_\varphi\varphi$, this leads to the system

$$\left(\frac{\dot{a}}{a}\right)^2 = \frac{4\pi G}{3}(\dot{\varphi}^2 + m_\varphi^2\varphi^2 + \dot{\psi}^2 - m_\psi^2\psi^2 + \frac{\lambda_\varphi}{2}\varphi^4 + \frac{\lambda_\psi}{2}\psi^4 + v\varphi^2\psi^2), \qquad (4.21)$$

$$\ddot{\varphi} + 3\frac{\dot{a}}{a}\dot{\varphi} + m_\varphi^2\varphi + \lambda_\varphi\varphi^3 + v\psi^2\varphi = 0, \qquad (4.22)$$

$$\ddot{\psi} + 3\frac{\dot{a}}{a}\dot{\psi} - m_\psi^2\psi - \frac{1}{a^2}\Delta\psi + \lambda_\psi\psi^3 + v\varphi^2\psi = 0. \qquad (4.23)$$

Here we take into account that - $\nabla a \ll \dot{a} \sim aH$.

For solving this system, put - $\psi = \psi(t,\vec{x}) = \Psi(t) \cdot \xi(\vec{x})$, where - $\xi(\vec{x}) = \xi_0 \exp(i\vec{k}\vec{x})$. The last expression indicates that perturbed potential is the plane wave with the time-variable amplitude, $\vec{k}$ its wave vector. Such a choice is analogous to Jeans' presentation of the perturbations in baryonic substrate.

Substitution of all of them into equation (4.23) will bring it to the following one

$$\ddot{\Psi} + 3\frac{\dot{a}}{a}\dot{\Psi} - \hat{m}_\psi^2 \Psi + \lambda_\psi \Psi^3 = \frac{1}{a^2}\frac{\Psi}{\xi}\Delta\xi \qquad (4.24)$$

with the non-zero right part. It is easy to see that - $\frac{\Delta\xi}{\xi} = \vec{k}^2$; hence the following equation leads from (4.24) one

$$\ddot{\Psi} + 3\frac{\dot{a}}{a}\dot{\Psi} + (-m_\psi^2 + \nu\varphi^2 + \frac{\vec{k}^2}{a^2})\Psi + \lambda_\psi \Psi^3 = 0. \qquad (4.25)$$

With the needed accuracy ($\frac{\Psi}{\varphi} \ll 1$), we have: $\frac{\dot{a}}{a} \approx \frac{\dot{a}_0}{a_0} = H = \sqrt{\frac{4\pi G}{3}} \cdot m_\varphi \varphi = const$

and $a_0 = A \exp Ht$ with arbitrary amplitude A. After neglecting the field self interaction, we get equation

$$\ddot{\Psi} + 3H\dot{\Psi} + (-m_\psi^2 + \nu\varphi^2 + \frac{\vec{k}^2}{A^2}\exp(-2Ht))\Psi = 0, \qquad (4.26)$$

whose exact partial solution, in accordance with [12], expresses as

$$\Psi(t) = C \exp\left(\frac{3}{2}Ht\right) \cdot J_n\left(2\frac{k}{A}\exp(-Ht)\right), \qquad (4.27)$$

where - $n = \sqrt{\frac{9}{4} - \frac{1}{H^2}(-m_\psi^2 + \nu\varphi^2)}$ is the index of Bessel function, C is an arbitrary constant.

Let $-m_\psi^2 + \nu\varphi^2 \sim \frac{9}{4}H^2$ for simplicity, then $n \sim 0$ and $\Psi(t) = C \exp\left(\frac{3}{2}Ht\right) \cdot J_0\left(2\frac{k}{A}\exp(-Ht)\right)$. Furthermore, assuming that - $2\frac{k}{A} \sim 1$, the field potential takes on the form

$$\psi(t,\vec{x}) = C \exp\left(\frac{3}{2}Ht\right) \cdot J_0(\exp(-Ht)) \cdot \exp(i\vec{k}\vec{x}). \qquad (4.28)$$

Now examine the behavior of this function in time. $Ht = \tau$ - is the dimensionless time. Using the numerical and the graphical representations of Bessel functions [13], we see that at large, argument ($\tau \gg 1$) function $J_0(\tau) \to 0$. However, for our purpose we must consider the opposite situation when $\tau \ll 1$.

For doing this, consider the series representation of Bessel function of zero order

$$J_0(z) = \sum_{k=0}^{\infty} (-1)^k \frac{z^{2k}}{2^{2k}(k!)^2}, \qquad (4.29)$$

and limit ourselves by first two terms only, because it rapidly ($\sim z^{-\frac{1}{2}}$) decreases as z increases. Hence, $J_0(z) = J_0(exp(-Ht)) \approx \frac{3}{4} + \frac{1}{2}Ht$ and function (45) reduces to the next one

$$\psi(t,\vec{x}) \approx \left(1 + \frac{8}{3}Ht\right) \cdot exp(i\vec{k}\vec{x}). \qquad (4.30)$$

This remarkable result – time-increasing amplitude – allows lighting the process of baryonic matter's macroscopic perturbations growth from a new side.

4.4. Influence of Field ψ on the Baryonic Matter's Density Perturbations Increase

Our next step is searching the influence of field ψ on the baryonic matter's density perturbations growing. Later on, we will base the Jeans equations on the adiabatic case.

In the usual designations (together with the equation of state for baryonic matter $P_m = P_m(\rho_m)$), they are

$$\frac{\partial \rho}{\partial t} + div(\rho \vec{V}) = 0, \qquad (4.31)$$

$$\frac{\partial \vec{V}}{\partial t} + \vec{V} grad\vec{V} + \frac{1}{\rho} grad P + grad U = 0, \qquad (4.32)$$

$$\nabla^2 U = 4\pi G \rho. \qquad (4.33)$$

For enriching our goal, it is necessary to use the well-known method of description the poly-components fluid dynamics. Namely, for searching the microscopic perturbations of baryonic matter, specify them in the standard manner [14]

$$\rho = \rho_0 + \delta\rho = \rho_0\left[1 + \gamma(t)\exp(i\vec{k}\vec{x})\right], \tag{4.34}$$

$$V = 0 + v = w(t)\exp(i\vec{k}\vec{x}), \tag{4.35}$$

$$U = U_0 + \delta U = U_0 + \Phi(t)\exp(i\vec{k}\vec{x}), \tag{4.36}$$

$$P = P_0 + \delta P = P_0 + \frac{\partial P}{\partial \rho}\cdot\delta\rho = P_0 + b^2\cdot\rho_0\cdot\gamma(t)\exp(i\vec{k}\vec{x}). \tag{4.37}$$

Besides, as we consider the influence of baryonic matter on the background of phantom field, it is needed to take into account that the last will contribute its additions at the pressure P and at the mass density ρ. Hence, - $\rho = \rho_m + \rho_f$, and - $P = P_m + P_f$. Moreover, according our assumption, we have

$$\rho_m = \rho_{m_0} + \delta\rho_m, \; \rho_f = \rho_{f_0} + \delta\rho_f, \tag{4.38}$$

$$P_m = P_{m_0} + \delta P_m, \; P_f = P_{f_0} + \delta P_f, \tag{4.39}$$

where $\delta\rho_m$, $\delta\rho_f$, δP_m and δP_f are the small perturbed additions to initials mass densities ρ_{m_0}, ρ_{f_0} and to initial pressures P_{m_0}, P_{f_0}. In our case, the main terms for mass density and for pressure, that are associated with vacuum and determined by the field φ, are

$$\rho_{f_0} = \frac{1}{2}m_\varphi^2\varphi^2 = const, \; P_{f_0} = -\frac{1}{2}m_\varphi^2\varphi^2 = const, \tag{4.40}$$

while the corresponding additions have the forms (we neglect the fields interaction, for simplicity)

$$\delta\rho_f \approx \frac{1}{2}\dot\psi^2 - \frac{m_\psi^2}{2}\psi^2 = \frac{1}{2}\left[\frac{64}{9}H^2 - m_\psi^2\left(1+\frac{8}{3}Ht\right)^2\right]\cdot\exp(i\vec{k}_f\vec{x})$$
$$= \rho_0\gamma_f(t)\cdot\exp(i\vec{k}_f\vec{x}) \tag{4.41}$$

$$\delta p_f \approx \frac{1}{2}\dot\psi^2 + \frac{m_\psi^2}{2}\psi^2 = \frac{1}{2}\left[\frac{64}{9}H^2 + m_\psi^2\left(1+\frac{8}{3}Ht\right)^2\right]\cdot\exp(i\vec{k}_f\vec{x})$$
$$= \rho_0\lambda_f(t)\cdot\exp(i\vec{k}_f\vec{x}) \tag{4.42}$$

That is why the variables in (4.54) – (4.55) may be represented as

$$\rho = \rho_0 + \delta\rho = \rho_0[1 + \gamma_m(t)\exp(i\vec{k}_m\vec{x}) + \gamma_f(t)\exp(i\vec{k}_f\vec{x})] \tag{4.43}$$

$$V = 0 + v = w_m(t)\exp(i\vec{k}_m\vec{x}) + w_f(t)\exp(i\vec{k}_f\vec{x}), \tag{4.44}$$

$$U = U_0 + \delta U = U_0 + \Phi_m(t)\exp(i\vec{k}_m\vec{x}) + \Phi_f(t)\exp(i\vec{k}_f\vec{x}), \tag{4.45}$$

$$P = P_0 + \delta P = P_0 + \frac{\partial P}{\partial \rho_f}\delta\rho_f =$$
$$P_0 + b^2\rho_0 \cdot [\gamma_m(t)\exp(i\vec{k}_m\vec{x}) + \gamma_f(t)\exp(i\vec{k}_f\vec{x})] \tag{4.46}$$

where we assume that speeds of sound in a baryonic matter and in the quintessence field are equal each other, i.e. $b_m = b_f = b \sim 1$. Besides, for next simplification, let $w_f(t) = 0$ (in [15] it was considered the bi-velocities type of fluid with special non-zero components) and for the case of dust-like baryonic substance, imply that - $P_{m_0} = 0$. It results in-
$\rho_0 = \rho_{m_0} + \rho_{f_0}$, $P_0 = P_{f_0}$.

Substituting (4.59) – (4.62) into (4.47) – (4.49) and making required transformations, we get the dynamical equation for two-component media

$$\frac{d^2\delta\rho_m}{dt^2} + (\vec{k}_m^2 - 4\pi G\rho_0)\delta\rho_m =$$
$$= -\frac{d^2\delta\rho_f}{dt^2} - [\Phi_f(t)\vec{k}_f(\vec{k}_f - \vec{k}_m) + (\vec{k}_m\vec{k}_f - 4\pi G\rho_0)]\delta\rho_f \cdot \exp(i(\vec{k}_f - \vec{k}_m)\vec{x}) \tag{4.47}$$

where - $\Phi_f(t) = 4\pi G\rho_0\gamma_f(t)k_f^{-2} = 4\pi G\delta\rho_f(t)k_f^{-2}$. Later on, assuming that $\vec{k}_m = \vec{k}_f = \vec{k}$ and accounting (4.47), the previous equation goes into

$$\frac{d^2\gamma_m}{dt^2} + (k^2 - 4\pi G\rho_0)\gamma_m + (k^2 - 4\pi G\rho_0) \cdot \rho_0^{-2} \cdot \left(\frac{32}{9}H^2 - \frac{1}{2}m_\psi^2\left(1 + \frac{8}{3}Ht\right)\right) = 0. \tag{4.48}$$

The most attractive consequence of this equation, that has the evolutionary nature, we get to omit the constant values in third term and replace $-\frac{4}{3}Ht$ by $\sin(-\frac{4}{3}Ht)$ (due to minuteness Ht here and after). Thus the standard equation of forced oscillations

$$\frac{d^2\gamma_m}{dt^2} + \Omega^2 \gamma_m = \alpha \sin\widetilde{\omega} t \qquad (4.49)$$

takes place, where $\Omega = (k^2 - 4\pi G\rho_0)^{\frac{1}{2}}$ is the basic internal frequency, $\widetilde{\omega} = \frac{4}{3}H$ and $\alpha = (k^2 - 4\pi G\rho_0)\frac{m_\psi^2}{\rho_0^2}$ are the externals frequency and amplitude, accordingly.

Now it should be pointed out that condition $\widetilde{\omega} - \frac{\pi}{2} \approx \Omega$ relates to the resonance case. Hence, the amplitude of baryonic matter's density perturbations gets the strong short-time splash. This result is very significant because the splash may be a real macroscopic mechanism of an initial matter density perturbations appearance. Moreover, the above-mentioned condition determines the resonance-case wave vector -

$$k_r = 2\sqrt{\left(\frac{H}{3} - \frac{\pi}{8}\right)^2 + \pi G\rho_0} \;.$$

In other – non-resonance – cases ($\widetilde{\omega} - \frac{\pi}{2} \neq \Omega$) the amplitude of oscillations, according to standard theory [16], at times $\tau \ll 1$ (see Section 4.3) becomes growth linearly, i.e. $\delta\rho_m \sim t$. So, it stimulates the process of matter's density perturbations increase (see next Section 4.5).

4.5. Effect of Stimulation of the Baryonic Matter Density Perturbations Increase

We point out that according to above-considered behavior of field ψ (equations (4.13) or (4.39)), it is necessary to generalize equation (4.47) in the similar manner. The result is obvious –

$$\frac{d^2\delta\rho_m}{dt^2} + 2H\frac{d\delta\rho_m}{dt} + \left(\vec{k}_m^2 - 4\pi G\rho_0\right)\delta\rho_m =$$
$$= -\frac{d^2\delta\rho_f}{dt^2} - 2H\frac{d\delta\rho_f}{dt} - [\Phi_f(t)\vec{k}_f(\vec{k}_f - \vec{k}_m) + (\vec{k}_m\vec{k}_f - 4\pi G\rho_0)]\delta\rho_f \cdot \exp(i(\vec{k}_f - \vec{k}_m)\vec{x}) \qquad (4.50)$$

where the Hubble constant is $H = \frac{\dot{a}}{a} \approx \frac{\dot{a}_0}{a_0} = \sqrt{\frac{4\pi G}{3}} \cdot m_\varphi \varphi = const$.

The "friction term" in right side of (4.50) will alter the amplitude and frequency of forced oscillations in (4.49) owing to previous results, but don't change the key conclusion of Section 4.4 – the linear time-growing of baryonic matter's perturbations. As for the "friction term" in left side of (4.50), it is necessary to say the following:

The problem of baryonic matter perturbations growing in the dust-like Universe with a non-variable Hubble constant was considered in a number of recent articles (see, for example [17]). But their common result adequate to those in [14] – the amplitude of perturbations increases as different powers of time, i.e. - $\delta\rho_m \sim t^l$, where, in particular, - $l = \frac{2}{3}, -1$, etc. More complicate results take place when the Hubble constant is the time-varying value [18]. For instance, in article [19], it was shown that - $\delta\rho_m \sim \frac{t}{\ln \chi t}$. In summary, in all of these cases, the "friction term" $2H \frac{d\delta\rho_m}{dt}$ at Jeans-like equation leads to a main consequence – the growing of baryonic matter perturbations in the expanding Universe.

Hence, for the baryonic Universe, the matter density perturbation will develop in time more rapidly, namely as

$$\delta\rho_m = C_1 \cdot t + C_2 t^{\frac{2}{3}}, \tag{4.51}$$

where C_1 and C_2 are any suitable constants.

That is why our result describes the effect of stimulation of the baryonic matter's density perturbations increase by phantom field.

4.6. Conclusions

Here it was shown that for deeper researching of the process of baryonic matter evolution in the expanding Universe, it is necessary to:

- know the physical property of the concrete field (or fields) that represents the background of non-baryonic substrate type of dark energy, and
- take into account the influence of such field on the continuous medium of baryonic matter.

In our article, these statements were realized for the quintessential field. As the result, we describe quintessence by two gravitating scalar fields. The first of them is the invariable field ($\varphi = const$), while the second evolves as the space-wave with linearly growing amplitude (see (4.30)). These fields give their contributions at the total pressure P and at the total mass density ρ of baryonic matter. As a consequence, the evolution of baryonic matter density perturbations obeys the standard equation of forced oscillations (4.49) and admits the resonance case, when amplitude of baryonic matter density perturbations gets the strong short-time splash. This splash was interpreted as the macroscopic mechanism of initial matter's density perturbations appearance.

In other – non-resonance - cases the amplitude of oscillations becomes growth linearly in time. That is why it also may stimulate the process of matter density perturbations increasing according to the expression (4.51).

As a result, it is possible to say that the quintessence field actively affects the baryonic matter's density perturbation growing in the Universe.

Some new results in this cosmological area have been discussed in articles [20].

5. ON THE STOCHASTIC DYNAMICS OF COSMIC STRINGS IN THE EARLY UNIVERSE

5.1. Introduction

The problem of galaxies creating correlates with the problem of primordial non-homogeneity of cosmological substrate. The standard reason of galaxies creation is the presence of cosmic strings that appeared at very early stages of the Universe evolution. Missing the fact of cosmic strings creation, we note that due to its giant linear mass density ($10^{22} g/cm$), they attract the substrate surrounding them.

The main goal of this article is to present the result that chaotic cosmic strings may also create the regular perturbations of substrate that lead to galaxies formation, in its turn.

The matter of this chapter has been partially published in [1].

5.2. The Energy-momentum Tensor of a Thread-like Substance with the Stochastic Perturbations

Let us consider a thread-like substance consisting of an infinite number of free strings. Due to its own movement, every string covers a two-dimensional hyper-surface that could be parameterized by two (τ - time-like and ρ - space-like) variables. Later on, we will consider the partial case when only parameter ρ is subjected to stochastic perturbations. This means that from parameter ρ, we must pass to such a parameter $\tilde{\rho}$ that

$$\tilde{\rho} = \rho + \rho'(x^0) = \rho\left(1 + \rho'(x^0)/\rho\right) = \rho(1 + z(x^0)), \tag{5.1}$$

where $z(x^0)$ is the dimensionless stochastic function of time. According to (5.1), we get the modified space-like vector with the linear accuracy

$$\tilde{l}^\alpha = dx^\alpha/d\tilde{\rho} \approx l^\alpha(1 - z(x^0)), \tag{5.2}$$

where $z(x^0) << 1$, and the standard time-like vector

$$u^\alpha = dx^\alpha/d\tau. \tag{5.3}$$

For searching the dynamics of thread-like substances, we will start from the well-known action principle

$$S = \int(\sqrt{-g}\mu)dV_4, \tag{5.4}$$

where g is the determinant of metric tensor, μ is the linear mass density. Calculating the variation of this action we get

$$\delta S = -\int \mu \left(u^\alpha u_\beta - \tilde{l}^\alpha \tilde{l}_\beta \right) \nabla_\alpha \delta x^\beta dV_4 = 0 . \tag{5.5}$$

It is easy to get the conservation law from (5.5)

$$\nabla_\beta \tilde{T}^{\alpha\beta} = 0, \tag{5.6}$$

where

$$\tilde{T}^{\alpha\beta} = \mu(u^\alpha u^\beta - \tilde{l}^\alpha \tilde{l}^\beta) \approx \mu(u^\alpha u^\beta - l^\alpha l^\beta) - 2\mu l^\alpha l^\beta z(x^0) \tag{5.7}$$

is the energy-momentum tensor of thread-like substance with stochastic addition.

5.3. Gravitational Field of Cosmic String Subjected to Stochastic Perturbations

Now it is easy to find the energy-momentum tensor of a solitary cosmic string from (5.7). In order to construct its expression, it is necessary to multiply (5.7) with the Dirac δ-function. So we get

$$\tilde{\mathrm{T}}^{\alpha\beta} = \tilde{T}^{\alpha\beta} \delta_3(\vec{x} - \vec{x}') \tag{5.8}$$

The linearized Einstein equations

$$\tilde{h}_{\alpha\beta} = -16\pi(\tilde{\mathrm{T}}^{\alpha\beta} - \tfrac{1}{2} \tilde{g}^{\alpha\beta} \tilde{\mathrm{T}}) \tag{5.9}$$

have the following stochastic solutions ($\tilde{h}_{\alpha\beta} = h_{\alpha\beta} + \hat{h}_{\alpha\beta}$)

$$\hat{h}_{00} = -4\gamma\mu \int \frac{z(x^0)}{|\vec{x}-\vec{x}'|} \delta_3(\vec{x}-\vec{x}') dV = 8\gamma\mu \cdot z(x^0) \cdot \ln \frac{r}{r_0} \tag{5.10a}$$

$$\hat{h}_{ab} = -8\gamma\mu \int \frac{z(x^0)}{|\vec{x}-\vec{x}'|} \delta_3(\vec{x}-\vec{x}') dV = -16\gamma\mu \cdot z(x^0) \cdot \ln \frac{r}{r_0} \tag{5.10b}$$

$(a,b = 1,2)$

$$\widehat{h}_{33} = -12\gamma\mu \int \frac{z(x^0)}{|\vec{x}-\vec{x}'|} \delta_3(\vec{x}-\vec{x}')dV = -24\gamma\mu \cdot z(x^0) \cdot \ln\frac{r}{r_0}. \qquad (5.10c)$$

Hence, the interval of gravitational field created by an infinite cosmic string which linear mass density is subjected to stochastic perturbations, may be written down as follows

$$ds^2 = \left(1 + 8\gamma\mu \cdot z(x^0) \cdot \ln\frac{r}{r_0}\right)(dx^0)^2 +$$
$$\left(1 - 8\gamma\mu \cdot z(x^0) \cdot \ln\frac{r}{r_0} + 16\gamma\mu \cdot z(x^0) \cdot \ln\frac{r}{r_0}\right)\left((dx^1)^2 + (dx^2)^2\right) - \qquad (5.11)$$
$$\left(1 + 24\gamma\mu \cdot z(x^0) \cdot \ln\frac{r}{r_0}\right)(dx^3)^2.$$

5.4. A Probe Thread's Equation of Motion in the Gravitational Field of Cosmic String Subjected to Stochastic Perturbations

Let us write out the thread's equation of motion at any external gravitational field

$$\frac{D^2 x^\alpha}{d\tau^2} - \frac{D^2 x^\alpha}{d\rho^2} = 0 \qquad (5.12)$$

and assume that

$$x^0 = x^0(\tau) \qquad (5.13\,a)$$
$$x^k = x^k(\tau, \rho), \qquad (5.13\,b)$$

for simplicity. For the solution of the equation (5.12), we will search by the post-Newtonian approximation method. According to this method, we may decompose vectors u^α and l^α as follows

$$u^0 = 1 + \underset{2}{u^0} + \underset{4}{u^0} + \cdots, \qquad (5.14\,a)$$

$$u^k = \underset{1}{u^k} + \underset{3}{u^k} + \cdots \qquad (5.14\,b)$$

and

$$l^0 = \underset{1}{l^0} + \underset{3}{l^0} + \cdots, \qquad (5.15\,a)$$

$$l^k = \underset{0}{l^k} + \underset{2}{l^k} + \cdots. \tag{5.15 b}$$

Then, in the lowest – zero – approximation for the null component, we have equation of motion from (5.12)

$$\frac{d^2 x^0}{d\tau^2} = 0 \tag{5.16}$$

with the simplest solution

$$x^0 = \tau. \tag{5.17}$$

For the spatial components of (5.12), the minimal order equals to two. Then, with account (5.17), we get the next equation

$$\frac{d^2 x^k}{dx^{0^2}} + 4\frac{\gamma\mu}{r}\eta^k - 4\frac{\gamma\mu}{r} \cdot z(x^o) \cdot \eta^k = 0. \tag{5.18}$$

Below we will give the possible cosmological application of this equation of motion. Therefore, for its research, it is enough to limit ourselves by the simplest type of the thread finite movement – the circular movement. In this case, $r = R_0 = const$ and equation of motion may be written down in the more habitual form

$$\frac{d^2 x^k}{dt^2} + 4\frac{\gamma m}{R_0^2} x^k = 4\frac{\gamma m}{R_0^2} \cdot z(t) \cdot x^k, \tag{5.19}$$

where $k = 1, 2$.

As the result, the equation of motion takes on the form of a stochastic differential equation of the second order.

5.5. Solution of the Average Stochastic Equation of Motion

By introducing two designations

$$\omega^2 = 4\frac{\gamma m}{R_0^2}, \quad f^k(x) = 3\omega^2 x^k, \tag{5.20}$$

we may write down equation (5.19) in the standard mathematical form

$$\frac{d^2 x^k}{dt^2} + \omega^2 x^k = 3 f^k(x) \cdot z(t). \tag{5.21}$$

Due to its linearity, this equation breaks up to the pair of identical stochastic equations for one-dimensional particle's oscillations

$$\frac{d^2 x}{dt^2} + \omega^2 x = 3 f(x) \cdot z(t). \tag{5.22}$$

Furthermore, we will assume that stochastic force describes the processes with independent increments for all of its arguments. They are named "white noise". Then, equation (5.21) could be rewritten as [2 - 4]

$$\frac{d^2 x}{dt^2} + \omega^2 x = f(x) \frac{d\zeta(t)}{dt}, \tag{5.23}$$

where $\zeta(t)$ - "white noise".

Solution of this equation equals to the solution of Ito's stochastic differential equations

$$dx(t) = y(t)dt, \tag{5.24a}$$

$$dy(t) = \omega^2 dt + f(x) d\zeta. \tag{5.24b}$$

Its solution is the Markov process $\{x(t), y(t)\}$ in the phase space of dynamical system. Distribution density of this process $W(x_0, y_0, t_0; x, y, t)$ satisfies the Kolmogorov-Fokker-Planck equation (KFP-equation)

$$\frac{\partial W}{\partial t} + \frac{\partial (yW)}{\partial x} - \frac{\partial (\omega^2 xW)}{\partial y} = \frac{1}{2} \frac{\partial^2 (f^2(x)W)}{\partial y^2} \tag{5.25}$$

with the corresponding initial data $W(x_0, y_0, t_0; x, y, t_0) = \delta(x - x_0)\delta(y - y_0)$ and normalizing condition

$$\int_{-\infty}^{\infty} \int_{-\infty}^{\infty} W(x_0, y_0, t_0; x, y, t) dx dy = 1. \tag{5.26}$$

After passing to the new coordinates

$$x = A\cos\varphi, \quad y = -\omega A \sin\varphi, \quad \varphi = \omega t + \theta \tag{5.27}$$

and using the average principle, the KFP-equation takes on the form

$$\frac{\partial W_0}{\partial t} + \frac{\partial}{\partial A}(A(A)W_0) + \frac{\partial}{\partial \theta}(B(A)W_0) = \frac{1}{2}\left\{\frac{\partial^2}{\partial A^2}(\Sigma(A)W_0) + 2\frac{\partial^2}{\partial A \partial \theta}(H(A)W_0) + \frac{\partial^2}{\partial \theta^2}(E(A)W_0)\right\}, \qquad (5.28)$$

where A, B, Σ, H, E are the totally calculating coefficients - diffusion, transport and mixed.

An important role the stationary distribution density plays is the analysis of the oscillatory system with stochastic perturbations. Therefore, we will consider this type of distribution density later on. After calculation of the needed coefficients, we have the resulted KFP-equation [5]

$$\frac{\partial}{\partial A}(\overline{A}(A)W_0) = \frac{1}{2}\left\{\frac{\partial^2}{\partial A^2}(\overline{\Sigma}(A)W_0) + \frac{\partial^2}{\partial \theta^2}(\overline{E}(A)W_0)\right\}. \qquad (5.29)$$

From (5.29), it follows that average stationary distribution density is possible to present as the sum of two items that depend on A and θ separately. Thus, we look for the solution of (5.29) as

$$W_0 = W'(A_0, A) + W''(\theta_0, \theta). \qquad (5.30)$$

This representation allows us to get the expressions

$$W'(A_0, A) = \sigma' \cdot (A - A_0), \qquad (5.31a)$$

$$W''(\theta, \theta_0) = \sigma'' \cdot (\theta - \theta_0), \qquad (5.31b)$$

where constant values σ' and σ'' can be determined from the normalizing conditions.

So, as the result, we have that distribution phase densities W' and W'' are linearly proportional to the amplitude A and phase θ, accordingly.

5.6. The Approximate Solution of the Stochastic Equation of Motion

By usage of designation $v^2 = 4\frac{\gamma}{R_0^2}$, we may write down equation (5.22) in the following form

$$\frac{d^2 x}{dt^2} + \omega^2 x = 3v^2 \cdot m(t) \cdot x, \qquad (5.32)$$

where $v^2 = 4\dfrac{\gamma}{R^2_0}$ and $m(t) = m \cdot z(t)$ - stochastic function of time. Let us put forward the assumption that $m(t)$ could be represented as the set of periodical stochastic function on time

$$m(t) = \sum_n (^A m_n \cos \Omega_n t + ^B m_n \sin \Omega_n t). \tag{5.33}$$

Moreover, we put that $^A m_n$ and $^B m_n$ are the stochastic values that have the identical dispersion σ_m and normal Gaussian distribution

$$f(^\Xi m_n) = \dfrac{1}{\sqrt{2\pi}\sigma_m} \exp\left\{-\dfrac{(^\Xi m_n - ^\Xi m_0)^2}{2\sigma_m^2}\right\}, \tag{5.34}$$

where $\Xi = A, B$.

In accordance with previous assumptions (5.2), we may look for the solution of (5.31) – (5.34) in the form $x = x_0 + x'$, where x_0 - non-perturbed and x' - small perturbation of it components $(x' \ll x_0)$. Thus, from (5.32), we get the partial non-perturbed solution

$$x_0(t) = R_0 \cos \omega t. \tag{5.35}$$

Substituting (5.34) into the (5.31) for the case of resonance $(\omega = \Omega)$ and in one-mode approximation, we get the equation for the perturbed variable

$$\dfrac{d^2 x'}{dt^2} + \omega^2 x' = \dfrac{1}{2} v^2 m R_0 [1 + 2 \cos \omega t]. \tag{5.36}$$

Its physically interesting solution (non-homogeneous part) is

$$\overset{*}{x}'(t) = \dfrac{1}{2} v^2 m \dfrac{R_0}{\omega^2}\left(1 - \dfrac{1}{3} \cos 2\omega t\right). \tag{5.37}$$

Due to (5.36), we may calculate the average magnitude of addition x' -

$$<\overset{*}{x}'(m)> = \int_{-\infty}^{\infty} \overset{*}{x}'(m) f(m) dm = \dfrac{v^2 R_0}{2\omega^2} m_0 (1 - \dfrac{1}{3} \cos 2\omega t) \tag{5.38}$$

and the average with respect to period T stochastic perturbation of x' -

$$\overline{<\overset{*}{x}'(m)>} = \frac{v^2 R_0}{2\omega^2 T} m_0 \int_0^T (1 - \frac{1}{3}\cos 2\omega t)dt = \frac{2}{3}\frac{m_0}{m} R_0 = \delta R_m. \quad (5.39)$$

5.7. Perturbations of Substrate Density Produced by Chaotic Motion of a Cosmic Thread

According to Smith-Vilenkin's model of the cosmic string network evolution [6], the typical distances between cosmic strings are $\xi \propto k^{-1/2} t$, where k - numerical factor of order in 20. Let $R_0 \propto \xi$. We also may introduce the value $\eta_m = \frac{\delta R_m}{R_0}$ that describes the measure of the cosmic strings' chaos due to stochastic changing of the mass "amplitude" distribution.

Earlier it was pointed out that at time scale $t_1 \approx 30\,\text{sec}$, the cosmic strings had the Brownian character, while at time scale $t_2 \approx 100\,\text{sec}$, they were straightened. Therefore, the value $\eta = \frac{\xi_1}{\xi_2} = \frac{t_1}{t_2} \sim 0.3$ can be chosen as measure of chaos in the Smith-Vilenkin model.

Let us put forward the assumption that

$$\eta_m \leq \eta. \quad (5.40)$$

This means that the initially dimensionless cosmic thread acquires the transversal size in virtue of its stochastic perturbations

$$\delta R \cong \eta R_0. \quad (5.41)$$

And due to this size, a thread will produce, in external substrate, the perturbations of wavelength ($\lambda_{pert} \propto \delta R$). It is easy to calculate that $\delta R \propto 10^{12}\,cm$. Comparison of this perturbing wavelength with the Jeans wavelength ($\lambda_J \propto 10^{11}\,cm$) leads to a result of $\frac{\lambda_{pert}}{\lambda_J} \sim 10$. It means that chaotic fluctuations of cosmic threads really can produce the long perturbing waves in external substrate. And such waves, in its turn, create the unstable state of substrate, i.e. split the primordially homogeneous medium to the number of separate clots and will form the galaxies from them later.

Some other results in this cosmological area have been discussed in articles [7].

6. Gravitational Interaction of Two Oscillating Cosmic Strings

6.1. Introduction

Alter the pioneer Vilenkin's paper [1] where the metric of a rectilinear massive cosmic string was deduced, many articles generalizing his result appeared. Really, the metric of rotating a massive cosmic string was founded in [2, 3]; non-rotating cosmic strings of finite size - in [4, 5]; rotating cosmic strings of finite size - in [6]. Moreover, it is necessary to mention that metrics of rectilinear [7, 8] and circular [9] charged cosmic strings, super-massive [10] and superconducting [11] cosmic strings, cosmic strings with kinks [12], hollow cosmic strings [13] and etc., also were derived.

But as it follows from what has been said above, these metrics are disregarded with one of the main features of cosmic strings behavior. We are talking about the fact that in its own evolution process, any cosmic string as an extent object will accomplish the oscillation. Especially in the case of cosmic strings with masses on it's ends [14]. That is why the real cosmic string on the early stages of the Universe evolution is called the oscillating string. These oscillations could be the reason of the plural processes of its re-commutation, intersection, string tears, closed strings origin and etc. Hence, the cosmic string oscillations will lead to the appearance of the corresponding additions to the components of metric tensor. And these additions as it follows from the general form of the solution of the string's equation of motion will contain only the periodical terms.

The special case of such type of metric was deduced by Vachaspati [15, 16]. It describes the space time interval in the vicinity of the cosmic string with weak disturbance waves traveling along it with the speed of light. Later, the simple physical process - light rays deflection - had been studied in this metric [17].

The main aim of our article is to deduce the interval of a weak-oscillating non-radiated cosmic string of another type, namely the interval of the massive cosmic string with applied standing flat disturbance waves on it. Moreover, we give the brief discussing of some physical processes occurring in the vicinity of these oscillating massive cosmic strings.

Among them, the dynamics of an open "probe" cosmic thread performing the constraint oscillations over its influence. The nature of these oscillations is drastically different from the previous mechanisms. Really, the cosmic string oscillations earlier were discussed in the context of dynamical friction under the string's moving in medium (radiation, non-relativistic particles [18], the intersections of cosmic strings [19, 20], its collapse [21] and etc.

The most detailed investigation of the oscillating cosmic string has been done under the consideration of its gravitational energy radiation [22 - 24]. In doing this, the solution of the string equation of motion was chosen as the traveling waves of different structure [25, 26]. This procedure had allowed us to get the time-periodical depended components of the metric tensor and to study the spectral distribution of the gravitational energy radiation.

In the present model, the strong oscillations arc occurring due to the external periodical force that tends not to the decreasing of the oscillations, but just the opposite - to their time-linear increasing. Then the gravitational radiation, emitted from the cosmic thread, will time-linear increase also.

The matter of this chapter has been partially published in [27].

6.2. Space time in the Vicinity of the Oscillating as Standing Waves Cosmic Thread

To realize our purpose, we start from the assignment of the stress-energy tensor due to the thread-like matter [28]

$$T^{\alpha\beta} = \sum_\alpha \overset{\alpha}{\mu}\left(\overset{\alpha}{u^\alpha}\overset{\alpha}{u^\beta} - \overset{\alpha}{l^\alpha}\overset{\alpha}{l^\beta}\right)\delta_3\left(\vec{x}-\overset{\alpha}{\vec{\xi}}\right) \qquad (6.1)$$

and the equation of motion of the α - th thread

$$\frac{D\overset{\alpha}{u^\alpha}}{d\overset{\alpha}{\tau}} - \frac{D\overset{\alpha}{l^\alpha}}{d\overset{\alpha}{\rho}} = 0, \qquad (6.2)$$

that follows from (6.1) in view of the conservation law. In (6.2), $\overset{\alpha}{u^\alpha} = \frac{d\overset{\alpha}{x^\alpha}}{d\overset{\alpha}{\tau}}, \overset{\alpha}{l^\alpha} = \frac{d\overset{\alpha}{x^\alpha}}{d\overset{\alpha}{\rho}}$ are the time-like and space-like vectors, accordingly, that characterize the motion and the orientation of the string as a whole. We put the usual orthonormal gauge $\left(\overset{\alpha}{u^\alpha} \pm \overset{\alpha}{l^\alpha}\right)^2 = 0$ on these vectors, also.

To the next investigation, we should be reminded that all calculations in [1] were carried out in the framework of the linearized Einstein's gravity, i.e. it was a setting that, $g_{\mu\nu} = \delta_{\mu\nu} + h_{\mu\nu}$, where $h_{\mu\nu}$ are the small corrections to the psevdoeuclidian background. Staying in this approximation, we can write the general relativity equations in the form

$$\Box h_{\mu\nu} = -16\pi\gamma\left(T_{\mu\nu} - \frac{1}{2}\delta_{\mu\nu}T\right). \qquad (6.3)$$

Here, the symbol □ means 4-dimensional D'Alambertian. Substituting the stress-energy tensor (6.2), here we get in components (for the N strings that are motionless as a whole)

$$\Box h_{00} = 0, \qquad (6.4)$$

$$\Box h_{kl} = 16\pi\gamma\sum_\alpha \overset{\alpha}{\mu}\left(\delta_{kl} + \overset{\alpha}{l_k}\overset{\alpha}{l_l}\right)\delta_3\left(\vec{x}-\overset{\alpha}{\vec{\xi}}\right). \qquad (6.5)$$

From (6.4), we have the partial solution $h_{00} = 0$. As for the equation (6.5), it is easy to see that the solution has the retarded potential form

$$h_{kl} = -4\gamma \sum_\alpha \int \overset{\alpha}{\mu} \frac{\delta_{kl} + \overset{\alpha}{l'_k} \overset{\alpha}{l'_l}}{|\vec{x} - \vec{x}'|} \delta_3\left(\vec{x} - \overset{\alpha}{\vec{\xi}}\right) dV', \tag{6.6}$$

where $\overset{\alpha}{l'_k} = \overset{\alpha}{l'_k}(\vec{x}', x'^0)$.

Accordingly to the problem organization, we should look for the metric tensor in the vicinity of the solitary cosmic string. Expanding the expression (6.6) into the power series of $\lambda = |\vec{x} - \vec{x}'|/x^0$, in the case $\alpha=1$ and with the main order accuracy, we have

$$h_{kl} = -4\gamma \int \mu \frac{\delta_{kl} + l'_k l'_l}{|\vec{x} - \vec{x}'|} \delta_3(\vec{x} - \vec{x}') dV', \tag{6.7}$$

where $l'_k = l'_k(\vec{x}', x^0)$ now. Then it is necessary to determine the vector l_k to finding the explicit form of (6.7).

In our previous paper [29], it was shown that

$$l^k = l_0^k + \sum_{n=1}^\infty \frac{\pi n}{L}\left(A_n^k \cos\frac{\pi n}{L}\tau + B_n^k \sin\frac{\pi n}{L}\tau\right)\cos\frac{\pi n}{L}\rho, \tag{6.8}$$

where A_n^k è B_n^k are the set of amplitudes, L is the string's length. Therefore, the potential (6.7) may be decomposed over two parts $\left(h_{kl} = {}^{(1)}h_{kl} + {}^{(2)}h_{kl}\right)$, that have the forms

$${}^{(1)}h_{kl} = -4\gamma \int \mu \frac{\delta_{kl} + l_0^k l_0^l}{|\vec{x} - \vec{x}'|} \delta_3(\vec{x} - \vec{x}') dV' \tag{6.9}$$

and

$${}^{(2)}h_{kl} = -4\gamma \delta_{kl} \int \mu \sum_{n=1}^\infty \frac{\pi^2 n^2}{L^2}\left(A_n^m \cos\frac{\pi n}{L}\tau + B_n^m \sin\frac{\pi n}{L}\tau\right)^2 \frac{\cos^2 \frac{\pi n}{L}\rho'}{|\vec{x} - \vec{x}'|} \delta_3(\vec{x} - \vec{x}') dV' \tag{6.10}$$

accordingly.

To calculate these integrals, we put that cosmic string lies along the z-axis while the disturbance waves are staying in the $\{y, z\}$ - plane. If $\hat{\mu} = \mu\delta(z - z')$ is the line mass density, we get the standard solution from (6.9)

$$^{(1)}h_{kl} = -8\ln\frac{\left(x^2+y^2\right)^{1/2}}{r_0}\delta_{kl}, \qquad (6.11)$$

where r_0 -constant value, having the sense of the massive string's cross-section size.

As for the expression (6.10), than in view of (6.8), the condition that $\rho = z$ along the cosmic string and by choice of the oscillations plane, it can be presented as follows

$$^{(2)}h_{22} = -8\gamma\hat{M}\ln\frac{y}{r_0}, \qquad (6.12)$$

with the time-variable line mass density

$$\hat{M} = \mu\sum_{n=1}^{\infty}\frac{\pi^2 n^2}{L^2}\left(\overset{*}{A}_n\cos\frac{\pi n}{L}wx^0 + \overset{*}{B}_n\sin\frac{\pi n}{L}wx^0\right)^2\cos^2\frac{\pi n}{L}z'\delta(z-z'). \qquad (6.13)$$

(We are marking $\overset{*}{A}_n = A_n^2$ and $\overset{*}{B}_n = B_n^2$ for simplify, here).

Hence, the space time interval of cosmic string due to the family of standing waves applied on it is

$$dS^2 = dx^{02} - \left[1 + 8\gamma\hat{\mu}\ln\frac{\left(x^2+y^2\right)^{1/2}}{r_0}\right]dx^2 -$$
$$-\left[1 + 8\gamma\hat{\mu}\ln\frac{\left(x^2+y^2\right)^{1/2}}{r_0} + 8\gamma\hat{M}\ln\frac{y}{r_0}\right]dy^2 - dz^2. \qquad (6.14)$$

Because the metric we have obtained above is drastically anisotropic, it is clear that all of the physical processes in it will be anisotropic, also. For example, the deflection angle will differ under it propagations along the x-axis and y-axis, accordingly.

The periodical character of this metric will radically influence on the oscillating processes near the oscillating cosmic string. Really, the "probe" thread in the space time (6.14) will perform the constrained oscillations without fail. Let us find the equation of these oscillations.

6.3. The Open "Probe" Thread in the Gravitational Field of the Massive Oscillating Cosmic String

Let us write out the equation of motion of a "probe" thread in the given spacetime in components

$$\left.\begin{array}{l}\dfrac{d^2x^0}{d\tau^2}-\dfrac{d^2x^0}{d\rho^2}+\Gamma^0_{00}\left(u^0u^0-l^0l^0\right)+\\ \quad +2\Gamma^0_{0m}\left(u^0u^m-l^0l^m\right)+\\ \Gamma^0_{mn}\left(u^mu^n-l^ml^n\right)=0\end{array}\right\}, \tag{6.15}$$

$$\left.\begin{array}{l}\dfrac{d^2x^k}{d\tau^2}-\dfrac{d^2x^k}{d\rho^2}+\Gamma^k_{00}\left(u^0u^0-l^0l^0\right)+\\ \quad +2\Gamma^k_{0m}\left(u^0u^m-l^0l^m\right)+\\ \Gamma^k_{mn}\left(u^mu^n-l^ml^n\right)=0\end{array}\right\}. \tag{6.16}$$

We look to the solution of equation (6.16) in the form

$$x^k = x_0^k + \xi^k, \tag{6.17}$$

where ξ^k are the small additions to the main non-perturbed displacement x_0^k. Substituting (6.17) into the (6.16) we get the homogeneous hyperbolic equation

$$\frac{d^2 x_0^k}{d\tau^2} - \frac{d^2 x_0^k}{d\rho^2} = 0, \tag{6.18}$$

that describes the thread free oscillations as a set of the standing waves.

To deduce the explicit form of the perturbed equation of motion, it is necessary to substitute the Riemann - Christoffel symbols, as well as the vectors u^m and ℓ^m into the (6.15) and (6.16).

Later on, we should take an interest to those terms in the metric tensor that have the periodical character only, i.e. the terms in g_{22}-component. In accordance to this remark, the required additions to the Riemann - Christoffel symbols are

$$\Gamma^0_{22}=\Gamma^2_{02}=-2\gamma\mu\sum_{m=1}^{\infty}\frac{\pi^3 m^3}{\Lambda^3}\left[\left(A_m^2-B_m^2\right)\sin\frac{\pi m}{\Lambda}\tau-\right.$$
$$\left.2A_m B_m\cos 2\frac{\pi m}{\Lambda}\tau\right]\cos^2\frac{\pi m}{\Lambda}z'\cdot\ln\frac{y}{r_0}, \tag{6.19}$$

$$\Gamma^2_{22}=2\gamma\mu\sum_{m=1}^{\infty}\frac{\pi^2 m^2}{\Lambda^2}\left[\left(A_m\cos\frac{\pi m}{\Lambda}\tau+B_m\sin\frac{\pi m}{\Lambda}\tau\right)^2\cos^2\frac{\pi m}{\Lambda}z'\right]\frac{1}{y}. \tag{6.20}$$

By virtue of the linearity of the Einstein's equations (6.3) and the solution of string's equation of motion [29], we may record the vectors u^μ and l^μ as its main terms

$$u^0 = \frac{1}{\omega},$$
$$u^2 = v_0^2 - \sum_{n=1}^{\infty} \frac{\pi n}{L}\left(a_n \sin\frac{\pi n}{L}\tau - b_n \cos\frac{\pi n}{L}\tau\right)\sin\frac{\pi n}{L}\rho; \quad (6.21)$$

$$l^0 = 0,$$
$$l^2 = \sum_{n=1}^{\infty} \frac{\pi n}{L}\left(a_n \cos\frac{\pi n}{L}\tau + b_n \sin\frac{\pi n}{L}\tau\right)\cos\frac{\pi n}{L}\rho. \quad (6.22)$$

Getting for simplification that $n=1$ and marking $\xi^2 = y, p = z$, the equation (6.15) provided (6.19), (6.21) and (6.22) acquires the form

$$\frac{d^2 x^0}{d\tau^2} = 2\gamma\mu \sum_{m=1}^{\infty} \frac{\pi^5 m^3}{\Lambda^3 L}\left[(A_m^2 - B_m^2)\sin 2\frac{\pi m}{L}\tau - 2A_m B_m \cos 2\frac{\pi m}{L}\tau\right]\cos^2\frac{\pi m}{\lambda}z$$
$$\left[2v_0\left(a\sin\frac{\pi}{L}\tau - b\cos\frac{\pi}{L}\tau\right)\sin\frac{\pi}{L}z + \frac{1}{L}\left(a\cos\frac{\pi}{L}\tau + b\sin\frac{\pi}{L}\tau\right)^2 \cos^2\frac{\pi}{L}\tau\right] \quad (6.23)$$
$$\left[\ln\frac{a}{r_0} + \ln\cos\frac{\pi}{L}\tau + \ln\sin\frac{\pi}{L}\tau\right].$$

In completely similar fashion, the equation (6.16) may be written as

$$\frac{d^2 y}{d\tau^2} - \frac{d^2 y}{dz^2} = 4\gamma\mu \sum_{m=1}^{\infty} \frac{\pi^4 m^3}{\Lambda^3 L}\left[(A_m^2 - B_m^2)\sin 2\frac{\pi m}{L}\tau - 2A_m B_m \cos 2\frac{\pi m}{L}\tau\right]$$
$$\cos^2\frac{\pi m}{\Lambda}z\left[\ln\frac{a}{r_0} + \ln\cos\frac{\pi}{L}\tau + \ln\sin\frac{\pi}{L}z\right]\frac{1}{w}\left(a\sin\frac{\pi}{L}\tau - b\cos\frac{\pi}{L}\tau\right)$$
$$\sin\frac{\pi}{L}z + 2\gamma\mu \sum_{m=1}^{\infty} \frac{\pi^4 m^3}{\Lambda^2 L}\left(A_m \cos\frac{\pi m}{\Lambda}\tau + B_m \sin\frac{\pi m}{\Lambda}\tau\right)^2 \cos^2\frac{\pi m}{\Lambda}z \cdot \quad (6.24)$$
$$\left[2v_0\left(a\sin\frac{\pi}{L}\tau - b\cos\frac{\pi}{L}\tau\right)\sin\frac{\pi}{L}z - \frac{\pi}{L}\left(a\sin\frac{\pi}{L}\tau - b\cos\frac{\pi}{L}\tau\right)^2 \sin^2\frac{\pi}{L}\tau + \right.$$
$$\left.\frac{\pi}{L}\left(a\cos\frac{\pi}{L}\tau + b\sin\frac{\pi}{L}\tau\right)^2 \cos^2\frac{\pi}{L}z\right]\frac{1}{\left(a\cos\frac{\pi}{L}\tau + b\sin\frac{\pi}{L}\tau\right)\sin\frac{\pi}{L}z}.$$

It is necessary to point out that we omit terms which are not describing the reciprocal influence of the cosmic string and the cosmic thread oscillations in (6.23) and (6.24). To avoid any misunderstandings, we are marking the *y*-th component of velocity as v_0, also. So, the equations of motion (6.23) - (6.24) are describing the oscillations of the "probe" thread that is moving along the *y*-axis to the massive cosmic string with the velocity v_0. But the cosmic string and the "probe" thread themselves are orientating along the *z*-axis as a whole.

Now it is necessary to estimate the orders of the terms that involve the right sides of (6.23) and (6.24). We see from (6.24) that one part of terms has the order $\dfrac{A^2}{\Lambda^2}$, while another $-\dfrac{A^2}{\Lambda^2}\dfrac{a}{L}$; from (6.23) that one part of terms has not only the order $\dfrac{A^2}{\Lambda^2}\dfrac{a}{L}$, but the order $\dfrac{A^2 a^2}{\Lambda^2 L^2}$, also. Discarding the terms of order $\dfrac{A^2}{\Lambda^2}\dfrac{a}{L}$ and higher in these equations it is easy to show that equation (6.23) can be lead to the trivial form

$$\frac{d^2 x^0}{d\tau^2} = 0 \qquad (6.25)$$

that gives the well-known solution

$$x^0 = \frac{\tau}{w}. \qquad (6.26)$$

As for the equation (6.24), that it will be unwieldy sufficiently, in spite of the above simplifications. That is why we put *m*=1 and *B*=*b*=0, also. Then, in view of (6.26), the (6.24) turns to be equal to

$$\frac{d^2 y}{dx^{0^2}} - w^2 \frac{d^2 y}{dx^2} = F(x^0; z), \qquad (6.27)$$

where the perturbed force is

$$F(x^0; z) = 4\gamma\mu \cdot v_0 \frac{\pi}{\Lambda^2 L} A^2 \cos^2 \frac{\pi w}{\Lambda} x^0 \cdot \cos^2 \frac{\pi}{\Lambda} z \cdot \sin \frac{\pi}{L} z \frac{\sin \frac{\pi w}{L} x^0}{\cos \frac{\pi w}{L} x^0} \qquad (6.28)$$

Furthermore, we will examine the case when the sizes of cosmic strings and cosmic threads satisfy the following condition

$$\Lambda = 2L. \qquad (6.29)$$

As the result, it is possible to write down the solution of the cosmic string's equation of motion as follows:

$$y(x^0, z) = y_1(x^0, z) + y_2(x^0, z) \tag{6.30}$$

with terms

$$y_1(x^0, z) = \frac{1}{2} \gamma \mu \cdot v_0 \frac{\pi w^2 A^2}{L^3} \cdot I(x^0) \cdot \sin \frac{\pi}{L} z, \tag{6.31}$$

$$y_2(x^0, z) = \frac{1}{4} \gamma \mu \cdot v_0 \frac{\pi w^2 A^2}{L^3} \cdot I(x^0) \cdot \sin 2\frac{\pi}{L} z, \tag{6.32}$$

$$I(x^0) = \int_0^{x^0} \cos^2 \frac{\pi \omega}{2L} \chi \cdot \frac{\sin \frac{\pi \omega}{L} \chi}{\cos \frac{\pi \omega}{L} \chi} \sin \frac{\pi w}{2l}(x^0 - \chi) d\chi. \tag{6.33}$$

Taking this integral, we verify that it turns to be equal to the sum of two periodic and two non-periodic terms. We have to deal with the most of the non-periodic item, further. Then omitting all other terms it is easy to get the suitable solution of the equation of motion

$$\left. \begin{array}{l} y_1(x^0, z) = \dfrac{1}{8} \gamma \mu \cdot v_0 \dfrac{wA^2}{L^2} \cdot x^0 \cdot \cos \dfrac{\pi w}{L} x^0 \cdot \sin \dfrac{\pi}{L} z, \\[2mm] y_2(x^0, z) = \dfrac{1}{16} \gamma \mu \cdot v_0 \dfrac{wA^2}{L^2} \cdot x^0 \cdot \cos \dfrac{\pi w}{L} x^0 \cdot \sin 2\dfrac{\pi}{L} z. \end{array} \right\} \tag{6.34}$$

This solution describes the periodical oscillations with the amplitude having the time-linear dependence. Therefore, the cosmic thread moving to the massive oscillating string will accomplish the constrained oscillations with the constantly increasing amplitude.

6.4. Gravitational Radiation from the Strong Oscillating Cosmic Thread

Let us estimate the full gravitational energy radiation, i.e. the energy less up to the time unit because of the strong cosmic thread oscillations that are occurring in the gravitational field of the massive oscillating cosmic string.

It is well-known that the full energy radiation over all directions is described by the formula

$$-\frac{d\varepsilon}{dx^0} = 4\pi d\bar{\Im}, \tag{6.35}$$

where $d\overline{\Im}$ - the average over all spatial directions radiation intensity into the solid angle dO is

$$d\overline{\Im} = |S|d\sigma = |S|R_0^2 dO, \qquad (6.36)$$

where $|S|$ - the Pointing vector module. The spatial components of the energy flux may be expressed over the component of the stress-energy pseudotensor as follows:

$$S^k = -it^{0k}. \qquad (6.37)$$

Keeping these remarks in mind, we may find the gravitational energy radiation from the strong oscillating cosmic thread. Marking the amplitudes that strongly depended from the time parameter as

$$\left.\begin{array}{l}\tilde{A}_1(x^0) = \dfrac{1}{8}\gamma\mu \cdot v_0 \cdot \dfrac{wA^2}{L^2} \cdot x^0 \\ \tilde{A}_2(x^0) = \dfrac{1}{16}\gamma\mu \cdot v_0 \cdot \dfrac{wA^2}{L^2} \cdot x^0\end{array}\right\}, \qquad (6.38)$$

we can rewrite the perturbation terms of the coordinates of the oscillating cosmic thread in the next form

$$\left.\begin{array}{c}\xi^0 = 0, \\ y = \xi^2 = \tilde{A}_1(x^0)\cos\dfrac{\pi w}{L}x^0 \cdot \sin\dfrac{\pi}{L}z + \tilde{A}_2(x^0)\cos\dfrac{\pi w}{L}x^0 \cdot \sin 2\dfrac{\pi}{L}z \\ \xi^1 = \xi^3 = 0\end{array}\right\} \qquad (6.39)$$

It is easy to find the suitable wave–type additions to the vectors u^μ and l^μ. But according to the general expression of the space components of the retard potential, that deduced by the solitary cosmic string, we are in need of the additions to the vector - l^k, only. Marking them as - ℓ^k, we find from (6.5), respectively,

$$\left.\begin{array}{c}\ell^0 = 0, \\ \ell^2 = \left(\tilde{A}_1(x^0)\dfrac{\pi}{L}\cos\dfrac{\pi}{L}z + 2\tilde{A}_2(x^0)\dfrac{\pi}{L}\cos 2\dfrac{\pi}{L}z\right)\cdot\cos\dfrac{\pi\omega}{L}x^0, \\ \ell^1 = \ell^3 = 0\end{array}\right\} \qquad (6.40)$$

We see that the suitable component of the gravitational potential is the potential of linear masses as before, that depends not only on z, but generally on x^0, also. Let us write it down in the form

$$\overset{*}{\delta h_{22}} = -4\gamma \overset{*}{\mu} \frac{\pi^2}{L^2} \int [\tilde{A}_1(x^{0'})\cos\frac{\pi}{L}z' + 2\tilde{A}_2(x^{0'})\cos 2\frac{\pi}{L}z']^2 \cdot \frac{\cos^2\frac{\pi w}{L}x^{0'}}{|x-x'|} \delta_3(\vec{x}-\vec{x}')dV', \quad (6.41)$$

where $\overset{*}{\mu}$ is the linear mass density of the "probe" cosmic thread. Marking as

$$\overset{*}{\rho}(x^{0'},z') = \overset{*}{\mu}\frac{\pi^2}{L^2}\left(\tilde{A}_1(x^{0'})\cos\frac{\pi}{L}z' + 2\tilde{A}_2(x^{0'})\cos 2\frac{\pi}{L}z'\right)^2 \cos^2\frac{\pi w}{L}x^{0'}, \quad (6.42)$$

the founded potential is reduced to the standard form

$$\overset{*}{\delta h_{22}} = -4\gamma \int \frac{\overset{*}{\rho}(x^{0'},z')}{|\vec{x}-\vec{x}'|}\delta_3(\vec{x}-\vec{x}')dV'. \quad (6.43)$$

Expanding this potential into the series over the parameter λ, we are holding the term that describes the gravitational field in the wave zone, only $(R_0 \rangle \rangle L)$ -

$$\tilde{\delta h}_{22} = -8\gamma \tilde{D}_{ij}(x^0;z)\frac{\partial^2}{\partial x_i \partial x_j}\ln\frac{y}{\Delta} = -8\gamma\Delta\ln\frac{y}{\Delta}\cdot\frac{d^2\tilde{D}_{ij}}{dx^{0^2}}\cdot n_i n_j, \quad (6.44)$$

where

$$\tilde{D}_{ij}(x^0,z) = \frac{1}{\tilde{\Delta}}\int \overset{*}{\rho}(x^0,z)\left(x^i x^j - \frac{1}{3}\delta_{ij}r^2\right)dV, \quad (6.45)$$

and $\tilde{\Delta}$ - the integer, characterizing the thread's own transversal size. Substituting (6.42) here, we get the expression

$$\tilde{D}_{ij}(x^0,z) = \overset{*}{\mu}\frac{\pi^2}{\tilde{\Delta}L^2}\left(\tilde{A}_1(x^0)\cos\frac{\pi}{L}z + 2\tilde{A}_2(x^0)\cos 2\frac{\pi}{L}z\right)^2\cdot\cos^2\frac{\pi w}{L}x^0\cdot D_{ij} = \frac{1}{\tilde{\Delta}}\overset{*}{D}_{ij}, \quad (6.46)$$

where

$$D_{ij} = \int \overset{*}{\mu}\left(x_i x_j - \frac{1}{3}\delta_{ij}r\right)^2 dV - \tag{6.47}$$

is the quadrupole moment.

To calculate the gravitational energy radiation, we use the Landau-Lifshitz pseudotensor. It is easy to verify then, that the energy momentum along the x-axis is [30]

$$t^{01} = \frac{1}{64\pi\gamma}\left(\frac{d\delta\tilde{h}_{22}}{dx^0}\right)^2. \tag{6.48}$$

Substituting the potential (6.44) here, we find for any point R_0 of the wave zone that

$$t^{01} = \frac{\gamma}{\pi}\tilde{\Delta}^2 \cdot \frac{1}{\tilde{\Delta}^2}\ln^2\frac{R_0}{\tilde{\Delta}} \cdot \frac{d^3\tilde{D}_{ij}}{dx^{0^3}} \cdot \frac{d^3\tilde{D}_{kl}}{dx^{0^3}} \cdot n_i n_j n_k n_l. \tag{6.49}$$

That is why the intensity of the gravitational energy radiation into the solid angle dO is

$$d\Im = \frac{\gamma}{\pi}\tilde{\Delta}^2 \cdot \frac{R_0^2}{\tilde{\Delta}^2} \cdot \ln^2\frac{R_0}{\tilde{\Delta}} \cdot \frac{d^3\tilde{D}_{ij}}{dx^{0^3}} \cdot \frac{d^3\tilde{D}_{kl}}{dx^{0^3}} \cdot n_i n_j n_k n_l \cdot dO. \tag{6.50}$$

For the next transformations, we note in passing that according to [31],

$$\frac{R_0}{\tilde{\Delta}}\ln\frac{R_0}{\tilde{\Delta}} = \sum_{k-1}^{\infty}\left(1 - \frac{\tilde{\Delta}}{R_0}\right)^k \cdot \sum_{m=1}^{k}\frac{1}{m}. \tag{6.51}$$

But in view of $\dfrac{\tilde{\Delta}}{R_0} \ll 1$ and correct to term of the first order, we have

$$-\frac{dE}{dx^0} = \frac{4\gamma}{45} \cdot \frac{d^3\overset{*}{D}_{ij}}{dx^{0^3}} \cdot \frac{d^3\overset{*}{D}_{ij}}{dx^{0^3}}. \tag{6.52}$$

And, at last, the expression (52) we may turn to the more well-known form

$$-\frac{dE}{dx^0} = G\gamma\overset{*}{\mu}^2, \tag{6.53}$$

where G is the full calculated numerical value. So, the founded expression in numerical coefficient is differing from the similar formulas deduced as for the gravitational energy radiation from the celestial bodies [30] and as the gravitational energy radiation from the massive cosmic strings [22, 23]. But in order, it is less than the last energy radiation, because of the chosen condition $\mu^* < \mu$. That is why it plays a part of the additional term to the full gravitational energy radiation from two oscillating cosmic strings. These additions, however, in view of (6.53), is linear increasing in time. Therefore, the gravitational energy radiation of the interacting cosmic strings may be more intensive that it has been supposed above in [22 – 24].

Some other results in this cosmological area have been discussed in articles [31].

7. THE STRETCHING OF COSMIC STRINGS IN THE PRESENCE OF A VACUUM DOMAIN WALL

7.1. Introduction

Gauge theories with a spontaneously broken symmetry predict the existence of topologically stable structures of three types – monopoles, strings, and domain walls whose space dimensions are zero, one, and two, respectively. Of these, the most dynamically evolving are cosmic strings which first appeared as Brownian trajectories and only at later stages of evolution straightened under tension and started to move as a whole at a velocity approaching that of light. This results in their intersections, overlaps, closed loops, high-intensity gravitational radiation, and other dynamic processes [1–6]. It follows from this scenario that straightening of cosmic strings plays the most important role in their evolution. Those straightening results from an increase in tension due to the expansion of the Universe should be pointed out. However, in the real Universe, the interaction occurs not only between a string and a Metagalaxy, among the cosmic strings themselves [7], but also between other topological vacuum defects – monopoles [8], walls [9], etc. Furthermore, we will examine the interaction between strings and domain walls alone, this interaction being discussed in [10]. Investigation of their interaction appears to be important not only for general theory but also for cosmology. The point is, this gravitational interaction turns out to be of repulsive character and therefore must lead to stretching of cosmic strings.

The matter of this chapter has been partially published in [11].

7.2. Oscillations of a Cosmic String in the Gravitational Field of a Vacuum Domain Wall

Let us consider the motion of a cosmic string in the gravitational field of a massive domain wall. Assuming it to be in the $\{y, z\}$ plane, the corresponding 4-interval can be written in the conformal flat form as [3]

$$ds^2 = \left(1 - 4\pi\frac{\gamma\sigma}{c^2}x^1\right)\left(dx^{0^2} - dx^{1^2} - dx^{2^2} - dx^{3^2}\right), \tag{7.1}$$

where σ - is the surface density of the domain wall.

Now let us write down the general relativistic equation of motion for a cosmic string

$$\frac{du^\mu}{d\tau} - \frac{dl^\mu}{d\rho} + \Gamma^\mu_{\alpha\beta}\left(u^\alpha u^\beta - l^\alpha l^\beta\right) = 0, \tag{7.2}$$

where we used the standard designations for the time-like $u^\mu = \dfrac{dx^\mu}{d\tau}$ and space-like $l^\mu = \dfrac{dx^\mu}{d\rho}$ vectors describing string dynamics. In addition, τ and ρ - are the values specifying parameterization of the hypersurface the string sweeps over when moving. Going to the coordinate relations, the initial equation of motion with allowance for the type of the external gravitational field must be rewritten as follows:

$$\frac{du^0}{d\tau} - \frac{dl^0}{d\rho} = 2\Gamma^0_{01}\left(l^0 l^1 - u^0 u^1\right), \tag{7.3}$$

$$\frac{du^1}{d\tau} - \frac{dl^1}{d\rho} = \Gamma^1_{00}\left(l^0 l^0 - u^0 u^0\right) + \\ \Gamma^1_{11}\left(l^1 l^1 - u^1 u^1\right) + \Gamma^1_{22}\left(l^2 l^2 - u^2 u^2\right) + \Gamma^1_{33}\left(l^3 l^3 - u^3 u^3\right), \tag{7.4}$$

$$\frac{du^2}{d\tau} - \frac{dl^2}{d\rho} = 2\Gamma^2_{12}\left(l^1 l^2 - u^1 u^2\right), \tag{7.5}$$

$$\frac{du^3}{d\tau} - \frac{dl^3}{d\rho} = 2\Gamma^3_{13}\left(l^1 l^3 - u^1 u^3\right). \tag{7.6}$$

Let us assume, for the following analysis of equations (7.3) – (7.6), that the string is located along the axis x^1, and the calculations will be performed by the method of successive approximations, considering the string to make free vibrations in the major approximation. Hereinafter, all of the values related to the free vibration will have the index «(0)».

Let $x^0 = x^0(\tau)$. Then, according to the definition l^α, we get $l^0_{(0)} = 0$. Thus, equation (7.3) becomes

$$\frac{du^0}{d\tau} - \frac{dl^0}{d\rho} = 0. \qquad (7.7)$$

Its solution $x^0 = \tau$, chosen for convenience sake, is a well-known temporal parameterization in string theory [12].

Let us analyze equations (7.3) – (7.6), bearing in mind that all of the following conditions specify parameters of motion of a string in the main approximation

$$\left.\begin{array}{l} u^0_{(0)} = u^0_{(0)}\left(1,0,0,u^3_{(0)}\right) \\ l^0_{(0)} = l^0_{(0)}\left(0,1,0,l^3_{(0)}\right) \end{array}\right\}. \qquad (7.8)$$

Thus, by virtue of equation (7.8), equation (7.5) takes on the form of a homogeneous hyperbolic equation

$$\frac{du^2}{d\tau} - \frac{dl^2}{d\rho} = 0. \qquad (7.9)$$

Its solution would be naturally chosen as $x^2 = 0$.

Further, with allowance for equation (7.8), equation (7.4) is, in fact,

$$\frac{du^1}{d\tau} - \frac{dl^1}{d\rho} = -\Gamma^1_{00} + \Gamma^1_{11} + \Gamma^1_{33}\left(l^3_{(0)}{}^2 - u^3_{(0)}{}^2\right) \qquad (7.10)$$

Finally, let us consider equation (7.6). Taking into account the parameters of motion of the string in the main approximation (8), we obtain

$$\frac{du^3}{d\tau} - \frac{dl^3}{d\rho} = 2\Gamma^3_{13}l^1_{(0)}l^3_{(0)}. \qquad (7.11)$$

Our further goal is the solution of equations (7.10) and (7.11). We rewrite them in the following way, performing for simplicity the substitution $x^1 \to x, x^3 \to z$ -

$$\frac{d^2x}{dx^{0^2}} = -\Gamma^1_{00} + \Gamma^1_{11} + \Gamma^1_{33}\left(l^3_{(0)}{}^2 - u^3_{(0)}{}^2\right), \qquad (7.12)$$

$$\frac{d^2z}{dx^{0^2}} - \frac{d^2z}{dx^2} = 2\Gamma^3_{13}l^3_{(0)}. \qquad (7.13)$$

First, let us analyze equation (7.13). Considering the approximate character of the desired solution, we assume $z = Z_{(0)} + \varsigma$, where the main displacement $Z_{(0)}$ is described by a solution to a homogeneous equation, and the addition ς – by a solution to an inhomogeneous one. Hence, an equation for $Z_{(0)}$ becomes

$$\frac{d^2 Z_{(0)}}{dx^{0\,2}} - \frac{d^2 Z_{(0)}}{dx^2} = 0. \tag{7.14}$$

Its solution in the one-mode approximation is a standing wave

$$Z_{(0)}(x, x^0) = \left(A \cos \frac{\pi}{L} x^0 + B \sin \frac{\pi}{L} x^0 \right) \cdot \sin \frac{\pi}{L} x, \tag{7.15}$$

where L is the length of the string.

The first unknown parameter describing the position of the string can be easily found as

$$l_{(0)}^3 = \frac{dZ_{(0)}}{dx} = \frac{\pi}{L} \left(A \cos \frac{\pi}{L} x^0 + B \sin \frac{\pi}{L} x^0 \right) \cdot \cos \frac{\pi}{L} x. \tag{7.16}$$

Since in our metrics $\Gamma_{13}^3 = -2 \frac{\pi \gamma \sigma}{c^2}$, the equation for correction ς becomes

$$\frac{d^2 \varsigma}{dx^{0\,2}} - \frac{d^2 \varsigma}{dx^2} = -4 \frac{\pi \gamma \sigma}{c^2} \cdot \frac{\pi}{L} \left(A \cos \frac{\pi}{L} x^0 + B \sin \frac{\pi}{L} x^0 \right) \cos \frac{\pi}{L} x. \tag{7.17}$$

To solve it, we consider the following equation

$$\frac{d^2 \varsigma}{dx^{0\,2}} - \frac{d^2 \varsigma}{dx^2} = f(x^0, x) = \Phi(x^0) \cdot \cos \frac{\pi}{L} x, \tag{7.18}$$

where

$$\Phi(x^0) = -4 \frac{\pi^2 \gamma \sigma}{c^2 L} \left(A \cos \frac{\pi}{L} x^0 + B \sin \frac{\pi}{L} x^0 \right). \tag{7.19}$$

For an unambiguous solution equation (7.18), we introduce the following initial conditions:

$$\varsigma(0,x) = \phi \cdot \cos\frac{\pi}{L}x,$$
$$(\varsigma)'_{x^0}(0,x) = \psi \cos\frac{\pi}{L}x \Biggr\}. \qquad (7.20)$$

and will seek the solution as a harmonic oscillation

$$\varsigma(x^0,x) = \varsigma(x^0) \cdot \cos\frac{\pi}{L}x \qquad (7.21)$$

with a time-dependent amplitude. According to [13], the function $\varsigma(x^0)$ is written as

$$\varsigma(x^0) = \varsigma_1(x^0) + \varsigma_2(x^0), \qquad (7.22)$$

where $\varsigma_1(x^0)$ is a general solution to a homogeneous linear equation, and $\varsigma_2(x^0)$ is a partial solution to an inhomogeneous one. Later, we are interested in the solution $\varsigma_2(x^0)$ which has the form

$$\varsigma_2(x^0) = \frac{L}{\pi}\int_0^{x^0}\cos\frac{\pi}{L}(x^0-\theta)\Phi(\theta)d\theta =$$
$$= -4\frac{\pi\gamma\sigma}{c^2}\int_0^{x^0}\cos\frac{\pi}{L}(x^0-\theta)\cdot\left(A\cos\frac{\pi}{L}\theta + B\sin\frac{\pi}{L}\theta\right)d\theta. \qquad (7.23)$$

Calculating the necessary integrals and leaving non-periodic terms only, we get the following solution for $\varsigma_2(x^0) = \varsigma(x^0)$ -

$$\varsigma(x^0) = -A(x^0)\cdot\cos\frac{\pi}{L}x^0 - B(x^0)\cdot\sin\frac{\pi}{L}x^0, \qquad (7.24)$$

where

$$A = A\frac{2\pi\gamma\sigma}{c^2}\cdot x^0, B = B\frac{2\pi\gamma\sigma}{c^2}x^0. \qquad (7.25)$$

As to a complete solution, $\varsigma(x^0,x)$ this is

$$\varsigma(x^0,\rho) = -\left(A(x^0)\cdot\cos\frac{\pi}{L}x^0 + B(x^0)\cdot\sin\frac{\pi}{L}x^0\right)\cdot\cos\frac{\pi}{L}x. \qquad (7.26)$$

Let us solve equation (7.10). Since the Riemann – Christoffel symbols appearing in the equation are approximately equal $\Gamma^1_{00} = -\Gamma^1_{11} = -\Gamma^1_{33} = -2\dfrac{\pi\gamma\sigma}{c^2}$, it takes on the form

$$\frac{d^2 x}{dx^{0^2}} = 2\frac{\pi\gamma\sigma}{c^2}\left(2 + l^{3\,2}_{(0)} - u^{3\,2}_{(0)}\right) \qquad (7.27)$$

To solve this equation, we must calculate the second unknown parameter determining the motion of the string. We find from equation (7.15) that

$$u^3_{(0)} = \frac{dz_{(0)}}{dx^0} = -\frac{\pi}{L}\left(A\sin\frac{\pi}{L}x^0 - B\cos\frac{\pi}{L}x^0\right)\cdot\sin\frac{\pi}{L}x. \qquad (7.28)$$

Substituting equation (7.28) into equation (7.27) and allowing for the determined equation (7.16), we derive

$$\frac{d^2 x}{dx^{0^2}} = 2\frac{\pi^3\gamma\sigma}{c^2}\left[2 + \frac{1}{L^2}\left(A\cos\frac{\pi}{L}x^0 + B\sin\frac{\pi}{L}x^0\right)^2\cdot\cos^2\frac{\pi}{L}x - \right.$$
$$\left. -\frac{1}{L^2}\left(A\sin\frac{\pi}{L}x^0 - B\cos\frac{\pi}{L}x^0\right)^2\right]\cdot\sin^2\frac{\pi}{L}x. \qquad (7.29)$$

A solution of this equation is also found by the method of successive approximations, assuming $x = X_{(0)} + \xi$, where $X_{(0)}$ is the solution to the equation of motion in the absence of domain wall. It is written as $\dfrac{d^2 X_{(0)}}{dx^{0^2}} = 0$. Since there is no motion of string along the x axis in the absence of domain wall, we assume that $X_{(0)} = 0$.

Equation for the addition ξ, hence, takes on the form

$$\frac{d^2\xi}{dx^{0^2}} = 2\frac{\pi^3\gamma\sigma}{c^2}\left[2 + \frac{1}{L^2}\left(A\cos\frac{\pi}{L}x^0 + B\sin\frac{\pi}{L}x^0\right)^2\right]. \qquad (7.30)$$

Its solution, as before, is a sum of periodic and non-periodic terms. Restricting ourselves by non-periodic term, as before, the simple integration results desired displacement

$$\xi = \left[4 + \frac{1}{2L^2}(A^2 + B^2)\right]\frac{\pi^3\gamma\sigma}{c^2}\cdot x^{0^2}, \qquad (7.31)$$

which has a quadratic dependence on time.

Let us analyze the results obtained. It follows from equation (7.26) that the amplitude of string vibrations along the z axis is decreased with time, since $\varsigma < 0$. On the other hand, it follows from expression (7.31) that $\xi > 0$ and, hence, the wavelength along the axis is increased with time. In summary, we can draw the following conclusion - the influence of the domain wall on the cosmic string oscillations is manifested in stretching which results in the straightening of the string. Although the solutions obtained are approximate and have meaning within limited time intervals, the fact of straightening of cosmic strings in the gravitational field of the domain wall is beyond question. This solution is a new physical mechanism of cosmic string straightening in addition to the well-known results [14].

7.3. The Tension Energy of a Cosmic String in the Gravitational Field of a Vacuum Domain Wall

Let us calculate the energy transmitted by the domain wall to the string, paying special attention to its dependence on time. An action for a relativistic string is known to have the form [12, 14]

$$S = -\mu \int_{\tau_1}^{\tau_2} L d\tau, \tag{7.32}$$

where the Lagrange function describes as

$$L = \int_{\rho_1(\tau)}^{\rho_2(\tau)} \sqrt{-g} \sqrt{(u^\alpha l_\alpha)^2 - (u^\alpha)^2 (l^\beta)^2} \, d\rho. \tag{7.33}$$

Bearing in mind that the length of the string is L, and allowing only for the additions of the first-order of smallness in equation (7.33), we get in the chosen calibration, that

$$L = L_0 + \widetilde{\delta L}, \tag{7.34}$$

where

$$L_0 = \int_0^L \sqrt{-g} \sqrt{(u_{(0)}^\alpha l_{\alpha(0)})^2 - (u_{(0)}^\alpha)^2 (l_{(0)}^\beta)^2} \, d\rho, \tag{7.35}$$

and

$$\delta\tilde{L} = \int_0^L \sqrt{-g} \left\{ u_{(0)}^\alpha l_{\alpha(0)} \left(u_{(0)}^\beta \lambda_{\beta(0)} + l_{(0)}^\beta v_\beta \right) - \right.$$
$$\left. \frac{\left(u_{(0)}^\alpha \right)^2 l_{(0)}^\beta \lambda_\beta - \left(l_{(0)}^\alpha \right)^2 u_{(0)}^\beta v_\beta}{\sqrt{\left(u_{(0)}^\alpha l_{\alpha(0)} \right)^2 - \left(u_{(0)}^\alpha \right)^2 \left(l_{(0)}^\beta \right)^2}^{1/2}} \right\} d\rho \qquad (7.36)$$

We used the following designations in the foregoing relations:

$$u^\alpha = u_{(0)}^\alpha + v^\alpha, \qquad (7.37)$$

where v^α and λ^α are the small corrections to the vectors describing the velocity and orientation of the string in space.

For our dynamic model with allowance for the parameters of motion (7.8) and adopted calibration, we get

$$u^\alpha = u^\alpha \left(1, v^1, 0, u_{(0)}^3 + v^3 \right)$$
$$l^\alpha = l^\alpha \left(0, 1, 0, l_{(0)}^3 + \lambda^3 \right). \qquad (7.38)$$

On the basis of equations (7.25), (7.26), and (7.30) and taking into account, as always, only non-periodic terms, we find

$$v^1 = \frac{d\xi}{dx^0} = \frac{\pi^3 \gamma \sigma}{L^2 c^2} \left(A^2 + B^2 \right) \cdot x^0, \qquad (7.39)$$

$$v^3 = \frac{d\varsigma}{dx^0} = 2 \frac{\pi^2 \gamma \sigma}{L c^2} \cos \frac{\pi}{L} x \cdot x^0 \left(A \sin \frac{\pi}{L} x^0 - B \cos \frac{\pi}{L} x^0 \right) \qquad (7.40)$$

and

$$\lambda^1 = \frac{d\xi}{dx} = 0, \qquad (7.41)$$

$$\lambda^3 = \frac{d\varsigma}{dx} = 2 \frac{\pi^2 \gamma \sigma}{L c^2} \sin \frac{\pi}{L} x \cdot x^0 \left(A \cos \frac{\pi}{L} x^0 + B \sin \frac{\pi}{L} x^0 \right). \qquad (7.42)$$

As to $u_{(0)}^3$ and $l_{(0)}^3$, they are describing by equations (7.28) and (7.16), respectively.

The obtained relations being cumbersome, we will consider the case of small oscillations, i.e., we will consider $\frac{A}{L} \ll 1$ and $\frac{B}{L} \ll 1$ and retain only the terms proportional to the first degrees of these relations. This results in the following values

$$u^\alpha_{(0)} l_{\alpha(0)} \left(u^\beta_{(0)} \lambda_\beta + l^\beta_{(0)} v_\beta \right) - \left(u^\alpha_{(0)} \right)^2 l^\beta_{(0)} \lambda_\beta - \left(l^\alpha_{(0)} \right)^2 u^\beta_{(0)} v_\beta \approx$$
$$\approx -\frac{\pi^3 \gamma \sigma}{L^2 c^2} \sin 2\frac{\pi}{L} x \cdot \left[\left(A^2 - B^2 \right) \cos 2\frac{\pi}{L} x^0 + 2AB \sin 2\frac{\pi}{L} x^0 \right], \quad (7.43)$$

and

$$\sqrt{\left(u^\alpha_{(0)} l_{\alpha(0)} \right)^2 - \left(u^\alpha_{(0)} \right)^2 \left(l^\beta_{(0)} \right)^2} \approx$$
$$\approx 1 + \frac{\pi^2}{2L^2} \left(A \cos \frac{\pi}{L} x^0 + B \sin \frac{\pi}{L} x^0 \right)^2 \cos^2 \frac{\pi}{L} x. \quad (7.44)$$

Since for metric (7.1) $\sqrt{-g} \approx 1 - 8\frac{\pi \gamma \sigma}{L^2 c^2} x$, then, substituting this relation together with equation (7.44) into equation (7.35), we have

$$L_0 = L + \delta L_1, \quad (7.45)$$

where

$$\delta L_1 = -8\frac{\pi \gamma \sigma}{c^2} \left[1 + \frac{\pi^2}{L^2} \left(A \cos \frac{\pi}{L} x^0 + B \sin \frac{\pi}{L} x^0 \right)^2 \right] \int_0^L x \cdot \cos^2 \frac{\pi}{L} x \, dx. \quad (7.46)$$

As to the addition $\widetilde{\delta L} = \delta L_2$, due to equations (7.36), (7.43) and (7.44), it has the form

$$\delta L_2 = -4\frac{\pi^2 \gamma \sigma}{L c^2} \cdot x^0 \left[\left(A^2 - B^2 \right) \cos 2\frac{\pi}{L} x^0 + 2AB \sin 2\frac{\pi}{L} x^0 \right] \int_0^{L/4} \sin 2\frac{\pi}{L} x d\left(\frac{\pi}{L} x \right). \quad (7.47)$$

Among the two obtained additions to the Lagrangian, the most interesting is term (7.47), since it is non-periodic function of time. Calculating expression (7.47), we get

$$\delta L_2 = 2\frac{\pi^2 \gamma \sigma}{L c^2} \cdot x^0 \left[\left(A^2 + B^2 \right) \cos 2\frac{\pi}{L} x^0 + 2AB \sin 2\frac{\pi}{L} x^0 \right]. \quad (7.48)$$

The corresponding addition to the energy is equal to the addition to the Hamiltonian, which, in turn, is equal to the correction to the Lagrangian with an opposite sign, that is,

$$\delta \varepsilon = \delta H = -\delta L_2 = 2\frac{\pi^2 \gamma \sigma}{Lc^2} \cdot x^0 \left[\left(B^2 - A^2 \right) \cos 2\frac{\pi}{L} x^0 - 2AB \sin 2\frac{\pi}{L} x^0 \right]. \qquad (7.49)$$

This result answers the question on the reasons for the stretching of cosmic strings. According to expression (7.49), the presence of domain wall results in increasing its tension energy by a quasi-periodic law. In other words, the domain wall pumps the string with an additional energy that increases its tension.

8. ON THE VACUUM-INDUCED GALAXY ROTATION

8.1. Introduction

The searching of galaxies dynamics in the anti-gravitational vacuum background and the baryonic matter perturbations increase in the presence of vacuum are the key problems of modern cosmology. Many articles dedicated to this problem appeared. In fact, the dynamics of Local Group in the vacuum background was searched in [1]; the forming of cosmic structures under action of dark energy – in [2]; the estimation of galaxies clusters' masses with the account of dark energy – in [3]; galaxy clustering under dark energy influence – in [4]; the estimation of dark energy parameter of state on the stellar dynamics examination – in [5]; the searching of dark energy nature with galaxy distribution – in [6]; etc.

It is necessary to point out that galaxies' dynamics were investigated on the basis of translational equations of motion in these articles. The last condition is equal to considering the galaxies as point masses that is possible to do on large scales – scales larger than ten mega parsecs.

Meanwhile, for the number of cosmic objects that locate at smaller mutual distances – double galaxies, narrow galaxies clusters, interacting and merging galaxies – such an approach is unacceptable from before. For searching these systems' evolution, it is necessary to take into account the proper sizes of galaxies. In its turn, this leads to the translational-rotational equations of motion usage for their dynamics description.

Such a type of equations of motion was argued in [7], at first. It was assumed that bodies posses their own momentum and interacting by gravitational force. But for the compact astronomical objects, the only gravitation interaction couldn't describe its real dynamics. For doing this, it is necessary to take into account the repelling force, also. For example, the braking force that appeared at the process of galaxies merging (see [8]). But in modern cosmology, the repelling forces are identifying with anti-gravitational phenomenon, [9].

In fact, it is recently established reliably that Universe dynamics is determined by vacuum. In doing this, the Universe expands, moreover it expands with acceleration, but the space-time becomes static at the same time. In spite of expansion, the vacuum density becomes constant ($\rho_V = const$); in units of crucial density ($\rho_c \approx 0.7 \cdot 10^{-29} g/cm^3$) it estimates as - $\Omega_V = \rho_V / \rho_c = 0.7 \pm 0.1$. Hence, for searching the dynamics of compact astronomical objects it is necessary to use the translational-rotational equations of motion that are added by vacuum repelling force.

But movement of above mentioned cosmological objects is possible to search and in the framework of limited problem organization. Really, as the boundary between solitary and double galaxies is conditionally often, than two-body astronomical problem in the reasonable approximation is possible to reduce to one-body problem with the known movement of its center of mass. Then the system of translational-rotational equations of movement decomposes into two independent parts.

The aim of present article is searching the rotational movement of galaxies caused by the antigravitational vacuum force, only. For doing this consider a galaxy as the rigid body and examine its rotation with respect to the center-of-mass point under the action of vacuum antigravitational force.

The outline of this article is as follows – in first section we calculate the vacuum force and find its potential; in second section we explore the rotational equation of motion for ellipsoidal galaxy and find the corresponding angular velocity; in third section we make some numerical estimation and find the minimal angular velocity that vacuum can create at arbitrary object. In conclusion we argue that the total Universe rotation or the Birch effect may be understand and in the framework of Newtonian limit of general relativity theory.

The matter of this chapter partially has been published in [10].

8.2. The Vacuum Perturb Force and Its Potential

Here we introduce three orthogonal systems of coordinates: absolute system - $\{X_O, Y_O, Z_O\}$, fixed at any space point - O; relative immovable system - $\{X, Y, Z\}$, situated at the galaxy center of mass - O; and mobile system - $\{\xi, \eta, \varsigma\}$, connected with the same point - O.

Now, set three Euler's angles in the standard manner: precession angle - ψ, formed by line OP, that crosses the planes OXY and $O\xi\eta$, with the direction of OX_o-axis; angle of the own rotation φ, formed by line OP and $O\xi$-axis; nutation angle θ, formed by OZ- and $O\varsigma$- axes.

And at last, choose inside of a galaxy arbitrary point M in a such way that its distance from the center of absolute coordinate system O is - $\vec{R}$, while its distance from the galaxy center O is $\vec{r}$. According to these designations, we mark the distance between points O and O as - $\vec{R}_0$.

Pick out in the galaxy an infinity small mass $dm_G = \rho_G d\Omega$ that concentrates around the point - M in volume - $d\Omega$. Let this mass be located at $\vec{R}$ distance from the absolute center - O. Then the gravitational force acts between unit mass, placed in this center, and small mass - dm is –
$$dF_G = -\frac{Gdm_G}{R^2} = -\frac{G\rho_G}{R^2}d\Omega.$$

As for the anti-gravitational - repelling - force, produced by mass of vacuum in the same volume - $d\Omega$, it is [9]

$$dF_V = -\frac{Gdm_G}{R^2} = -\frac{G\rho_G d\Omega}{R^2} = 2\frac{G\rho_V d\Omega}{R^2}. \tag{8.1}$$

In deducing (1), we take into account the vacuum equation of state - $p_V = -\rho_V$ and the fact, that according to Friedmann's cosmological model, the gravitation produces by generalized mass density - $\rho_G = \rho + 3p$, where p is the substrate pressure.

Let γ be the angle between vectors $\vec{r}$ and - $\vec{R}_0$. Then,

$$R = R_0 \left(1 - 2\frac{r}{R_0}\cos\gamma + \frac{r^2}{R_0^2}\right)^{\frac{1}{2}}. \tag{8.2}$$

Suppose that condition $r/R_0 \ll 1$ is acquiring. Therefore, with the first order accuracy, we find the element of vacuum repelling force

$$dF_V = \frac{2G\rho_V\left(1 + 2\frac{r}{R_0}\cos\gamma\right)}{R_0^2} dxdydz. \tag{8.3}$$

Later on, we will be interested the part of vacuum potential force that depends on galaxy sizes, only. Thus, after integration, we get the following perturb item:

$$f_V = \frac{\partial U_V}{\partial R} = 4\frac{G\rho_V}{R_0^3}\int_\Omega r\cos\gamma dxdydz, \tag{8.4}$$

where integral must be taken over the volume Ω of galaxy.

Express vector $\vec{r}$ in components of the immovable coordinate system - $\{X,Y,Z\}$, i.e. put - $\vec{r} = l\vec{i} + m\vec{j} + n\vec{k}$. But then its projections necessary transform to the components of mobile coordinate system. Thus, we get

$$\left.\begin{array}{l}l = a_{11}\xi + a_{12}\eta + a_{13}\varsigma \\ m = a_{21}\xi + a_{22}\eta + a_{23}\varsigma \\ n = a_{31}\xi + a_{32}\eta + a_{33}\varsigma\end{array}\right\}. \tag{8.5}$$

In doing this, the direction cosines will have the standard form [11].

Having in mind (8.5), and taking into consideration the following expression - $\cos\gamma = (lX + mY + nZ)/\sqrt{l^2 + m^2 + n^2} \cdot \sqrt{X^2 + Y^2 + Z^2}$ - we find the perturb force

$$f_V = 4\frac{G\rho_V}{R_0^4} \times$$
$$[\ (a_{11}X + a_{21}Y + a_{31}Z)\int \xi d\xi \int d\eta d\varsigma +$$
$$(a_{12}X + a_{22}Y + a_{32}Z)\int \eta d\eta \iint d\xi d\varsigma + \qquad (8.6)$$
$$(a_{13}X + a_{23}Y + a_{33}Z)\int \varsigma d\varsigma \iint d\xi d\eta\ ]$$

Introduce the following denotations:

$$I_\xi = \rho_V \int \xi d\xi,$$
$$I_\eta = \rho_V \int \eta d\eta, \qquad (8.7)$$
$$I_\varsigma = \rho_V \int \varsigma d\varsigma$$

for galaxy's vacuum moments along coordinate axes ξ, η, ς;

$$S_{\eta\varsigma} = \iint d\eta d\varsigma,$$
$$S_{\xi\varsigma} = \iint d\xi d\varsigma\ ; \qquad (8.8)$$
$$S_{\xi\eta} = \iint d\xi d\eta$$

for the corresponding cross-section areas. Moreover, let - $S_\xi, S_\eta, S_\varsigma$ are the dual vectors, and $e_1 = X/R_0$, $e_2 = Y/R_0$, $e_3 = Z/R_0$ are the relative projections of vector $\vec{R}_0$ on the absolute coordinate system.

For writing down the vacuum force in abbreviate form, introduce the following table -

$i, j, k \Rightarrow \begin{Bmatrix} \xi = 1 \\ \eta = 2 \\ \varsigma = 3 \end{Bmatrix}$, where every index in the left side runs all of three values in the right side.

Finally, the vacuum perturb force takes on the form

$$f_i = 4\frac{G}{R_0^3} a_{ik} \cdot I_j S_j \cdot e_k = -\frac{\partial U_V}{\partial R_0} e_i. \qquad (8.9)$$

Hence, from the last expression, it is easy to find the needed vacuum potential -

$$U_V = -4Ga_{ik} \cdot I_j S_j \cdot \int \frac{e_i e_j}{R_0^3} dR_0 .$$ (8.10)

8.3. Equations of Rotational Motion

Later on, as the model of examining type of galaxy, we choose the elliptical galaxy. This leads not only to simplify the process of the analytical solving of our problem, but allows to compare theoretical results with the real astronomical observations. Such types of galaxies – E-galaxies – contains the compact star component (stellar gas) and possesses by the right shape – ellipsoids of different power contraction.

Consider an elliptical galaxy with two equal main axes, i.e. its moments are - $A = B \neq C$. Then, according to [11], the potential will not depend on angle - φ. Hence, it may be arbitrary. Let it equal to zero, i.e. $\varphi = 0$. After calculation, the direction cosines for this case and substitution them into (8.10), we find the vacuum potential

$$U_V = -4G \times I_j S_j \times \begin{pmatrix} \cos\psi \int \frac{e_1^2}{R_0^3} dR_0 + \sin\psi \cos\theta \int \frac{e_1 e_2}{R_0^3} dR_0 - \sin\psi \cos\theta \int \frac{e_1 e_2}{R_0^3} dR_0 + \\ + \cos\psi \cos\theta \int \frac{e_2^2}{R_0^3} dR_0 + \sin\psi \sin\theta \int \frac{e_1 e_3}{R_0^3} dR_0 - \cos\psi \sin\theta \int \frac{e_2 e_3}{R_0^3} dR_0 + \\ + \sin\theta \int \frac{e_2 e_3}{R_0^3} dR_0 + \cos\theta \int \frac{e_3^2}{R_0^3} dR_0 \end{pmatrix}$$ (8.11)

For the chosen shape of galaxy, its equations of rotational motion were argued in [11]. They are

$$\begin{aligned} \frac{d\theta}{dt} &= \frac{1}{C\omega \sin\theta} \cdot \frac{\partial U}{\partial \psi}, \\ \frac{d\psi}{dt} &= -\frac{1}{C\omega \sin\theta} \cdot \frac{\partial U}{\partial \theta} \end{aligned}$$ (8.12)

It is necessary to make some comments concerning this system. First, ω is the first integral of the rotational motion here, i.e. $\omega = const$. It describes the component of angular velocity with respect to the specific momentum - C. Second, at deducing (8.12), the condition that galaxy angular velocity is very small was put forward. This allowed neglecting the squared angular velocities and the angular accelerations. And third, we assume that arbitrary potential equals to the vacuum potential, i.e. we put - $U = U_V$.

Now based on (8.11), calculate its derivatives with respect to angles ψ and θ, and plug them into the system (8.12). Due to the minuteness of angular velocities, we may put small the suitable angles, also. Hence, neglecting squared angles and taking into account that value of any integral in (8.11) is not larger the expression $-1/R_0^2$, with the needed accuracy, we obtain

$$\theta \frac{d\theta}{dt} = \kappa \cdot (\theta - 2\psi)$$
$$\theta \frac{d\psi}{dt} = \kappa \cdot (2\theta - \psi)$$ (8.13)

where coefficient - $\kappa = 4G \cdot I_j S_j / C\omega R_0^2$. The system (8.13) easily transforms to the differential equation of the first order

$$\frac{d\theta}{d\psi} = F\left(\frac{\theta}{\psi}\right),$$ (8.14)

where the right side is - $F\left(\frac{\theta}{\psi}\right) = 2\left(1 - \frac{1}{2}\frac{\theta}{\psi}\right) \Big/ \left(1 - 2\frac{\theta}{\psi}\right)$.

The solution of this equation expresses by quadrature - $\ln \psi = \int \frac{du}{F(u) - u} + C$ with the new variable $u = \frac{\theta}{\psi}$ and constant C [12]. This quadrature consists of two items. First, that proportional - $arctg\left(\frac{8}{3}u - \frac{5}{6}\right)$, is limited from above. Thus, it cannot lead to results of evolutional character. But the second item $\frac{1}{8}\ln\left|1 - \frac{5}{4}u + 4u^2\right|$ allows us to realize this possibility. Hence, we get - $\psi = \left(1 - \frac{5}{4}u + 4u^2\right)^{\frac{1}{8}}$.

From the above admitted condition about the minuteness of angles θ and ψ, we obtain that their relation approximately equals to one, i.e. $u \approx 1$. Hence, - $\psi^4 \propto u$, and - $\theta \propto \psi^5$.

Putting our last dependency into the second equation of (8.12), we have - $\psi^5 \frac{d\psi}{dt} = \kappa(-\psi + 2\psi^5)$. Analysis of this equation shows that its general solution consists of increasing and decreasing parts. From the physical viewpoint, we will be interested in the first of them, only. So, the needed solution for the evolving of precession angle is

$$\psi(t) = 8 \frac{GI_j S_j}{C\omega R_0^2} \cdot t .$$ (8.15)

Based on this result, it is easy to calculate the angular velocity of the elliptical galaxy around OZ axis. Naturally, as for this case, we have - $\omega = -\dot{\psi}$, then its module equals

$$\omega = \omega_V = \sqrt{8\frac{GI_j S_j}{CR_0^2}} = const. \quad (8.16)$$

This expression describes the angular velocity that a galaxy acquires due to the vacuum anti-gravitation force.

8.4. The Numerical Estimations

Consider the giant elliptical galaxy NGC 4486 (M 87) that is located at distance - $R_0 = 15 Mps$; its proper sizes l equals - $60 Kps$. With account of above-cited vacuum density, it is possible to estimate the magnitude of expression (8.16) now. So, admitting $I_j S_j \sim 3 IS \sim \rho_V l^4$ and assuming that - $C \sim l^2$, we find - $\omega_V \propto \sqrt{8 G \rho_V} \cdot \frac{l}{R_0} \propto \omega_0 \cdot \frac{l}{R_0}$.

Substitution of the needed values into this expression give the following estimation - $\omega_V = 6,4 \cdot 10^{-21} \sec^{-1}$. Another result we will get if we take the frame of reference connected with galaxy center-of-mass.

Indeed, the most large-scale elliptical galaxies are joining into a special group of cD-galaxies. They have the compact stellar system and giant rarefied shell consisting from stars, also. Scales of shell are tens and hundreds kilo parsecs of order. In fact, for the same system NGC 4486 M 87, its radius is about - $8 Kps$, while its shell stretches on $60 Kps$ from the center. So, putting $l = 8 Kps$ and $R_0 = 60 Kps$ we get - $\omega_V = 2.1 \cdot 10^{-19} \sec^{-1}$.

The very important consequence results from this – the larger galaxy locates from us, the smaller is its angular velocity. At least, this relates to part of angular velocity that depends on vacuum anti-gravitation force. But the subsequent collection of angular momentum up to the observable magnitudes will determine by mechanisms based on the gravitational interaction.

At the same time, analyses (8.16) shows that its maximal magnitude will be under the condition - $l \propto R_0$. Then, expression for the vacuum angular velocity simplifies and takes on the form

$$\omega_V = \omega_0 \propto \sqrt{G \rho_V}. \quad (8.17)$$

This expression is possible to interpret as the minimal angular velocity in the Universe that possesses an arbitrary object due to the vacuum presence, independently on its shape. Its numerical value is - $10^{-19} \sec^{-1}$. Hence, the vacuum itself creates the identical initial angular velocity for all of the cosmic objects. It may say that cosmic vacuum not only creates the microscopic baryonic substance (particles born from vacuum), but pre-determines its macroscopic and global characteristics, also. Really, the vacuum is the moving force of the Universe's total evolution [9].

8.5. Conclusion

So, the rotational movement of elliptical galaxies, that created by vacuum antigravitational force, is searched in articles. The estimation of corresponding angular velocity for the real elliptical galaxies gives a magnitude of about - $10^{-19} \sec^{-1}$.

At the same time, the modern observational data gives the essentially larger value. Really, for elliptical galaxy NGC 1600, according to [13], its angular velocity equals - $10^{-14} \sec^{-1}$. Other E-galaxies have the angular velocities of the same order [14]. As it seems, the reason of divergence is in the fact that these galaxies already collect the corresponding angular momentum during their evolution. And this collection is determined by different mechanisms of gravitational nature (see, for example, [15].) Thus, (8.17) is possible to be considered as the primordial angular velocity, that is produced by vacuum.

And at last, the expression (8.17) is invariant, that doesnt depend on choice neither the frame of reference nor the geometrical scales of the Universe at the given cosmological epoch. Hence, at the earliest stages of the Universe evolution, for instance, at the baryonic asymmetry epoch when vacuum density was $10^{-15} g/cm^3$ of order, the angular velocity occurs equal - $\omega_V \propto 10^{-11} \sec^{-1}$.

But it is hardly possible to talk about acquiring the "mechanical" rotation of baryons. It will say correctly about the basing of Birch effect [16] - the total Universe rotation - in the framework of Newtonian gravitation theory.

Our Universe, according to this effect, rotates with angular velocity $\omega \propto 10^{-20} \sec^{-1}$ that is practically equal to effect (8.17) at our epoch. Moreover, for the very early Universe when vacuum density was - $10^{90} g/cm^3$, the Universe angular velocity is - $\omega_V \propto 10^{42} \sec^{-1}$. This magnitude is also practically equal to the result of article [17], that was written in the framework of general relativity theory ($\omega \propto 10^{43} \sec^{-1}$).

Now, the question arises – why our results, which we get in Newtonian gravity, coincide with results of [17]? The answer, as it seems, is as follows – because the density of baryonic matter in the relativistic cosmological models with rotation was chosen in such a way, that it coincides with the density of non-baryonic matter (vacuum, in particular). But the latest estimations of the baryonic matter density gives the magnitude on two orders smaller than non-baryonic matter density [9]. That is why the numerical results of [17] are too overstated from a modern viewpoint.

Some other results in this cosmological area have been discussed in articles [18].

REFERENCES

Section 1

[1] Weinberg S. Gravitation and Cosmology, New-York – London – Sydney – Toronto, 1972.

[2] Zel'dovich Ya.B., Novikov I.D. Relativistic Astrophysics, v. II, Chicago Press, 1983.

[3] Padmanabhan T. Structure Formation in the Universe, Cambridge University Press, 1993.
[4] Linde A.D. Particle Physics and Inflationary Cosmology, Harwood Academic Publishers, Chur, Switzerland, 1990.
[5] Peebles P.J.E. Physical Cosmology, Princeton University, 1971.
[6] Komatsu E., Smith K.M., Dunkley J., et all // [arXiv:1001.4538v3], 2010; Gorbynov V.S., Rybakov V.A. Introduction to the Theory of Early Universe. Cosmological Perturbations. Inflationary Theory. Moscow, URSS, 2009 (in Russian).

Section 2

[1] Riess A.G., et al. // Astron. J., 1998, 116, 1009; Perlmutter S., et al. // *Astrophys. J.*, 1999, 517, 565; Chernin A.D. // *Physics-Uspekhi (Advances in Physics Sciences).* 2001, 44, 1099.
[2] Sadoulet B. *Current topics in Astrofundamental Physics*: Primordial Cosmology. / NATO ASI Series. Kluwer Academic Publishers. 1998, 517; Chiba T., Okabe T., Yamaguchi M. // *Phys. Rev.*, 2000, D62, 023511; Gudmundsson E.H., Björnsson G. // Astron. J., 2002, 565, 1; Daly R.A., Djorgovski S.G. Astron. J. // 2004, 612, 652.
[3] Ponman T.L. et al. // *Nature*, 1994, 369, 462; Jones L.R. et al. // MNRAS, 2000, 312, 139; Sun M. et al. // *Astrophys. J.,* 2004, 612, 805; D'Onghia E. et al. // *Astrophys. J.*, 2005, 630, L109.
[4] Zel'dovich Ya.B., Novikov I.D. *Relativistic Astrophysics*, v. II, Chicago Press, 1983.
[5] Weinberg S. *Gravitation and Cosmology*. New-York – London – Sydney – Toronto, 1972.
[6] Minz A., Orlov V. / *ASP Conference Series*, 2004, 316, 291.
[7] Percival W.J. // *A&A*, 2005, 443, 819.
[8] Dolgachev V.P., Domozhilova L.M., Chernin A.D. // *Astronomy Reports*, 2003, 47, 728.
[9] Coble K., Dodelson S., Friemann J.A. // http//xxx.lanl.gov./abs/astro-ph/9608122; Primac J.R. // http//xxx.lanl.gov./abs/astro-ph/9707285; Aleamaiz J.Z., Lima J.A.S. // *Astronomical J.*, 2001, 550, L133.
[10] Chechin L.M. // *Chinese Physics Letters*, 2006, 23, 2344.
[11] Gibbons G.S., Hawking S.W. // *Phys. Rev.* 1977, D15, 2738; Qin Y.-P. // *Chinese Physics Letters*, 2006, 23, 758.

Section 3

[1] Linde A. *Particle Physics and Inflationary Cosmology*. Harwood Academic Publishers, Chur, Switzerland, 1990.
[2] Weinberg S. *Gravitation and Cosmology*. New-York – London – Sydney – Toronto, 1972; Gurevich L.E., Chernin A.D. *Introduction to the cosmogony*. Moscow, Nauka, 1978; Dolgov A.D., Zel'dovich Ya.B., Sazhin M.V. *Cosmology of the early Universe*. Moscow, Moscow University Press, 1988. (in Russian).
[3] Solov'eva L.V., Nurgaliev I.C. // *Astronomicheskij Zhyrnal*, 1985, 62, 459. (in Russian).
[4] Chernin A.D. // *Physics-Uspekhi (Advances in Physics Sciences).* 2001, 44, 1099.

[5] Viana P.T.P., Liddle A.R. // *MNRAS*, 1996, 281, 323; Wang L., Steinhardt P. // Astrophys. J., 1998, 508, 483; Linder E.V., Jenkins A. // *MNRAS*, 2003, 346, 573; Persival W.J. // A&A, 2005, 443, 819; Nunes N.J., da Silva A.S., Aghanim N. // A&A, 2006, 450, 899.

[6] Chechin L.M. // *Chinese Physics Letters*, 2006, 23, 2344; Chechin L.M. // Doklady NAS Republic of Kazakhstan. // 2006, № 4, 31 (in Russian).

[7] Chevallier M., Polarski D. // Int. J. Mod Phys., 2001, D10, 213; Linder E.V. // *Phys. Rev. Lett.* 2003, 90, 091301.

[8] Riess A.G., et al. // Astron. J. 1998, 116, 1009; Perlmutter S., et al. // *Astrophys. J.* 1999, 517, 565.

[9] Sahni V., Starobinsky A. // *Int. J. Mod. Phys.*, [arXiv:astro - ph/0610026], 2006; Knop R. et al. // *Astrophys.* J. 2003, 598, 102; Linder E.V., Jenkins A. // MNRAS. [arXiv:astro - ph/0305286], 2003; Hannestad S., Mörtsell E. // JCAP, 2004.

[10] Chechin L.M., Myrzakul Sh.R. // *Russian Physical Journal*, 2009, 52, 286.

[11] Jassal H.K.,.Bagla J.S, Radmanabhan, T. // MNRAS, 2004, [arXiv:astro - ph/0404378], 2004; Liberato L., Rosenfeld R. // JCAP, 2006, 07, 009; Lazkoz R., Nesseris S., Perivolaropoulos L. // *JCAP*, 2005, 11, 010.

[12] Kamke E., *Differentialgleichungen Lösungsmethoden und Lösungen.* Leipzig, 1959.

[13] de-Vega H.J., Siebert, J.A [arXiv: astro - ph/0305322], 2003.

[14] Ciardi B., Ferrara A. // Space Science Reviews, 2005, 116, 625; Tsagas Ch.,G.; Challinor A., Maartens R. // *Physics Reports*, 2008, 465, 61.

Section 4

[1] Ginzburg V.L. Physics –Uspekhi. (Advances in Physics Sciences), 2002, 45, 205; Sandage A. *The Universe at Large.* / Eds. G. Munch, A.Mampaso, F.Sanchez. Cambridge University Press, 1997.

[2] Zel'dovich Ya.B., Novikov I.D. *Relativistic Astrophysics*, v. II, Chicago Press, 1983; Padmanabhan T. *Structure Formation in the Universe*, Cambridge University Press, 1993.

[3] Dvali G., Gruzinov A., Zaldarriaga M. // *Phys. Rev.*, 2003, D69, 023505; Polarsky D., Gannouji R. // Phys. Lett., 2008, B660, 439.

[4] Verheest F., Shukla P.K., Jacobs G., Yaroshenko V.V. // Phys. Rev., 2003, E68, 027402; Johnson M.C., Kamionkowski M. // *Phys. Rev.*, 2008, D78, 063010; Thompson T. A., *Astrophysical Journal*, 2008, 684, 212; Dunsby P.K.S. // *Classical and Quantum Gravity*, Vol. 8, Issue 10, 1991, p. 1785.

[5] Chyornyj L.T. *The Relativistic Models of Continuous Media.* Moscow, Nauka, 1983 (in Russian); Serov S.A., Serova S.S. // E-print arXiv:physics / 0503215.

[6] Bartolo N., Corasaniti P. – S., Liddle A., Malquarti M. // *Phys. Rev.*, 2004, D70, 043532.

[7] Macorra de la. // *JCAP*, 2008, 01, 030.

[8] Podol'sky D.I. // *Astronomy Letters*, 2002, 28, 434.

[9] Copeland E.J., Sami M., Tsujikawa Sh. // [arXiv: hep-th/0603057v3].

[10] Chernin A.D. // *Physics-Uspekhi, (Advances in Physics Sciences)*, 2001, 44, 1099.

[11] Linde A.D. *Particle Physics and Inflationary Cosmology*. Harwood Academic Publishers, Chur, Switzerland, 1990.
[12] Kamke E.*Differentialgleichungen Lösungsmethoden und Lösungen*. Leipzig, 1959.
[13] Watson G.A. *Treatise on the Theory of* Bessel Functions, 2 ed. CUP, 1944.
[14] Peebles P.J.E. Physical Cosmology. 1971, Princeton University; Weinberg S. *Gravitation and Cosmology*. New-York – London – Sydney – Toronto, 1972.
[15] Brenner H. // *Physica*, 2009, A388, 3391.
[16] Landau L.D., Lifshitz, E.M. *Mechanics*, M, Nauka, 1958, (in Russian).
[17] Chernin A.D., Nagirner D.I., Starikova S.V., // *Astronomy and Astrophysics*, 2003, 399, 19; Nunes N.J., da Silva A.S., Aghanim N., // *Astronomy and Astrophysics*, 2006, 450, 899.
[18] Munshi D., Porciani C., Yun Wang. // MNRAS, 2004, 349, 281; Linder E. V, Jenkins A. // *MNRAS*, 2003, 346, 573; Persival W. J // *Astronomy and Astrophysics*, 2005, 443, 819. Wang L., Steinhardt P. // *Astrophys. J.*, 1998, 508, 483; Basilakos S.// arXiv:0903.0452v1 [astro-ph. CO], 2009.
[19] Chechin L.M. // *Chinese Physics Letters*, 2006, 23, 2344.
[20] Perrotta F., Baccigalupi C., Matarrese S. // *Phys. Rev.* 2000, D61, 023507; Sahni Varun. // *Classical and Quantum Gravity*, 2002, 19, 3435; Copeland E.J.; Sami M.; Tsujikawa Sh. // *International Journal of Modern Physics* 2006, D15, 1753.

Section 5

[1] Chechin L.M. Omarov T.B. // Vestnik KazNY named after al-Farabi, 2002, No. 2, 136. (in Russian).
[2] Adomian G. *Stochastic Systems. Mathematics in Science and Engineering*. Orlando, FL: Academic Press Inc., 1983.
[3] Øksendal B.K. *Stochastic Differential Equations: An Introduction with Applications*. Berlin: Springer, 2003.
[4] Mitropolskyi Yu.A., Kolomietz V.G. // *Matematicheskaya fizika i nelinejnaya mekhanika*, Kijev, 1986, No. 5(39), 23.(in Rusian).
[5] Baxendale P.H., Lototsky S.V. *Stochastic Differential Equations: Theory and Applications*. World Scientific Publishing Co., Pte. Ltd., 2007.
[6] Smith A.G., Vilenkin A. // *Phys.Rev.* 1987, D36, 987.
[7] Albrecht A., Turok N. // *Phys. Rev.* 1989, D40, 973.

Section 6

[1] Vilenkin A // *Phys. Rev.*, 1981. 24, 2082.
[2] Deser S., Jackiw R., t'Hooft G. // *Ann. Phys.* (N.Y.); 1984. 152, 220.
[3] Clement G. // *Int. J. Theor. Phys.*, 1985, 24, 267.
[4] Allen B.,Ottewill A. C. // *Phys. Rev.*, 1990, D42, 2669.
[5] Jensen B.,Soleng H. // *Phys.Rev.*, 1992, D45, 3528.
[6] Gluschenko G.N. // *JETF*, 1995. 107, 273. (in Russian).

[7] Carter B. In "Formation and Evolution of Cosmic Strings", Cambridge University Press, Cambridge. 1990.
[8] Peter P., Puy D. // *Phys. Rev.*, 1993, D48, 5546.
[9] Larsen A. L. // *Phys. Lett.*, 1992, A170, 174.
[10] Mignel O. // *Phys. Rev.*, 1991, D43, 2521.
[11] Letelier P.S.,Verdaguer E. // *Phys. Rev. Lett.*, 1985.60, 2228.
[12] Garfinkle D., Vachaspati T // *Phys. Rev.*, 1988, D37, 257.
[13] Clement G., Zouzou I. // *Phys. Rev.*, 1995, D50, 7271.
[14] Letelier P. // *Phys. Rev.*, 1983, D28, 2414.
[15] Vachaspati T. // *Nucl. Phys.*, 1986, B277, 593.
[16] D.Garfinkle. // *Phys. Rev.*, 1990, D41, 1112.
[17] Vollick D.N.,Unruh A. // *Phys. Rev.*, 1990, D42, 2621.
[18] Garfinkle D., Will C. // *Phys. Rev.*, 1987, D35, 1124.
[19] Vilenkin A. // *Phys. Rev. Lett.*, 1981, 46, 1169.
[20] Thompson Ch. // *Phys. Rev.*, 1988, D37, 283.
[21] Garfinkle D., Duncan G.C. // *Phys. Rev.*, 1994, D49, 2752.
[22] Vachaspati T., Vilenkin A. // *Phys. Rev.*, 1985, D31, 3052.
[23] Sakellariadou M. // *Phys. Rev.*, 1990, D42, 354.
[24] Anzhong W., Nilton S. // *Class. Quant. Grav.*, 1996, 13, 715.
[25] DeLaney D., Engle K., Scheich X. // *Phys. Rev.*, 1990, D41, 1775.
[26] Turok N. // *Nucl. Phys.*1984. B242, 520.
[27] Letelier P.S. // *Phys. Rev.*, 1979, D20, 1294.
[28] Omarov T.B., Chechin L.M. // *General Relativity and Gravitation*, 1999, 31, 443; *Astronomical and Astrophysical Transactions*. 2003, 22, 155.
[29] Omarov T.B., Chechin L.M. // *Doklady MS-AS RK*, (1998).N2, 43. (in Russian).
[30] Landau L.D., Lifshitz E.M. *The Classical Theory of Fields*. (2d ed.), Pergamon Press. Oxford. London. Paris. Frankfurt, 1962.
[31] Gradshteyn I.S., Ryzhik I.M. *Table of Integrals, Series, and Products*. Academic, New York, 1980.
[32] Geroch R., Traschen J. // *Phys. Rev.*, 1987, D36, 1017; Vilenkin A., Shellard E.P. *Cosmic strings and other topological defects*. 1994, Cambridge, Cambridge University Press; Hindmarsh M.B.; Kibble T.W.B. // *Reports on Progress in Physics*, 1995, 58, 477.; Kibble T.W.B. / E-print [arXiv:astro-ph/0410073]; Damour T., Vilenkin A. // *Phys. Rev.*, 2005, D71, id. 063510; Kallosh R., Linde A. // JCAP, 2007, 04, 017; Sakellariadou M. // *Nuclear Physics, Proceedings Supplements*, 2009, B192, 68.

Section 7

[1] Kibble T.W.B. // *J. Phys.*, 1976, A9, 1387.
[2] Zel'dovich Ya.B. // *MNRAS*, 1980, 192, 663.
[3] Vilenkin A. // *Phys. Rev.*, 1981, D23, 852.
[4] Garfinkle D. // *Phys. Rev.*, 1985, D32, 1323.
[5] *The Formation and Evolution of Cosmic Strings*. / Ed. Gibbons G.W., Hawking S.W., Vachaspati T., Cambridge, Cambridge University Press, 1990.
[6] Mejerovich B.E. // *Physics-Uspekhi (Advances in Physics Sciences)*. 2001, 121, 1033.

[7] Letelier P. // *Phys. Rev.*, 1983, D28, 2414; Shellard E.P.S. // *Nucl. Phys*, 1987, B283, 624; Omarov T.B., Chechin L.M. // *Reports NAS Kazakhstan*, 1995, No.1, 4; Omarov T.B., Chechin L.M. // GRG, 1999, 31, 443.

[8] Drechsler W., Havas P., Rosenblum A. // *Phys. Rev.*, 1984, D29, 658; Chechin L.M. // *Russian Physical Journal*, 1995, No.2, 59; Nabuyuki Sakai. // *Phys. Rev.*, 1996, D54, 1548.

[9] Vilenkin A. // *Phys. Rev.*, 1981, D23, 852; Wu Z.Ch. // *Phys. Rev.*, 1983, D28, P.1898; Widrow L.M. // *Phys. Rev.*, 1989, D40, 1002; Larsen A.L. // *Phys. Rev.*, 1994, D49, 4154.

[10] Linde A.D. *Particle Physics and Inflationary Cosmology*, Harwood Academic Publishers, Chur, Switzerland, 1990.

[11] Chechin L.M. // *Russian Physics Journal*, 2005, 48, 20.

[12] Garriga J., Vilenkin A. // *Phys. Rev.* D., 1991, 44, 1007.

[13] Barbashov B.M., Nesterenko V.V. *The Relativistic String Model in Hadrons' Physics*. Moscow, Energoatomizdat, 1987. (in Russian).

[14] Tikhonov A.N., Samarskij A.A. *Equations of Mathematical Physics*. Moscow 1977 (in Russian); Courant R., Gilbert D. *Methods of Mathematical Physics*, N.Y., Interscience, 1962.

[15] Turok N., Bhattacharjee. // *Phys. Rev.*, 1984, D29, 1557; Bennet D. // *Phys. Rev.*, 1986, D34, 3592; Austin D., Copeland E.J., Kibble T.W.B. // *Phys. Rev.* 1993, D48, 5.

Section 8

[1] Dolgachev V.P., Domozhilova L.M., Chernin A.D. // *Astronomy Reports*, 2003, 47, 728.

[2] Linder E.V., et al. // [arXiv:astro-ph/0305286], 2003.

[3] Munshi D., et al. // *MNRAS*, 2004, 349, 281.

[4] Grillio C., et al. // *A&A*, 2008, 477, 397.

[5] Solevi P, et al. // *MNRAS*, 2006, 336, 1346.

[6] Doubochine G.N. // *Astronomitcheskyi Zhyrnal* 1958, 35, 265. (in Russian).

[7] Vorontzov-Vel'yaminov B.A. *Extra-galaxy Astronomy*. 1987, M., Nauka, (in Russian).

[8] Cooper R G, et al. // *Astrophys. J.*, 1982, 254, 16.

[9] Chernin A.D. // *Advances in Physical Sciences*, 2001, 171, 1153; Chernin A.D. // *Advances in Physical Sciences*, 2008, 178, 267. (in Russian).

[10] Chechin L.M. // *Astronomical Reports*, 2010, 78, 784.

[11] Abalakin V.K., et al. *Handbook on Celestial Mechanics and Astrodynamics*. 1976, M., Nauka, (in Russian).

[12] Stepanov V.V. *Course of Differential Equations*. M., GIFML, 1958. (in Russian).

[13] IAU Symposium 127, *Structure and Dynamics of Elliptical Galaxies*, de Zeeuw P. T. (ed.), Dordrecht, Reidel, 1987.

[14] de Zeeuw P.T., et al. // *Annual Review on Astronomy and Astrophysics*, 1991, 29, 239.

[15] Gurevitch L.E., Chernin A.D., *Introduction to Cosmogony*. 1978, M., Nauka. (in Russian); Chechin L.M. // *Cosmology and Gravitation*, 2003, 9, 281.

[16] Birch P. // *Nature*, 1982, 298, 451.

[17] Panov V.F. // Izvestiya VUZov, *Fizika*, 1989, 30, № 6, 67. (in Russian).
[18] Shi Chun Su, Chu M.-C. // [arXiv:astro-ph./0902.4575v2], 2009; Godlowski W., Szydlowski M. // [arXiv: astro-ph/0409073v1], 2004.

In: The Big Bang: Theory, Assumptions and Problems
Editors: J. R. O'Connell and A. L. Hale

ISBN: 978-1-61324-577-4
© 2012 Nova Science Publishers, Inc.

Chapter 4

TEMPORAL TOPOS AND U-SINGULARITIES

Goro C. Kato
Mathematics Department, California Polytechnic State University,
San Luis Obispo, CA, U. S.

ABSTRACT

Several papers and books by C. Isham, C.Isham-A. Doering, F. Van Oystaeyen, A. Mallios-I. Raptis, C. Mulvey, and Guts and Grinkevich, have been published on the methods of categories and sheaves to study quantum gravity. Needless to say, there are well-written treatises on quantum gravity whose methods are non-categorical and non-sheaf theoretic. This paper may be one of the first papers explaining the methods of sheaves with minimally required background that retains experimental applications.

Temporal topos (t-topos) is related to the topos approach to quantum gravity being developed by Prof. Chris Isham of the Oxford-Imperial research group (with its foundations in the work of F. W. Lawvere). However, in spite of strong influence from papers by Isham, our method of t-topos is much more direct in the following sense. Our approach is much closer to the familiar applications of the original algebraic geometric topos where little logic is involved.

The distinguishable aspects of this paper "Temporal Topos and U-Singularity" from other topos theorists' approaches are the following. For a particle, we consider a presheaf associated with the particle. By definition, a presheaf is a contravarinat functor; however, in the t-topos theory, such a presheaf need not be defined for every object of a t-site over which the topos of presheaves are defined. When such an associated presheaf is not defined (or non-reified), we say that the presheaf (the particle) is in ur-wave state. Therefore, the duality is already embedded in our t-topos theory. We also have the notion of a (micro) decomposition of a presheaf (a particle) to obtain microcosm objects. Another important aspect of our approach is the associated space and time sheaves for a given particle-presheaf. The sheaves associated with space, time, and space-time are treated differently from a particle associated presheaf. Namely, Yoneda Lemma and its embedding are crucial for formulating and capturing the nature of space-time. In this formulation, the space and time sheaves would not exist unless a particle (presheaf) exists. Such a non-locality nature as the EPR type non-locality is also embedded in t-topos. Applications to singularities (a big bang, black holes, and subplanck objects) are formulated in terms of universal mapping properties of direct limit and inverse limit in category theory. Furthermore, the uncertainty principle is formulated through the concept of a micro-morphism in t-site. Our t-topos theoretic approach enables us to formulate a light cone in macrocosm and also in microcosm. However, such a light cone in microcosm has non-reified space-time regions because of the uncertainty principle (a miro-morophsim).

PROLOGUE

Introducing a categorical approach to quantum field theory will avoid divergent expressions, e.g., for the total amplitude of a quantum process. One may also take categorical and sheaf theoretic methods as avoidance of the Dedekind-Cantor continuum approach to physical entities. The Dedekind-Cantor type continuum is one of the sources of infinites in physical theory.

The concept of a sheaf has been effectively used for the foundations of quantum physics and quantum gravity especially among people in the C. Isham school at Imperial College as in [1], [2], [3], Mallios' school as in [4], [5], and Penrose as twistor cohomology of sheaves in [6]. Even though direct connections to our temporal topos method are not known, a few names should be mentioned: Mulvey, Heller, and Sasin. In particular, the noncommutative geometry approach, called virtual topology of F. Van Oystaeyen, seems to be quite relevant to our work (See the treatise *Virtual Topology and Functor Category*, Tayler and Francis Group, 2007).

See [7], [8], [9] for developments and the history of sheaf theory in the theory of holomorphic functions in several complex variables, algebraic analysis, and algebraic geometry.

In this article we will summarize what we have obtained in the series on the fundamentals of the theory of temporal topos following [10], [11], [12]. Our method of temporal topos, referred to as t-topos for short, differs from Isham's and Mallios' schools, and also from the Russian school directed by A. K. Guts and E. B. Grinkevich. However, we should acknowledge the motivational influence coming especially from the paper [1] by C. Isham. As we have mentioned earlier, compared with other approaches to quantum gravity via sheaves, our method is a more direct and straightforward application of commonly used familiar algebraic geometric (categorical-cohomological) methods. That is, in order to express the changing state of a particle over a time period, the associated presheaf representing the particle is "parameterized" by an object in a t-site. We call such an object in t-site a *generalized time period*. Namely, we introduce such a state controlling parameter as a generalized time period-object in the t-site to keep track of varying states of a particle. (See below for more on t-site.)

Our goals include studying the topos of presheaves (t- topos) defined on a t-site and its applications to quantum gravity. However, in t- topos theory, a presheaf is not always defined on every object in a t-site. When it is defined, a presheaf in t-topos theory satisfies the properties of a contravariant functor. This is one of the issues relevant to the Kochen-Specker theorem in [2] and [3]. The t-topos theory is a background independent theory and also a scale independent theory * (See (*) below.) in the following sense: all the concepts are defined in our theory in terms of presheaves associated with a macro or micro particle together with the associated space, time, and space-time sheaves. For a particle state in the usual sense, we associate a presheaf m so that each particle state of the particle is represented by the reified pair of the presheaf m and an object V (which is called a generalized time period V) in a site S. [At the Second International Conference on Theoretical Physics and Topos, held at Imperial College, London, 2003, (*) C. Isham said (In the definitions in t-topos theory) " --- a particle can be replaced by an elephant."] Such a site as used in t-topos theory is called the t-site.

Recall that a site in general is a category with a Grothendieck topology as defined in [9], [14], [15]. An ur-particle state of the presheaf m associated with a particle is expressed as $m(V)$ as an object in a product category

$$\prod_{\alpha \in \Delta} C_\alpha \tag{0.0}$$

(See [10], [11], [14].) One of the reasons for introducing the product category indexed by a finite set is that for each physical quantity possibly measured, we need a category where such a measurement (interpreted as a morphism in t-topos) can take place. Following the terminology used among topos theorists, the category $\hat{S}$ of presheaves on a site S (with a restricted sense as follows) is said to be a temporal topos or simply t-topos. Namely, $\hat{S}$ is the category of contravariant functors from the t-site S to $\prod_{\alpha \in \Delta} C_\alpha$. However, such a t-topos theoretic presheaf is more restricted than the usual definition of a presheaf. That is, $m(V)$ may not be defined for every pair of an object m of $\hat{S}$ and an object V of S.

Definition: A presheaf m, an object of $\hat{S}$, and a generalized time period V, i.e. an object of S, are said to be reified (or compatible) when $m(V)$ is defined.

Hence, an object of the t-topos $\hat{S}$ may be more appropriately called an ur-presheaf (or t-presheaf) rather than just a presheaf. Let m and P be presheaves, i.e., objects of $\hat{S}$. We say that m is observed (measured) by P over a generalized time period V (i.e. an object of the t-site S), when there exists a morphism from $m(V)$ to $P(V)$. For a presheaf m associated with a particle, there are the space, time, and space-time (pre)sheaves κ_m, τ_m and ω_m associated with m. The associated (pre)sheaves with space, time and space-time do not exist without the particle. (See the forthcoming [17] for a complete description of t-topos theory, especially the treatment of space-time sheaf $\omega = (\kappa, \tau)$.)

Also recall that a presheaf m is said to be in a *particle ur-state (or ur-particle state)* if there exists an object V in S such that $m(V)$ is defined. Otherwise, m is said to be in a *wave ur-state* (or ur-wave state). For example, when such an object V in the t-site cannot be specified between the two as in the case of double slit experiment, m is said to be in a wave ur-state. (See [15] for the application of t-topos to a double slit experiment.) Recall also that m and m' are ur-entangled when presheaves m and m' are defined always on the same objects of S. (See [10], [16] for connections to EPR type non-locality.)

In this paper, for a presheaf m representing a particle and for an object V in the t-site, a decomposition of m and a covering of V play major roles in defining a notion of entropy.

For a presheaf m in $\hat{S}$, consider a (micro)decomposition of m by subpresheaves m_j:

$$m = \prod_{j \in J} m_j, \tag{0.1}$$

and let

$$\{V \longrightarrow V_k\} \tag{0.2}$$

be a covering of V by a family of objects in the sense of [9], [13], [14]. We will define the various concepts of entropies as the numbers of defined (reified) pairs of those m_j and V_k. Among all of the decompositions and coverings of m and V as in (0.1) and (0.2), respectively, we have compatible pairs $m_j(V_k)$. We will define a notion of entropy of the state $m(V)$ as a number of such compatible pairs in the next section. For a microdecomposition, see [11].

1. METHODS OF TEMPORAL TOPOS

We have introduced notions of a microdecomposition and a micromorphism. For example, the concept of a t-topos theoretic light cone is viewed as a light cone with holes where non-reified states occur. This is because the notion of a micromorphism gives the impossibility of factorization between two states corresponding to two generalized time periods. Together with a microdecomposition and a further refinement of a covering in what will follow, we get similar "unreified" pairs of particle-decomposed presheaves and covering-decomposed objects in a t-site. Such a state as unmatched pairs of particle presheaves and objects in the t-site is considered as an ultra-microcosm, and the state is closer to "u-singularity". Even though the method of t-topos is a more kinematical and qualitative theory, the dynamical aspect is embedded in the space and time sheaves. Namely, space-time sheaf $\omega = (\kappa, \tau)$ is associated with a particle. Hence, for example, when the curvature of space-time $\omega_m = (\kappa_m, \tau_m)$ caused by m (representing a particle with mass) is measured, the fundamental composition principle can be used to assign a value. (See what will follow.) Another view of a dynamical aspect of t-topos is the following. When two particles, represented by presheaves m and m' are close enough to influence space-time in the common "region" of two space-time sheaves, then one can associate the two gravitationally interacting particles with the "product space-time" of the associated space-time sheaf induced by m and m'. (See [17] for details.)

Let a presheaf m associated with a particle be observed twice over generalized time periods V and U. Consider the case where m is observed over V first and then over U. That is, time $\tau_m(V)$ precedes time $\tau_m(U)$ in the usual classical linearly ordered sense. Then there exists a morphism g from V to U in the t-site S. Such a morphism g is said to be a *linearly t-ordered morphism*. Note that not every morphism from V to U in S represents such a linear temporal order in the above sense. This is one of the reasons for introducing the concept of a site rather than just a topological space. Recall that an inclusion is the only morphism, if it exists, between two open sets of a topological space.

Suppose that m is measured (or observed) by P over V, then there exists a morphism s_V from $m(V)$ to $P(V)$ which is the definition of an observation (or measurement). Categorically speaking, this means that s is a natural transformation (a morphism of functors) from m to P. Then we have the following diagram:

$$m(V) \xleftarrow{m(g)} m(U)$$
$$\downarrow \quad [\quad ^{s_V \, \text{O} m(g)}$$
$$P(V) \tag{1.1}$$

where the composition $s_V \, \text{O} m(g)$ in the above diagram should be understood as the measurement of $m(U)$ by measuring $m(V)$ by $P(V)$. Namely, the image of the composite morphism $s_V \, \text{O} m(g)$ is the amount of information P can obtain on the future state $m(U)$ by measuring the state of m over V. According to the quantum mechanical way of thinking, the phrase that a particle can *"be"* in several different locations at *"the same time"* has been used. However, such expressions need to be examined more carefully since t-topos gives more precise descriptions of such issues. That is, for a particle to be at a place, an object of t-site must be chosen. Then the particle must not be in a wave ur-state since an object in S has been specified. Namely, an expression as "An electron moves from point A to point B taking all available paths simultaneously" assumes the following. If such an electron were observed in addition to the two states corresponding to A and B, then there would be a non-trivial factorization of $V \xrightarrow{g} U$, i.e., $g = g_2 \circ g_1$ via $\{W\}$ in the t-side, corresponding to A and B. Then in the diagram

$$V \xrightarrow{g} U$$
$$] \, ^{g_1} \quad Z \, _{g_2}$$
$$W \quad , \tag{1.2}$$

$V \xrightarrow{g_1} W$ and $W \xrightarrow{g_2} U$ would become non-trivial linearly t-ordered morphisms. In particular, if such a $V \longrightarrow U$ is a micromorphism, then there does not exist such a proper factorization. The number of such paths between A and B (linearly t-ordered) are precisely equal to the number of non-trivial factorizations by linearly t-ordered morphisms of $V \longrightarrow U$.

For a given state $m(V)$ of m over V, assume that there exists an object V' in the t-site S so that $\tau_m(V')$ precedes $\tau_m(V)$. Namely, $V' \longrightarrow V$ is linearly t-ordered. Continue the process to obtain a sequence of objects, generalized time periods, of S. That is, we get

$$--- \to V'' \longrightarrow V' \longrightarrow V. \tag{1.3}$$

A definition of a t-topos theoretic light cone is given in [11]. We will give another definition of a light cone using the presheaf associated with a photon.

Definition 1.1. Let γ be a photon presheaf which is observed over a generalized time period V. Then consider all the light cone sequences with respect to the state $\gamma(V)$, or going through V

$$---\to \gamma(V_2)\to \gamma(V_1)\to \gamma(V)\to \gamma(V^1)\to \gamma(V^2)---, \qquad (1.4)$$

where

$$---\leftarrow V_2 \leftarrow V_1 \leftarrow V \leftarrow V^1 \leftarrow V^2 \leftarrow --- \qquad (1.5)$$

is an arbitrary sequence of linearly t-ordered objects in S. In terms of space-time sheaf, we have

$$---\to \omega(V_2)\to \omega(V_1)\to \omega(V)\to \omega(V^1)\to \omega(V^2)--- \qquad (1.6)$$

associated with γ.

Note that sequence (1.4) emphasizes the states of γ, and sequence (1.6) emphasizes the corresponding space-time.

2. Entropy and Limits

We will define the notion of an entropy for a decomposition as in (0.1) of m and for a covering as in (0.2) of V of objects in the t-topos $\hat{S}$ and t-site S, respectively. Furthermore, we can continue decomposing to get a sequence of decompositions as

$$\prod_{j\in J} m_j \longrightarrow \prod_{k\in K} m_{jk} \longrightarrow ---. \qquad (2.1)$$

Definition 2.1· The t-entropy of the state $m(V)$ for a micro-decomposition $\prod_{j\in J} m_j$ and a covering $\{V \longrightarrow V_k\}_{k\in K}$ of V is defined by the number of compatible (reifiable) pairs $\{m_j(V_k)\}_{j\in J, k\in K}$.

Definition 2.2· The formal entropy of $m(V)$ for the decomposition and the covering is defined by the product of cardinalities of index sets J and K.

Definition 2.3. The absolute entropy of the state $m(V)$ is defined as the maximum number of compatible pairs for all decompositions and coverings of m and V, respectively.

Note 2.4. Among the compatible pairs in the definitions of entropies, the corresponding generalized time periods need not be linearly t-ordered. Note also that the rest of the pairs between the decomposition and the covering are the collection of non-reified (non-measurable) particle associated presheaves.

Since a covering $\{V_k \longleftarrow V_{k,l}\}_l$ of V_k of a covering $\{V \longleftarrow V_k\}_k$ of V is a covering $\{V \longleftarrow V_{k,l}\}_{k,l}$ (See [9, 13, 14].), we get a sequence for V,

$$\{V \longleftarrow V_k\}_k \longrightarrow \{V \longleftarrow V_{k,l}\}_{k,l} \longrightarrow ---. \tag{2.2}$$

Next, we will consider limits of such sequences as in (2.1) and (2.2) and sequences as in (1.5, 1.6).

Definition 2.5. A presheaf m is said to be a *fundamental* presheaf when a decomposition in the sense of the sequence (2.1) becomes stable. That is, further decompositions consist of only several isomorphic objects. Namely, all the components of a decomposition are isomorphic presheaves.

Definition 2.6. An object V of the t-site S is said to be *fundamental* when a covering of V consists of all isomorphic objects to V itself.

Remarks 2.7. (1) A fundamental presheaf should be associated with elementary particles. Such a pair of a fundamental presheaf and a fundamental object of t-site is said to be a *fundamental pair*.

(2) The notion of a direct limit (inverse limit) is defined by a universal mapping property as in [9, 13, 14]. In this sense, such a notion as a direct (or invese) limit is an ultimate and universal object. Therefore, we propose the following definitions. The direct limit of the sequence (2.1) is said to be an *ur-subplanck decomposition* of m since sequence (2.1) is obtained by decomposing each presheaf in each step. Similarly, the direct limit of sequence (2.2) is said to be an *ur-subplanck covering* of V.

(3) As such a decomposition in (2.1):

$$\prod_{j \in J} m_j \longrightarrow \prod_{k \in K} m_{jk} \longrightarrow ---$$

approaches the direct limit of this sequence, more fundamental presheaves appear. Then a fundamental presheaves m_α associated with short-lived particles cause severe curvatures of space-time in microcosm. Note that being short-lived means the smallness of the assigned value via *FUNC* of the corresponding time sheaf $\tau_{m_\alpha}(V_{\underset{\rightarrow}{\lim}})$ evaluated at the corresponding $V_{\underset{\rightarrow}{\lim}}$ after sufficient refinements of the covering with more fundamental objects of S:

$$\{V \longleftarrow V_k\}_k \longrightarrow \{V \longleftarrow V_{k,l}\}_{k,l} \longrightarrow ---)$$

as in (2.2). Note also that the entropy of such a condition is also small, because the number of reified fundamental pairs of presheaves and objects of S decreases, i.e., there are more isomorphic objects in $\hat{S}$ and S, respectively. One might associate this t-topos theoretic interpretation of the ultra-microscopic state of short-lived fundamental presheaves $\{m_\alpha\}_\alpha$ and fundamental objects $\{V_{\underset{\rightarrow}{\lim}}\}$ of the t-site (i.e., generalized time periods) with the foam-like condition. For the connections to other singularities, see Remark 3.1(2) in the next section and the Epilogue, where we introduce the notions of ur-big bangs of the 0th stage and $(-1)^{st}$ stage, respectively.

3. U-SINGULARITIES

Let be m_Ω a fundamental presheaf of $\hat{S}$. Assume that m_Ω was observed at a generalized time period V of S. Then we can consider such a situation as we have considered earlier. Namely, for this given state of m_Ω over V, i.e., $m_\Omega(V)$, assume that there exists an object V' in the t-site S so that $\tau_{m_\Omega}(V')$ may precede $\tau_{m_\Omega}(V)$. Namely, $V' \longrightarrow V$ is linearly t-ordered. Continue this process successively to obtain a sequence of objects, i.e., generalized time periods of S. That is, as we evaluate at the (contravariant) fundamental presheaf m_Ω, we get the following sequence

$$--- \leftarrow m_\Omega(V'') \longleftarrow m_\Omega(V') \longleftarrow m_\Omega(V). \tag{3.1}$$

The direct limit $m_\Omega(V_{\underset{\leftarrow}{\lim}}) \overset{def}{=} \underset{\longrightarrow}{\lim}(--- \leftarrow m_\Omega(V'') \longleftarrow m_\Omega(V') \longleftarrow m_\Omega(V))$ of (3.1) is said to be the *inverse u-singularity* of the state $m_\Omega(V)$. On the other hand, in Remark (3), $\{m_\alpha(V_{\underset{\rightarrow}{\lim}})\}$ is said to be the *direct u-singularity* of the state $m(V)$. By the very definition of a direct limit, the corresponding space-time $\omega_{m_\Omega}(V_{\underset{\leftarrow}{\lim}})$, and in particular $\tau_{m_\Omega}(V_{\underset{\leftarrow}{\lim}})$, can not be preceded by any usual classical time. We are assuming the existence of a particle which has survived from the earliest universe. As a candidate for such a particle represented by m_Ω, we can consider cosmic background radiation.

At the inverse and direct u-singularity states, only isomorphic fundamental presheaves and fundamental generalized time periods (appearing in the direct limit of a covering sequence as in (2.2)) are available to be reified.

Remarks 3.1. (1) One may like to associate the "Ancestor's Rule" with the further decompositions of a given state in (2.1) to obtain (or to reach) fundamental presheaves in the following sense. Each individual has 2^n ancestors in the n generations back. After a certain number of generations, individuals have only several common fundamental ancestors.

(2) The inverse u-singularity and the direct u-singularity correspond to a big bang type singularity and to a microcosm quantum fluctuation, respectively. Notice that the definition of

the inverse u-singularity is defined as the direct limit of the sequence (3.1) induced by the linearly t-ordered sequence of generalized time periods (i.e., objects of the t-site). Namely, it is possible to have reified pairs of fundamental presheaves and fundamental generalized time periods *without* having linearly t-ordered fundamental objects *beyond* the inverse u-singularity. That is, beyond the inverse u-singularity, there is no distinction of the time notion of "before and after" in the linearly t-ordered sense and hence also in the classical sense. One might phrase such a state as: possible reified pairs simply exist without the past-future notion. Then note also that space sheaf κ and time sheaf τ can not be distinguished since there is no particular role difference without a linearly t-ordered concept. Further beyond such a state as no past-future, our t-topos approach does not tell whether there can exist such a situation as no reified pairs of fundamental presheaves and the t-site objects at all or not.

EPILOGUE

Our basic approach toward quantum behavior of a particle (elementary or not) is to capture an ur-particle state as a reified pair of the associated presheaf m and an object V of t-site. Generally speaking, for a macrocosm presheaf, there are more decompositions, and for a macrocosm covering, there are further refinement coverings. The t-topos theory is a scale independent theory not only because all the concepts are defined independently of the scales, but also in the following sense. For example, for a given morphism in the t-site, the shorter the sequence of the factorization of the morphism is, the more microscopic the morphism. A similar statement can be asserted for a presheaf.

When presheaf m does not have an object to be reified, m is said to be in an ur-wave state. This ur-wave state includes the case of the double slit experiment because such a choice of one object of the t-site cannot be determined. When applied to the notion of a light cone of a particle in a microcosm, such a t-topos theoretic microscopic light cone is a light cone with missing states where the associated particle presheaves do not have objects from t-site to be reified. One of the missing elements in our approach to t-topos is the aspect of dynamics. However, in the t-topos theory, there is a notion for such relativistic dynamics in terms of the space and time presheaves depending upon a particle (locally defined). For a full-scale description of the space-time sheaf as a measuring device of a particle, see the forthcoming [17]. However, further study is needed to develop the t-topos theory to treat more applications. The development of t-topos methods is still at the early stage. The t-topos aspect of the time delay effect, for example near a black hole, is yet to be formulated. In the near future, our plan is to investigate the t-topos theoretic interpretations of Hawking radiation and quantum tunneling. See our forthcoming papers, e.g., [17]. Our theory may belong to a hidden variable approach (with direct experimental applications) as indicated in [19].

A similarity between a back hole type singularity and a big bang type singularity is the concept of u-singularity, i.e., the categorical notion of a limit (inverse or direct). Namely, for a compatible pair of a presheaf and an object (generalized time period) of the t-site, a quantum fluctuation type singularity is described as limits of microdecompositions of the given presheaf and of micro coverings of the object of the t-site. Meanwhile, a big bang type singularity is given as a limit of a linearly t-ordered sequence for such a compatible object of the t-site with an arbitrary fundamental presheaf. In Remark 3.1, we have considered that fundamental pairs may exist, but no linearly t-ordered time, and furthermore, the totally incompatible (non-reified) state of fundamental presheaves and fundamental objects of the t-site. We may call such states the *ur-big bang* of the 0th *stage* and the *ur-big bang* of $(-1)^{st}$ *stage*, respectively. The ur-big bang of stage $(-1)^{st}$ is the "unmatched melting pot" of presheaves of t-topos and t-site objects without any compatible pairs. Notice that there are similarities between the singularity type of a big bang and ultra microcosm in the sense of the direct u-singularity in Remark 2.7. The difference is also clear as well. The inverse u-singularity is induced by a linearly t-ordered sequence, but the direct u-singularity is induced by the coverings. However, at the level of ur-big bang state of the 0th stage and the subplanck covering level, there is a similarity since the both cases are consisting of fundamental pairs without the usual space-time notion. Note also that some results from particle physics may tell us how many fundamental objects are in t-topos and t-site and in particular at the big bang. In order to make the t-topos theory into a quantitative theory, we may be able to use the so called the fundamental composition principle as in [2] and [3] for V defined for an operator in a Hilbert space H corresponding to a physical quantity. Namely, the following diagram consisting of the vertical morphism of Hilbert space H induced by a function from $\square$ of real numbers to $\square$

is commutative. See [2], [3], [10] for details. For the mathematical foundations for t-topos theory, see the forthcoming [17]. In this paper, sheaf cohomology per se does not appear. However, sheaf cohomology via coverings is crucial for Penrose's work as mentioned in Prologue. The papers [4] and [5] by Mallios and Rptis, De Rham cohomology, i.e., the hypercohomolgy with coefficient in the complex of sheaves of differential forms, plays an important role. The methods of sheaf cohomology also appears in [20]. Namely, in order to obtain the Veneziano amplitude, Volovich's p-adic string theory requires the computation zeta function obtained from the 1st p-adic cohomology group of Fermat curve over a finite field in characteristic p. (See the references in [20].) More general treatments of cohomologies can be found in [9] and [14]. As for cohomologies of sheaves for physics, see the forthcoming [17].

ACKNOWLEDGMENT

During the last several years the author has been given opportunities to scheme the methods of temporal topos at the following institutions. The author expresses his gratitude toward the individuals and the institutions for the invitations and the hospitality shown during these visits: Antwerp University (Antwerp, Belgium), Lund University (Lund, Sweden), Imperial College (London, U.K.), Darmstadt University (Darmstadt, Germany), Calabria University (Arcavacata di Rende, Italy), George Mason University (Fairfax, U.S.A.), the

Institute for Advanced Study (Princeton, U.S.A.), Indian Statistical Institute (Kolkata, India), Kyoto University (Kyoto, Japan), Okayama University (Okayama, Japan), Tsukuba University (Tsukuba, Japan), and Shizuoka University (Shizuoka, Japan).

REFERENCES

[1] Butterfield J. and Isham C. J., *Spacetime and the Philosophical Challenge of Quantum Gravity*, arxiv:gr-qc/9903072 v1 18 Mar 1999.
[2] Butterfield J. and Isham C. J., *Int. J. Theor. Phys.* **37** (1998) 2669, quantum ph/980355.
[3] Butterfield J. and Isham C. J., *Int. J. Theor. Phys.* **37** (1999) 827, quantum ph/9808067.
[4] Mallios, A. and Raptis, I., Finitary, Causal and Quantal Vaccum Einstein Garvity, *Int. J. Theor. Phys.* **42**, 1479 (2003), gr-qc/0209048.
[5] Rapitis, I., Finitary-Algebraic 'Resolution' of the Inner Schwarzschild Singularity, *Int. J. Theor. Phys.*; gr-qc/0408045.
[6] Penrose, R., The Road to Reality Alfred A. Knopf, N.Y., (2005).
[7] Grauert, H. and Remmert, R., Coherent Analytic Sheaves, *Grundlehren der Mathematischen Wissenschaften*, **265**, Springer-Verlag, 1984.
[8] Kato, G. and Struppa, D. C., *Fundamentals of Algebraic Microlocal Analysis, Pure and Applied Math.*, No. 217, Marcel Dekker Inc., 1999.
[9] Kato, G., The *Heart of Cohomology*, Springer, 2006.
[10] Kato, G., *Elemental Principles of t-Topos, Europhysics Letters*, Vol. 68, No. 4, pp467-472, (2004).
[11] Kato, G., Elemental t.g. Principles of Relativistic t-Topos, *Europhysics Letters*, Vol. 71, No. 2, pp. 172-178, (2005).
[12] Kato, G., u-Singularity and t-Topos Theoretic Entropy, *Int. J. Theor. Phys* (2010) 49: 1952 – 1960.
[13] Gelfand, S. I. and Manin, Y. I., *Methods of Homological Algebra*, Springer, 1996.
[14] Kashiwara H. and Schapira, P., *Sheaves and Categories*, Springer, 2006.
[15] Kato, G. and Tanaka, T., Double-slit Interference and Temporal Topos, *Foundations of Physics*, Vol. 36, No. 11, Nov. 2006.
[16] Kato,G.,Kafatos, M., Roy, S., Tanaka, T., Sheaf Theory and Geometric Approach to EPR *Nonlocality*, submitted.
[17] Kato, G., *Elements of Temporal Topos*, Springer, in preparation.
[18] Kato, G. and Takemae, S. A., *Hawking radiation and t-topos, in preparation*.
[19] Genovese, M., *Research on hidden variable theories: A review of recent progresses, Physics Reports* **413** (2005) 319-396.
[20] Vladimirov, V. S., Volovich, I. V., and Zelenov, E. I., p-Adic *Analysis and Mathematical Physics*, World Scientific, 1994.

Chapter 5

FULLY QUANTUM STUDY OF THE FRW MODEL WITH RADIATION AND CHAPLYGIN GAS

Sergei P. Maydanyuk[1] and Vladislav S. Olkhovsky[2]
Institute for Nuclear Research, National Academy of
Science of Ukraine, Ukraine

ABSTRACT

We give an introduction into quantum cosmology with emphasis on its conceptual fully quantum consideration in framework of formalism based on minisuperspace Wheeler–DeWitt equation. After a general review of different approaches of quantum cosmology, we study the closed Friedmann-Robertson-Walker model with quantization in presence of the positive cosmological constant, radiation and Chaplygin gas. For analysis of tunneling probability for birth of an asymptotically deSitter, inflationary Universe as a function of the radiation energy we introduce a new definition of a "free" wave propagating inside strong fields. On this basis we correct Vilenkin's tunneling boundary condition, define and calculate penetrability and reflection in fully quantum stationary approach (without application of any semiclassical and perturbation theory approximations).

Keywords: physics of the early universe, inflation, Wheeler-De Witt equation, Chaplygin gas.

1. INTRODUCTION

There are some reasons why one should try to set up a quantum theory of gravity. The key motivations come from quantum field theory and general relativity. A unification of all possible fundamental interactions is an important aim. The quantum field theory should provide this with a fundamental cut-off scale. So, quantum gravity is only one aspect of the ambitious venture to unify all interactions. Here, string theory has become in the highest level of perspectives.

---
[1] E-mail address: maidan@kinr.kiev.ua.

From a general-relativistic point of view, quantization of gravity is key point to supersede general relativity. This is due to the fact that general relativity predicts its own break-down. This happens when quantities occuring in the theory itself diverge. The famous divergence here is consideration of the birth of Universe, called as the big-bang singularity.

Quantum gravity is expected to provide a quantum theory of the gravitational field. This could be done by different ways in dependence on the considered structure which should be quantized. The most conservative way is quantization of Einstein's general relativity theory. Such approach is separated into covariant and canonical techniques. While the covariant approaches use covariance of space-time in formalism, the canonical ones begin from separation of space-time into space and time and further work with Hamiltonian formalism. In the latter case, the initial metric is already considered as evolution of a three-dimensional metric in time. Different ways of covariant approaches are developed in the path-integral technique and perturbation theory (with implication of Feynman diagrams). On the other side, the canonical approaches are developed in frameworks of loop quantum gravity and quantum geometrodynamics.

The string theory is the main alternative to the direct quantization of Einstein's theory. The aim of this theory is to construct the unified quantum theory of all possible interactions. In this direction, the quantum gravity is only a simple aspect in region where the gravity could be studied as independent interaction. Up to-date, different string-based quantum cosmological models have been actively developed and cause an increased interest.

2. BOUNDARY CONDITIONS IN COSMOLOGY IN THE FULLY QUANTUM CONSIDERATION

Among all variety of models of early Universe one can select two prevailing and alternative approaches: these are the Feynman formalism of path integrals in multidimensional spacetime, developed by the Cambridge group and other researchers, called the *"Hartle-Hawking method"* [1], and a method based on direct consideration of tunneling in 4-dimensional Euclidian spacetime called the *"Vilenkin method"* [2]. In the quantum approach we have the following picture of the Universe creation: a closed Universe with a small size is formed from "nothing" (vacuum), where by the word "nothing" one refers to a quantum state without classical space and time. A wave function is used for a probabilistic description of the creation of the Universe and such a process is connected with transition of a wave through an effective barrier.

In majority of models tunneling is studied in details in the semiclassical approximations (for example, see [3–6]). Here, a *tunneling boundary condition* [3] could seems to be natural, where the wave function should represent an outgoing wave at the large scale factor a. However, whether is such a wave free in asymptotic region? If to draw attention on increasing gradient of potential, used with opposite sign and having a sense of force, acting ``through the barrier" on this wave, then one come to contradiction: *influence of the potential on this wave is increased strongly at increasing a* [7]. Now a new question has been appeared: what should the wave represent in the cosmological problem? So, we come to necessity *to define "free" wave inside strong fields*.

---

[2] E-mail address: olkhovsky@mail.ru, olkhovsk@kinr.kiev.ua.

The problem of correct definition of the wave in cosmology is reinforced more, if to calculate incident and reflected waves before the barrier. Even with known exact solution for the wave function there is uncertainty in determination of these waves. But penetrability is based on them.

In order to estimate probability of formation of Universe accurately, we need in the fully quantum basis. In order to construct it, we should realize the following steps:

(1) to give definition of the wave in strong fields;
(2) to construct the fully quantum stationary method of determination of the penetrability and reflection using the definition of the wave above;
(3) to estimate how much the semiclassical approach is differed from the fully quantum one.

3. A MODEL IN THE FRIEDMANN-ROBERTSON-WALKER METRIC WITH RADIATION AND GENERALIZED CHAPLYGIN GAS

Let us start from a case of a closed ($k = 1$) FRW model in the presence of a positive cosmological constant $\Lambda > 0$, radiation and component of the Chaplygin gas. The minisuperspace Lagrangian has the form [8]:

$$L(a,\dot{a}) = \frac{3a}{8\pi G}\left(-\dot{a}^2 + k - \frac{8\pi G}{3}a^2 \rho(a)\right), \tag{1}$$

where a is scale factor, $\dot{a}$ is derivative of a with respect to time coordinate t, $\rho(a)$ is a generalized energy density. In order to connect the stage of Universe with dust matter and its another accelerating stage, in Ref. [9] a new scenario with the *Chaplygin gas* was applied to cosmology. A quantum FRW-model with the Chaplygin gas has been constructed on the basis of equation of state instead of $p(a) = \rho(a)/3$ (where p (a) is pressure) by $p_{Ch}(a) = -A/\rho_{Ch}^\alpha$, where A is positive constant and $0 < \alpha \leq 1$. Solution of equation of state gives

$$\rho_{Ch}(a) = \left(A + \frac{B}{a^{3(1+\alpha)}}\right)^{1/(1+\alpha)}, \tag{2}$$

where B is a new constant. This model through one phase α connects the stage of Universe where dust dominates and DeSitter stage. At limit $\alpha \to 0$ eq. (2) is transformed into ρ_{dust} plus ρ_Λ. From such limit we find $A = \rho_\Lambda$, $B = \rho_{\text{dust}}$ and write the following generalized density [7]:

$$\rho(a) = \left(\rho_\Lambda + \frac{\rho_{dust}}{a^{3(1+\alpha)}}\right)^{1/1+\alpha} + \frac{\rho_{rad}}{a^4}, \qquad (3)$$

where ρ_{rad} is component describing the radiation (equation of state for radiation is $p = \rho_{rad}/3$, p is pressure) and $\rho_\Lambda = \Lambda/8\pi G$.

The passage to the quantum description of the evolution of the Universe is obtained by the procedure of canonical quantization in the Dirac formalism for systems with constraints. We obtain the Wheeler-De Witt (WDW) equation, which after multiplication on factor and passage of component with radiation ρ_{rad} into right part transforms into the following form [7]:

$$\left\{-\frac{\partial^2}{\partial a^2} + V(a)\right\}\varphi(a) = E_{rad}\,\varphi(a), \quad E_{rad} = \frac{3\rho_{rad}}{2\pi G},$$

$$V(a) = \left(\frac{3}{4\pi G}\right)^2 ka^2 - \frac{3}{2\pi G}a^4 \cdot \left(\rho_\Lambda + \frac{\rho_{dust}}{a^{3(1+\alpha)}}\right)^{1/3+\alpha}, \qquad (5)$$

where $\varphi(a)$ is wave function of Universe. For the Universe of closed type (at $k = 1$) at $8\pi G \equiv M_p^{-2} = 1$ we have:

$$V(a) = 36a^2 - 12a^4\left(\Lambda + \frac{\rho_{dust}}{a^{3(1+\alpha)}}\right)^{1/3+\alpha} \qquad (6)$$

and $E_{rad} = 12\rho_{rad}$.

Let us expand the potential (6) close to arbitrary point $\bar{a}$ by powers $q = a - \bar{a}$ and restrict ourselves by linear item:

$$V_{Ch}(q) = V_0 + V_1 \cdot q. \qquad (7)$$

For coefficients V_0 and V_1 I find:

$$\begin{aligned}V_0 &= V_{Ch}(a = \bar{a}), \\ V_1 &= 72a + 12a^3\left\{-4\Lambda - \frac{\rho_{dust}}{a^{3(1+\alpha)}}\right\}\left(\Lambda + \frac{\rho_{dust}}{a^{3(1+\alpha)}}\right)^{-\alpha/1+\alpha}\end{aligned} \qquad (8)$$

and eq. (4) obtains form:

$$-\frac{d^2}{dq^2}\varphi(q)+(V_0-E_{rad}+V_1 q)\varphi(q)=0.\tag{9}$$

After change of variable

$$\xi=\frac{E_{rad}-V_0}{|V_1|^{2/3}}-\frac{V_1}{|V_1|^{2/3}}q\tag{10}$$

we have

$$\frac{d^2}{d\xi^2}\varphi(\xi)+\xi\cdot\varphi(\xi)=0.\tag{11}$$

4. MOTIVATIONS TO CORRECT VILENKIN'S BOUNDARY CONDITION OF TUNNELING

Let us analyze how much a choice of the boundary condition in the asymptotic region is motivated.

- In tasks of decay in nuclear and atomic physics the potentials of interactions tend to zero in the asymptotic region. Here, an application of the boundary condition at limit of infinity does not give questions. In cosmology we deal with another, principally different type of the potential: with increasing of the scale factor a modulus of this potential increases. A gradient of the potential, used with opposite sign and having a sense of force acting on the wave, increases also. So, there is nothing mutual with free propagation of the wave in the asymptotic region [7].
- Results of Ref. [8] reinforce a seriousness of such a problem: the scale factor a in the external region is larger, the period of oscillations of the wave function is smaller. This requires with increasing a to decrease step of calculations of the wave function. This increases time of calculations, increases errors. So, boundary condition in the asymptotic region has no practical sense in cosmology. In contrary, in nuclear and atomic physics calculations of the asymptotic wave are the most stable.
- It has not been known whether Universe expands at extremely large scale factor a. Just the contrary, it would like to clarify this, imposing Universe to expand in initial stage.

So, we redefine the boundary condition as [7]:

the boundary condition should fix the wave function so that it represents the wave, interaction between which and the potential barrier is minimal at such a value of the scale factor a where action of this potential is minimal.

To give a mathematical formulation for this definition, we are confronted with two questions:

1. What should the free wave represent at arbitrary point inside cosmological potential?
2. At which coordinate is imposition of this boundary condition the most corrected?

At first, let us solve the second question. Which should this point be: or this is a turning point (where the potential coincides with energy of radiation), or this is a coordinate where a gradient of the potential (having a sense of *force of interaction*) becomes zero, or this is a coordinate where the potential becomes zero?

We define this coordinate where the force acting on the wave is minimal. We define the force as the gradient of the potential used with opposite sign.

5. DEFINITION OF THE WAVE MINIMALLY INTERACTING WITH THE POTENTIAL

Definition 1. [strict definition of the wave]:
The wave is such a linear combination of two partial solutions of the wave function that the change of the modulus ρ of this wave function is closest to constant under variation of a:

$$\left. \frac{d^2}{da^2} \rho(a) \right|_{a=a_{tp}} \to 0. \tag{12}$$

For some types of potentials it is more convenient to define the wave less strongly.

Definition 2 [weak definition of the wave]:
The wave is such a linear combination of two partial solutions of wave function that the modulus ρ changes minimally under variation of a:

$$\left. \frac{d}{da} \rho(a) \right|_{a=a_{tp}} \to 0. \tag{13}$$

Now we shall look for the function $\varphi(\xi)$ as

$$\varphi(\xi) = T \cdot \Psi^{(+)}(\xi), \tag{14}$$

$$\Psi^{(\pm)}(\xi) = \int_0^{u_{max}} \exp \pm i\left(-\frac{u^3}{3} + f(\xi)u\right) du, \tag{15}$$

where T is an unknown factor, $f(\xi)$ is an unknown continuous function satisfying $f(\xi) \to$ const at $\xi \to 0$, and u_{max} is the unknown upper limit of integration. The real part of $f(\xi)$ gives a contribution to the phase of the integrand function while the imaginary part of $f(\xi)$ deforms modulus.

At small enough values of $|\xi|$ we represent $f(\xi)$ in the form of a power series:

$$f(\xi) = \sum_{n=0}^{+\infty} f_n \xi^n, \qquad (16)$$

where f_n are constants. Substituting formula (15) for $\Psi(\xi)$ into equation (11), we find (see Ref. [7]):

$$f_2^{(\pm)} = \pm \frac{f_1^2}{2i} \cdot \frac{J_2^{(\pm)}}{J_1^{(\pm)}},$$

$$f_3^{(\pm)} = \pm \frac{4 f_1 f_2^{(\pm)} J_2^{(\pm)} - J_0^{(\pm)}}{6i J_1^{(\pm)}}, \qquad (17)$$

$$f_{n+2}^{(\pm)} = \pm \frac{\sum_{m=0}^{n}(n-m+1)(m+1) f_{n-m+1}^{(\pm)} f_{m+1}^{(\pm)}}{i(n+1)(n+2)} \cdot \frac{J_2^{(\pm)}}{J_1^{(\pm)}},$$

where

$$J_0^{(\pm)} = \int_0^{u_{max}} \exp\pm i\left(-\frac{u^3}{3} + f_0 u\right) du,$$

$$J_1^{(\pm)} = \int_0^{u_{max}} u \cdot \exp\pm i\left(-\frac{u^3}{3} + f_0 u\right) du, \qquad (18)$$

$$J_2^{(\pm)} = \int_0^{u_{max}} u^2 \cdot \exp\pm i\left(-\frac{u^3}{3} + f_0 u\right) du.$$

In order to be the solution $\Psi(\xi)$ closer to the well-known Airy functions, $\mathrm{Ai}(\xi)$ and $\mathrm{Bi}(\xi)$, we choose $f_0 = 0, f_1 = 1$.

6. CALCULATIONS OF THE WAVE FUNCTION

In order to provide a linear independence between two partial solutions for the wave function effectively, I look for the first partial solution increasing in the region of tunneling and the second one decreasing in this tunneling region. At first, I define each partial solution and its derivative at a selected starting point, and then I calculate them in the region close enough to this point using the *method of beginning of the solution*. Here, for the partial solution which increases in the barrier region, as the starting point I use arbitrary point $\bar{a}$ inside well with its possible shift at non-zero energy $E_{\rm rad}$ or equals to zero $a=0$ at null energy $E_{\rm rad}$, and for the second partial solution which decreases in the barrier region, I select the starting point to be equal to external turning point $a_{\rm tp,\,out}$. Then both partial solutions and their derivatives I calculate independently in the whole required range of a using the *method of continuation of the solution*, which is improvement of the Numerov method with a constant step. By such a way, I obtain two partial solutions for the wave function and their derivatives in the whole studied region.

Having obtained two linearly independent partial solutions $\varphi_1(a)$ and $\varphi_2(a)$, we make up a general solution (prime is for derivative with respect to a):

$$\varphi(a) = T \cdot (C_1 \varphi_1(a) + C_2 \varphi_2(a)), \tag{19}$$

$$C_1 = \left.\frac{\Psi \varphi_2' - \Psi' \varphi_2}{\varphi_1 \varphi_2' - \varphi_1' \varphi_2}\right|_{\bar{a}}, \quad C_2 = \left.\frac{\Psi' \varphi_1 - \Psi \varphi_2'}{\varphi_1 \varphi_2' - \varphi_1' \varphi_2}\right|_{\bar{a}}, \tag{20}$$

where T is normalization factor, C_1 and C_2 are constants found from the boundary condition.

7. PROBLEM OF INTERFERENCE BETWEEN THE INCIDENT AND REFLECTED WAVES

Rewriting the wave function $\varphi_{\rm total}$ in the internal region through a summation of incident $\varphi_{\rm inc}$ wave and reflected $\varphi_{\rm ref}$ wave:

$$\varphi_{\rm total} = \varphi_{\rm inc} + \varphi_{\rm ref}, \tag{21}$$

we consider the total flux:

$$j(\varphi_{\rm total}) = j_{\rm inc} + j_{\rm ref} + j_{\rm mixed}, \tag{22}$$

where

$$\begin{aligned} j_{\rm inc} &= i \left(\varphi_{\rm inc} \nabla \varphi_{\rm inc}^* - {\rm h.c.} \right), \\ j_{\rm ref} &= i \left(\varphi_{\rm ref} \nabla \varphi_{\rm ref}^* - {\rm h.c.} \right), \\ j_{\rm mixed} &= i \left(\varphi_{\rm inc} \nabla \varphi_{\rm ref}^* + \varphi_{\rm ref} \nabla \varphi_{\rm inc}^* - {\rm h.c.} \right). \end{aligned} \tag{23}$$

The j_{mixed} component describes interference between the incident and reflected waves in the internal region (let us call it as *mixed component of the total flux* or simply *flux of mixing*). From constancy of the total flux j_{total} we find flux j_{tr} for the wave transmitted through the barrier, and:

$$j_{inc} = j_{tr} - j_{ref} - j_{mixed}, \quad j_{tr} = j_{total} = \text{const}. \tag{24}$$

Now one can see that *the mixed flux introduces ambiguity in determination of the penetrability and reflection for the same known wave function.*

In the radial problem of quantum decay definition of penetrability and reflection looks to be conditional as the incident and reflected waves should be defined inside internal region from the left of the barrier. In order to formulate these coefficients, we shall include into definitions coordinates where the fluxes are defined (denote them as x_{left} and x_{right}):

$$T = \frac{j_{tr}(x_{right})}{j_{inc}(x_{left})}, \quad R = \frac{j_{ref}(x_{left})}{j_{inc}(x_{left})}, \quad M = \frac{j_{mixed}(x_{left})}{j_{inc}(x_{left})}. \tag{25}$$

From eqs. (24) and (25) we obtain [7] (j_{tr} and j_{ref} are directed in opposite directions, j_{inc} and j_{tr} in the same directions):

$$|T| + |R| - M = 1. \tag{26}$$

Now we see that condition $T + R = 1$ has sense in quantum mechanics only if there are no any interference between incident and reflected waves which we calculate, and it is to use $j_{mixed} = 0$.

8. THE PENETRABILITY IN THE FULLY QUANTUM AND SEMICLASSICAL APPROACHES

Now we shall estimate by the method described above the coefficients of penetrability and reflection for the potential barrier (6) with parameters $A = 36$, $B = 12\Lambda$, $\Lambda = 0.01$ at different values of the energy of radiation E_{rad}. We shall compare the found coefficient of penetrability with its value, which the semiclassical method gives. In the semiclassical approach we shall consider two following definitions of this coefficient:

$$P_1^{WKB} = \frac{1}{\theta^2},$$

$$P_2^{WKB} = \frac{4}{\left(2\theta + 1/(2\theta)^2\right)^2}, \tag{27}$$

where

$$\theta = \exp \int_{a_p^{(int)}}^{a_p^{(ext)}} |V(a) - E| \, da. \qquad (28)$$

One can estimate also *duration of a formation of the Universe*, using by definition (15) in Ref. [10]:

$$\tau = \frac{2 a_{tp,int}}{P}. \qquad (29)$$

Table 1. The penetrability, *P*, of the barrier and duration τ of the formation of the Universe defined by eq. (29) in the FRW-model with the Chaplygin gas obtained in the fully quantum and semiclassical approaches (minimum of the hole is -93.579 and its coordinate is 1.6262, maximum of the barrier is 177.99 and its coordinate is 5.6866): the fully QM method 1 is calculations by the fully quantum approach for the boundary located in the coordinate of the minimum of the internal hole (i. e. coordinate is 1.6262), the fully QM method 2 is calculations by the fully quantum approach for the boundary located in the internal turning point $a_{tp,\,in}$ (coordinates of the turning points are in Table 2)

Energy	Penetrability, *P*			Time, τ	
	Full QM method 1	Full QM method 2	Method WKB	Full QM method 1	Method WKB
20	7.6221×10^{-30}	7.7349×10^{-29}	1.4692×10^{-29}	9.4313×10^{29}	4.8928×10^{29}
40	3.5680×10^{-26}	6.5169×10^{-25}	1.2298×10^{-25}	2.1257×10^{26}	6.1670×10^{25}
60	2.0591×10^{-22}	4.3423×10^{-21}	8.1523×10^{-22}	3.8719×10^{22}	9.7797×10^{21}
80	1.5530×10^{-18}	2.3850×10^{-17}	4.4346×10^{-18}	5.3862×10^{18}	1.8862×10^{18}
100	3.2922×10^{-14}	1.1053×10^{-13}	2.0304×10^{-14}	2.6622×10^{14}	4.3167×10^{14}
120	8.6052×10^{-11}	4.4005×10^{-10}	7.9523×10^{-11}	1.0678×10^{11}	1.1555×10^{11}
140	2.2012×10^{-8}	1.5460×10^{-6}	2.7128×10^{-7}	4.3888×10^{8}	3.5612×10^{7}
160	2.9685×10^{-5}	4.9980×10^{-3}	8.1663×10^{-4}	3.4471×10^{5}	1.2530×10^{4}
170	3.4894×10^{-3}	2.6078×10^{-1}	4.2919×10^{-2}	3.0460×10^{3}	2.4820×10^{2}

Results are presented in Table 1. P_2^{WKB} is not included in the table because it coincidence with P_1^{WKB} to the first 8 digits. One can see that the fully quantum approach gives the penetrability close to its semiclassical value, which differs from results [10].

In the next Table 2 the coefficients of the penetrability, reflection and mixing calculated in the fully quantum method are presented for the energy of radiation E_{rad} close to the height of the barrier. One can see that summation of all such values for coefficients allows to reconstruct the property (26) with accuracy of the first 11–18 digits. Now it becomes clear that the approach proposed in Ref. [10] and the semiclassical methods do not give such an accuracy.

Table 2. The coefficients of the penetrability, reflection and mixing calculated by the fully quantum method and test on their summation for the FRW-model with the Chaplygin gas density component (the fully quantum approach 1 is used at the internal boundary located in the coordinate of the minimum of the internal hole)

Energy	Fully quantum method				Turning points	
	Penetrability	Reflection	Interference	Summation	$a_{tp,in}$	$a_{tp,out}$
20	$7.6221543404 \times 10^{-30}$	1.00000000000000	8.55×10^{-20}	1.00000000000000	3.59	7.05
40	$3.5680158760 \times 10^{-30}$	1.00000000000000	2.34×10^{-19}	1.00000000000000	3.79	6.97
60	$2.0591415452 \times 10^{-22}$	1.00000000000000	4.86×10^{-20}	1.00000000000000	3.98	6.88
80	$1.5530040238 \times 10^{-18}$	1.00000000000000	2.08×10^{-19}	1.00000000000000	4.18	6.78
100	$3.2922846164 \times 10^{-14}$	0.99999999999996	3.13×10^{-20}	1.00000000000000	4.38	6.67
120	$8.6052092530 \times 10^{-11}$	0.99999999991394	1.06×10^{-19}	1.00000000000000	4.59	6.55
140	$2.2012645564 \times 10^{-8}$	0.99999997798735	1.60×10^{-19}	1.00000000000000	4.83	6.39
160	$2.9685643504 \times 10^{-5}$	0.99997031435611	6.94×10^{-20}	0.99999999999961	5.11	6.18
170	$3.4894544195 \times 10^{-3}$	0.99651054553176	2.02×10^{-19}	0.99999999995131	5.31	6.02

CONCLUSIONS

We have presented the fully quantum formalism for study of the evolution of Universe in frameworks of the closed Friedmann-Robertson-Walker model with quantization in the presence of a positive cosmological constant, radiation and Chaplygin gas. Note the following:

1. A fully quantum definition of the wave propagating inside strong field and interacting with them minimally has been formulated. The tunneling boundary condition has been corrected.
2. A quantum stationary method of determination of penetrability and reflection relatively the barrier has been developed. Here, non-zero interference between the incident and reflected waves has been taken into account and for its estimation the coefficient of mixing has been introduced.

In such an approach the penetrability of the barrier for the studied FRW-model has been estimated. Note the following:

1. The probability for birth of asymptotically deSitter Universe is close to results obtained by the semiclassical approach, but it differs from results obtained by non-stationary approach [10] (see Tables 1 and 2).
2. The reflection from the barrier has been determined at the first time. It differs essentially from 1 at the energy of radiation close enough to the barrier height (see Table 2).
3. The modulus of the coefficient of mixing (indicating the interference between the incident and reflected waves) is less 10^{-19}.
4. A property (26) is reconstructed to the first 11–18 digits (see Table 2).

Analysis of fully quantum estimates of rates for the Universe evolution in its first stage, role of fully quantum approaches and their physical contents are important task of quantum cosmology, which is planned to be studied in closest future in details.

REFERENCES

[1] J. B. Hartle, S. W. Hawking, *Wave function of the Universe*, Phys. Rev. **D28**, 2960–2975 (1983).
A. Vilenkin, *Creation of Universes from nothing*, Phys. Lett. **B117**, 25–28 (1982).
B. Vilenkin, *Approaches to quantum cosmology*, Phys. Rev. **D50**, 2581–2594 (1994), gr-qc/9403010.

[2] V. A. Rubakov, *Quantum cosmology*, Proceedings: Structure formation in the Universe (Edited by R. G. Crittenden and N. G. Turok, Kluwer, 1999), p. 63–74, gr-qc/9910025.

[3] R. Casadio, F. Finelli, M. Luzzi, G. Venturi, *Improved WKB analysis of cosmological perturbations*, Phys. Rev. D71, 043517 (2005) [12 pages], gr-qc/0410092.

[4] M. Luzzi, *Semiclassical Approximations to Cosmological Perturbations*, Ph. D. thesis (Advisor: Prof. Giovanni Venturi, University of Bologna, 2007), p. 148, arXiv:0705.3764.

[5] S. P. Maydanyuk, A fully quantum method of determination of penetrability and reflection coefficients in quantum FRW model with radiation, *Int. Journ. Mod. Phys.* **D19**, 395–435 (2010), arXiv:0812.5081.

[6] S. P. Maydanyuk, Wave function of the Universe in the early stage of its evolution, *Europ. Phys. Journ.* **C57**, 769–784 (2008), arxiv.org:0707.0585.
A. Y. Kamenshchik, U. Moschella, V. Pasquier, *Phys. Lett.* **B511**, 265 (2001), gr-qc/0103004.

[7] G. A. Monerat, G. Oliveira-Neto, E. V. Correa Silva, L. G. Ferreira Filho, P. Romildo, Jr., J. C. Fabris, R. Fracalossi, S. V. B. Goncalves, and F. G. Alvarenga, Dynamic of the early universe and the initial condition for inflation in a model with radiation and a Chaplygin gas, *Phys. Rev.* D76, 024017 (2007) [11 pages].

In: The Big Bang: Theory, Assumptions and Problems
Editors: J. R. O'Connell and A. L. Hale

ISBN: 978-1-61324-577-4
© 2012 Nova Science Publishers, Inc.

Chapter 6

A BRIEF INTRODUCTION REVIEW ON THE PROBLEMS OF THE UNIVERSITY ORIGIN

V. S. Olkhovsky
Institute for Nuclear Research, National Academy of Science of Ukraine, Ukraine

ABSTRACT

In the presented brief introduction review there are briefly analyzed various cosmologic and quantum cosmologic Big-Bang hypothesis and theories. It is made the general conclusion that namely in the *beginning* is arising the general problem of the world origin and a lot of related problems. The schematic description of some problems between them are presented. And it is added the problem of the inevitable choice meta-theoretical dilemma: the beginning of the Universe formation from vacuum ("nothing") is *either* a result of the irrational randomness after passing from other space-time dimensions or from other universe, caused by some unknown process, *or* a result of the creation of the expanding Universe (together with the laws of its functioning) by the supreme intelligent design from *nigilo*. And to the first doctrine there was adjoined in the XXI c the meta-physical doctrine of the parallel other universes with some kind of interaction between them or with an irrational spontaneous passage of the matter from them to our Universe – those hypothetical universes are or the exactly same as ours (in quantum mechanics), or with other space-time dimensions (in some Big-Bang theories), or with other values of the physical constants (in other Big Bang theories).

INTRODUCTION

In the science history and in the science philosophy of XX-XXI cc (especially in the field of the natural sciences, beginning from physics) there has been a lot of interesting things, which had not obtained a sufficiently complete elucidation and analysis yet. Firstly, under the influence of scientific and technological progress a great attention has been paid to the justification of such direction in the science philosophy as the scientific realism (i.e. the correspondence of the science to the reality), which has successively acquired three forms: the naïve realism, the usual realism and the critical science realism. Secondly, some new important problems of physics (especially the problem of the essentially probabilistic description of the reality of the microscopic world, the problem of the essential influence of the observer on the reality, the collapse of the wave function) had been revealed in the development of quantum mechanics, *the continuously complicated interpretation of the Universe origin and the expansion after the Big Bang,* and also *no* succeeded attempt in explaining the origin of the biological life in terms of physics and other natural sciences, all being with a variety of interpretation versions, connected with the world-views of the researchers.

As to "great" and "grand" problems of natural sciences: There is an extensive introduction in the large number of open problems in many fields of physics, published by the Russian physicist V.Ginzburg in [1], which is rather interesting to study. Inside this large list of open problems of modern physics (and in a certain degree of modern natural sciences), represented by V.Ginzburg repeatedly in Russian editions, some of them are marked him "great" or "grand" problems. Between namely these problems I would like to underline three of them:

a) The problem of interpretation and comprehension of quantum mechanics (even of the non-relativistic quantum theory) remains still topical. The majority of critics of quantum mechanics are unsatisfied with the probabilistic nature of its predictions. One can add here also the questions and paradoxes of the theory of quantum measurements theory, especially like the wave-function reduction. The appearance of quantum mechanics, and, in particular, the discussion of N.Bohr with A.Einstein (lasting many years), had seriously undermined the traditional forms of the naïve realism in the philosophy of the scientific realism and had strongly influenced (and are continuating to influence) not only on physics but also on other kinds of knowledge in the sense of the dependence of the reality on the observer and, moreover, on our understanding of the human knowledge at all. More lately the new interpretation of quantum mechanics is appeared: in it the hypothesis of many universes, which are the exactly same as ours, permits to avoid the wave-function reduction.

b) The relationship between physics and biology and, specifically, the problem of reductionism. The main problem, according to V.Ginzburg, is connected with the explanation of the origin of the biologic life and the origin of the human abstract thinking (but the second one, as to me, is connected not with biology but with the origin of the human spiritual life which is far beyond natural sciences). V.Ginzburg assumes that for a possible explanation of the origin of the biologic life one can naturally imagine a certain jump which is similar to some kind of phase transition (or, may be, certain synergetic process). But there are other points of view too.

c) *The cosmological problem* (in other words, *the problem of the Universe origin*). According to V.Ginzburg, it is also a grand problem, or strictly speaking, a great complex of cosmic problems many of which is far from the solution.

d) I did also analyzed in [2] these three problems in the context of other aspects, first of all regarding the increasing discussions between the supporters of two different meta-theoretical, meta-philosophical doctrines: *either* the beginning of the Universe formation from vacuum ("nothing") is *either* a result of the irrational randomness after passing from other space-time dimensions or from other universe, caused by some unknown process, *or* a result of the creation of the expanding Universe (together with the laws of its functioning) by the supreme intelligent design from *nigilo*.

SCHEMATIC DESCRIPTION OF THE PROBLEMS CONNECTED WITH THE UNIVERSE ORIGIN AND EXPANSION

Earlier, after Enlightenment till approximately 1920, scientists in the natural sciences did usually consider the Universe as *eternally existing and eternally moving*.

Now the most convincing arguments against the model of the eternally existing Universe are:

(a) *the second law of thermodynamics which does inevitably bring to the heat death of the Universe,*
(b) *the observed cosmic microwave background.*

The most surprising conclusion of the revealed non-stationary state of the Universe is *the existence of the "beginning"*, under which the majority of physicists understand *the beginning of the Universe expansion.*

The cosmologic problem as the problem of the origin and evolution of the Universe has initiated to be analyzed by A.Einstein (after 1917) and now it is connected with papers of many other physicists. The first several authors had been G.Lemaître (who proposed what became known as the Big Bang theory of the origin of the Universe, although he called it his "hypothesis of the primeval atom"), A.Friedman and G.Gamow.

And what namely had been in the "beginning"? Gamow had assumed in 1921 that the expansion had initiated from the super-condensed hot state as a result of the *Big Bang*, to which he and others had ascribed the time moment $t=0$, i.e. the beginning of the Universe history. The initial state in this model is in fact *postulated*. *The nature* of the initial super-condensed hot Universe state *is not known*. Such initial point (or super-small region), in which the temperature, pressure, energy density etc had reached the anomalous huge (almost infinite) values, can be considered as a particular point, where the "physical" processes cannot be described by physical equations and in fact are excluded from the model analysis. Under these conditions the theory of grand unification (or superunification) of all four known interactions (strong, weak, electromagnetic and gravitational) is assumed to act. But no satisfactory superunification has yet been constructed. The superstring theory claims the role of such superunification, but this goal has not yet been achieved [1].

Strictly speaking, namely in the region of this point (from $t = 0$ till $\sim t_0 = 10^{-44}$ sec, where t_0 is the Planck time) is arising the general problem of the world origin and also the choice dilemma: the beginning of the Universe formation from vacuum ("nothing") is *either* a result of the irrational randomness after passing from other space-time dimensions or from other universe, caused by some unknown process, *or* a result of the creation of the expanding Universe (together with the laws of its functioning) by the supreme intelligent design from *nigilo*.

The framework for the standard cosmologic model relies on Einstein's general relativity and on simplifying assumptions (such as homogeneity and isotropy of space). There are even non-standard alternative models. Now there are many supporters of Big Bang models. The number of papers and books on standard versions of the cosmologic Big Bang models is too enormous for citing in this short paper (it is possible to indicate, only for instance, [3-6] for the initial reading in cosmology of the Universe and in the different quasi-classical and quantum approaches in cosmology for description of the creation and the initial expansion of the Universe). However, there is no well-supported model describing the Universe history prior to 10^{-15} seconds or so. Apparently a new unified theory of quantum gravitation is needed to break this barrier but the theory of quantum gravitation is only schematically constructed in the quasilinear approximation. Understanding this earliest era in the history of the Universe is currently one of the most important unsolved problems in physics. Further, over the time interval $\sim 10^{-35}$ s, which is much larger than the Planck time and so can still be considered classically, the Universe was expanding (inflating) much more rapidly than in the known Friedman models. After the inflation, the Universe had been as though developing in accord with the Friedman's scenario [1]. It may be possible to deduce what happened before inflation through observational tests yet to be discovered, and a crucial role at the inflation stage could be played the so-called Λ-term added to the Einstein equations of the General Relativity.

A lot of observations testify that there is exists non-luminous matter in the Universe which manifests itself owing to its gravitational interaction and is present everywhere – both in the galaxies and in the intergalactic space. And what is the nature of dark mass? According to the very popular hypothesis, the role of dark matter is played by the hypothetical WIMPs (Weakly Interacting Massive Particles) with masses higher than protons [1]. There are also exist some other candidates for the role of dark matter (for instance, pseudoscalar particles – axions) [7]. Cosmic strings can be also mentioned [1].

The possibility of the existence of the above-mentioned Λ-term in equations of the General Relativity is now frequently referred to as "dark energy" or quintessence. For $\Lambda > 0$ it "works" as "antigravity" (against the normal gravitational attraction) and testifies to the acceleration of the Universe expansion in our epoch [1, 8].

Moreover, it is worth to underline that many physicists consider that the second law of thermodynamics is universal for all closed systems, including also our Universe as a whole (which is closed in naturalistic one-world view). Therefore the heat death is inevitable (see, for instance, [1] and especially [9]).

There are also versions of the non-standard versions of the cosmologic Big Bang models [10-14]. We shall shortly refer to these models, noting that at least one of them (by B.Setterfield and T.Norman [10]) clearly speaks on the young Universe: They indicate that after the Big Bang the light speed had been gradually decreased approximately 10^6-10^7 times and it was deduced that the velocities of the electromagnetic and radioactive decays had been gradually decreased near 10^7 times too. In [10, 11] it was deduced that after the inflation the Universe had not been really expanding.

ON THE ANTHROPIC PRINCIPLE

From 1973 (and particularly after eighties) the term *"anthropic principle"*, introduced by B. Carter, has become to acquire in the science and out of the science a certain popularity [15, 16]. Carter and other authors had been noted that physical constants must have values in the very narrow interval in order the existence of the biologic life can become possible, and that the measured values of these constants are really found in this interval. In other words, the Universe seems to be exactly such as it is necessary for the origin of the life. If physical constants would be even slightly other, then the life could be impossible.

After meeting such testimonies, a number of scientists had formulated several interpretations of anthropic principle each of which brings the researchers to the worldview choice in its peculiar way. We shall consider here two of them.

According to *the weak anthropic principle* (WAP), the observed values of physical and cosmological constants caused by the necessary demand that the regions, where the organic life would be developed, ought to be possible. And in the context of WAP there is the possibility of choice between two alternatives:

(1) *Either* someone does irrationally believe that there are possible an infinity of universes, in the past, in the present and in the future, and we exist and are sure in the existence of our Universe namely because the unique combination of its parameters and properties could permit our origin and existence.
(2) *Or* someone does (also irrationally) believe that our unique Universe is created by Intelligent Design of a Creator (God) and the human being is also created by Creator in order to govern the Universe.

According to *the strong anthropic principle* (SAP), the Universe has to have such properties which permit earlier or later the development of life. This form of the anthropic principle does not only state that the universe properties are limited by the narrow set of values, compatible with the development of the human life, but does also state that this limitation is *necessary* for such purpose. So, one can interpret such tuning of the universe parameters as the testimony of the supreme intelligent design of a certain creative basis. There is also a rather unexpected interpretation of SAP, connected with the eastern philosophy, but it is not widely known.

CONCLUSIONS

We have presented here only some main cosmological problems, referred to the Universe origin and history and may be their number will increase with time.

Then, there was started to increase the discussion between the supporters of various meta-theoretical (semi-scientific and semi-philosophic) approaches to the problems of the origin of the whole Universe (see, for instance,[2] and also interviews of many researchers of the Big Bang and the Universe history) – between the hypothesis of the Supreme Intelligent Design (the creationism) and the hypothesis of the self-organization of the matter from the lower levels (beginning from the 0-th level, i.e. vacuum) into the higher levels. And to the last doctrine there was adjoined in the XXI c the meta-physical doctrine of the parallel other universes with some kind of interaction between them or with an irrational spontaneous passage of the matter from them to our Universe – those hypothetical universes are or the exactly same as ours (in quantum mechanics), or with other space-time dimensions (in quantum cosmology), or with other values of the physical constants (in other approaches to the co-existing universes, as the naturalistic interpretation of WAP).

The interpretation questions in the considered here grand and great problems of physics and also of all natural sciences are practically inevitably connected with the world-views of the researchers. Therefore, it is quite clear that the strong divergences in the various interpretations and even paradigms of various researchers, especially relating to these grand and great open problems, can be caused by the incompatibility of their world-views.

REFERENCES

[1] Ginzburg, V.L. (1999). What problems of physics and astrophysics seem now to be especially important and interesting (30 years later, already on the verge of XXI century), *Physics – Uspekhi*, **42**, 353-272; (2002) *On some advances in physics and astronomy over the past 3 years*, **45**, 205-211.

[2] Olkhovsky, V.S. (2010) A retrospective view on the history of the natural sciences in XX-XXI, *Natural Science*, v.2, N3, 228-245.

[3] Hartle, J.B. and Hawking, S.W., (1983) Wave function of the Universe, *Phys. Rev. D* **28**, 2960-2975.

[4] Vilenkin, A., (1994) Approaches to quantum cosmology, *Phys. Rev. D* **50**, 2581-2594; gr-qc/9403010.

[5] Kragh H., (1996) *Cosmology and Controversy*, Princeton (NJ), Princeton University Press; ISBN 0-691-02623-8.

[6] Peacock J., (1999) *Cosmological Physics*, Cambridge University Press; ISBN 0521422701.

[7] Ellis, J. (1998) Particle components of dark matter, *Proc. Natl. Acad. Sci. USA*, **95**, 53.

[8] Armendariz-Picon C., Mukhanov V., Steinhardt P.J., Dynamical Solution to the Problem of a Small Cosmological Constant and Late-Time Cosmic Acceleration, *Phys. Rev. Lett.*, **85**, 4438 (2000)

[9] Adams F.C. and Laughlin G., (1997) A Dying Universe: the Long-Term Fate and Evolution of Astrophysical Objects, *Rev. of Mod. Phys.*, **69**, 337-372.

[10] B. Setterfield and T. Norman, *The Atomic Constants, Light and Time,* paper for SRI International, 1987.

[11] V.S. Troitskii, 1987 {in Russian: В.С.Троицкий, *Астрофизика и космическая наука*, 1987, т.139, стр.389-411}.

[12] J.P. Petit (1988), An interpretation of cosmological model with variable light velocity, *Mod. Phys. Lett. A* **3**(16), 1527–1532. doi:10.1142/S0217732388001823; J.P. Petit (1988), Cosmological model with variable light velocity: the interpretation of red shifts, *Mod. Phys. Lett. A* **3**(18), 1733–1744. doi:10.1142/S0217732388002099; J.P. Petit, M. Viton (1989), Gauge cosmological model with variable light velocity. Comparizon with QSO observational data, *Mod. Phys. Lett. A* **4**(23): 2201–2210. doi:10.1142/S0217732389002471.

[13] J. Moffat (1993), Superluminary Universe: A Possible Solution to the Initial Value Problem in Cosmology, *Int.J. Mod. Phys. D* **2**(3): 351–366. doi:10.1142/S0218271893000246; arXiv:gr-qc/9211020.

[14] A. Albrecht, J. Magueijo (1999), A time varying speed of light as a solution to cosmological puzzles, *Phys. Rev. D* **59**: 043516; arXiv:astro-ph/9811018.

[15] Carter, B., (1974) "Large Number Coincidences and the Anthropic Principle in Cosmology", *IAU Symposium* **63**: *Confrontation of Cosmological Theories with Observational Data*, Dordrecht: Reidel, 1974.

[16] Barrow J.D., and Tipler F.J., (1986) *The Anthropic Cosmological Principle*, Clarendon Press, Oxford, 1986.

In: The Big Bang: Theory, Assumptions and Problems ISBN: 978-1-61324-577-4
Editors: J. R. O'Connell and A. L. Hale © 2012 Nova Science Publishers, Inc.

Chapter 7

THE QUANTUM THEORY OF THE BIG BANG: EFFECTIVE THEORY OF QUANTUM GRAVITY

Subodha Mishra
Department of Physics, Institute of Technical Education and Research,
Siksha O Anusandhan University, Jagamohan Nagar,
Bhubaneswar-751030, India

Abstract

Considering the fact that our universe might have been born out of the *Big Bang* and formed subsequently out of a system of fictitious self-gravitating particles, fermionic in nature, we formulate a quantum mechanical theory of the universe which turns out to be an effective theory of quantum gravity. This theory is the special relativized quantized Newton-Cartan theory which is the nonspecial relativistic limit of Einstein's General Theory of Relativity. We are able to obtain a compact expression for the radius R_0 of the universe by using a model density distribution $\rho(r)$ for the particles which is singular at the origin. This singularity in $\rho(r)$ can be considered to be consistent with the so called Big Bang theory of the universe. By assuming that Mach's principle holds good in the evolution of the universe and taking the age of the present universe to be $\tau_0 \simeq 14 \times 10^9 yr$, we obtain the total mass of the universe ($M \simeq 1.0 \times 10^{23} M_\odot$), $M_\odot$ being the solar mass and the value of ratio of the variation of the universal gravitational constant with time to G in extremely good agreement with the results obtained by others. Our theory reproduces the ratio of the number of neutrinos to that of nucleons and density of neutrinos in the present universe correctly. We also study quantum mechanically our expanding universe which is made up of gravitationally interacting particles such as particles of luminous matter, dark matter and dark energy as a self-gravitating system using a well-known many-particle Hamiltonian, but only recently shown as representing a soluble sector of quantum gravity. Describing dark energy by a repulsive harmonic potential among the points in the flat 3-space and incorporating Mach's principle to relativize the problem, we derive a quantum mechanical relation connecting temperature of the cosmic microwave background radiation, age, and cosmological constant of the universe. When the cosmological constant is zero, we get back Gamow's relation with a much better coefficient. Otherwise, our theory predicts a value of the cosmological constant 2.0×10^{-56} cm^{-2} when the present values of cosmic microwave background temperature of 2.728 K and age of the universe 14 billion years are taken as input.

The current theory for the origin of the universe is the Big Bang theory[1, 2] according to which the universe is considered to have started with a huge explosion from a superdense and superhot state. Theoretically, it means that the universe has started from a mathematical singularity with infinite density. the expansion of the universe is being described by the Hubble's law[3] which is given by the relation

$$v = H_0 d \qquad (1)$$

where v denotes the radial velocity at which each distant galaxy is receding from us, d is the distance of the galaxy and H_0 is the Hubble constant. The time which would have elapsed since the scale factor $R = 0$ to the present stage of the universe is given by $t_0 = 1/H_0$, where t_0 is called the Hubble time. The Hubble time represents the maximum age of the universe.

In spite of the fact that the Big Bang theory proves to be very successful description of the universe for the whole range of times starting almost from Big Bang to the present age, it has a few serious shortcomings in the sense that using this theory, a number of very obvious questions have been left yet unanswered. The first question involves the ratio of the number of protons and neutrons in the universe to the number of photons. Photons are found mainly in the cosmic back ground radiation, while protons and neutrons form the atomic nuclei of matter that make up the galaxies. From the calculated abundances of the nuclides mentioned earlier, it is concluded[4] that the observed universe contains $\sim 10^9$ photons for every proton or neutron. The standard Big Bang theory (SBBT) does not explain this ratio but instead assumes that the ratio is given as a property of the initial condition. There has been an estimate of the average density of photons in the universe considering the fact that most of the diffuse photons form the microwave background radiation[5, 6, 7]. From this it follows that there should be about 500 photons per cubic centimeter of the universe. It has also been established that there would be an equal number of neutrinos too, per cubic centimeter in the universe. This has not been explained using SBBT. The third question which has not been answered yet by the SBBT is related to the large scale homogeneity of the observed universe. Although the universe we observe is, in many ways, quite inhomogeneous because of the presence of stars, galaxies and cluster of galaxies within it, when looked over very large scales, it apperas to be very homogeneous. The last thing for which there is no explanation in the SBBT is related to the mass density ρ_0 of the universe, which is measured relative to some critical density denoted by ρ_c. If ρ_0 exceeds ρ_c, then the gravitational pull of everything over everything else will be strong enough to halt the expansion, eventually without considering accelerated expansion. This would cause the universe to collapse, resulting in what is sometimes called a Big crunch. If on the otherhand, $\rho_0 < \rho_c$, the universe will go on expanding forever. Cosmologists introduce a quantity, designated by a symbol Ω, which is defined by the ratio of the mass density to the critical mass density (ρ_0/ρ_c). The correct value is $\Omega \approx 1$, where $\Omega = \Omega_m + \Omega_\Lambda$ and $\Omega_m = \rho/\rho_c$, $\Omega_\Lambda = \rho_\Lambda/\rho_c$ and the critical density $\rho_c = 3H^2/8\pi G$.

Assuming the best value of the Hubble constant to be equal to $\approx 70 \ km/s \ /Mpc$, which corresponds to an age of the universe as $t = 13.7 \ Gyr$, we calculate[3] ρ_c which gives $\rho_c = 6.8 \times 10^{-30} \ g/cm^3$. Measurement of the average density of universe ρ_0 is extremely difficult, because the universe consists of all sorts of matter and all of them contribute to ρ_0. Actually now it is known that 70% of total mass comes from the vacuum energy or cosmological constant and baryonic matter amounts to only 5% and the nonbaryonic dark matter constitutes 25% of the critical density of the universe, that is $\rho_{DM} = 1.7 \times 10^{-30} \ g/cm^3$. The neutrinos which are hot or relativistic constitute only 1% of the dark matter. The large scale structure formation implies that most of the dark matter should be cold or nonrelativistic. However, the average density of matter in a galaxy can be determined using the relation $\rho_G = n_G M_G$, where M_G is the mean mass of a galaxy and n_G, the number of galaxies in a unit volume. For some fixed region of the sky, we count the number of galaxies and find n_G. In order to determine M_G, we consider the fact that the stars in a galaxy revolve around the center of galaxy. This is equivalent to the motion of planets around the sun. Hence once we know the rotating velocity of some of the stars in a galaxy and distances from the galactic center, M_G can be found out[4]. Such measurements show that $\rho_G \approx 3 \times 10^{-31} \ g/cm^3$. assuming that the cosmic matter is mainly in galaxies we have $\rho_0 \approx \rho_G$. This means that $\rho_0 < \rho_c$ and hence, the universe will be infinite, open and ever expanding.

The very assumption that matter in the universe is mainly concentrated in galaxies need not be correct. It is very likely that the space between the galaxies is not a vacuum but contains gases or extinguished stars[8]. The question is, just how much of such unknown matter is there in the universe ? Looking at the work of Zwicky[9] relating to the measurement of the mass of coma cluster, we find that the dynamical mass of coma cluster is almost four hundred times its luminosity mass. The luminosity mass of a cluster is determined by measuring the luminosities of the member galaxies of a cluster and then adding up all the masses of the members. As regards the dynamical mass is concerned, this is determined by measuring the relative velocities among the galaxies. In this case, the mean relative velocity is related to the whole cluster. The great discrepancy one notices in the two measured values for the masses of the coma cluster can only be interpreted by the fact that the mean mass of the coma cluster is not only contributed by the visible galaxies, but by a large amount of invisible matter within the cluster. The mass measured from the luminosities includes only the mass in the light emitting regions and does not include the mass that exists in regions not emitting light. If non light emitting region contains a large amount of matter, then only the luminosity mass would be much less than dynamical mass. Thus, from Zwicky's work, we get an idea about the missing mass or about the Dark Matter (DM) of the present universe.

Since light emitting bodies must have a lot of baryons and the dominant contribution to the cosmic mass comes from the DM, it is therefore believed that the particles constituting the DM can not be baryons. Besides, particles of DM are expected not to participate in electromagnetic interactions. If they are governed by weak interactions, then only they can not be detected in today's laboratory experiments. Out of the list of non-baryonic kind of particles that have been suggested to constitute the DM of the universe, the neutrinos which are hot or relativistic constitute only 1% of the dark matter. Other dark matter particle candidates are weakly interacting massive particles (WIMP)[10] which may be massive with mass in order of GeV considered to be the most suitable candidates to form the particles for the DM. The neutrinos have a non-zero rest mass. A recent study[11] by Schramm and Steigmann suggests that mass of the neutrino must be between $(4 \ to \ 20 \ eV/c^2)$. Actually the recent[12] upper bound on the neutrino mass is $m_\nu = 0.20 \ eV$ and the lower bound is $m_\nu > 10^{-3} \ eV$.

Though the cosmological constant can contribute 70% to the critical density, the recent realization is that since the scale of cosmological constant and neutrino masses are of the same order of magnitude, some kind of radiative models of neutrino masses which require Higgs scalar, could be a dark matter candidate. Though the present upper limit on neutrino mass does not permit standard neutrinos to be dark matter candidates, right-handed sterile neutrinos could be a dark matter candidate which again depend on their mass mixing with active neutrinos[13]. Considering the fact that the major constituents of the present universe are dark matter particle, which might have been formed out of some fictitious particles which represent effectively the DM and Cosmological constant contribution, and these particles are of certain mass that had filled the early universe at a time some two minutes after the Big Bang, we, in this chapter, have tried to calculate the mass, mean density and radius of the universe by treating the universe just as a system made out of these particles which are self gravitating. The present calculation is based on making an intuitive choice for the distribution of particles in the universe, characterized by a distribution function having a singularity at the origin. Such a form of single particle density seems to be consistent with the concept like the Big Bang theory of the Universe. The present calculation is the result of our earlier study[14, 15, 16] where such a form of singular density distribution has been used to calculate the binding energy of a system of self gravitating particles like the neutron stars and white dwarfs etc., having no source for nuclear power generation present at their cores. In all these cases our theory has proved to be of a great success due to the fact that it has correctly reproduced the results for the binding energies and the radii of the neutron stars etc, that agree with those obtained by other workers. In the present work, we assume the fictitious particles to be fermionic in nature, carrying a tiny mass m, which interact among themselves through gravitational forces. Considering the fact that the present age[4] of the universe is about 14×10^9 yr, we have been able to make an estimate for the radius of the universe by adjusting the mass of these fictitious particles. The age of the Universe is determined using the fact that the horizon of the universe has been expanding with a speed equal to the speed of light $'c'$. With this, we not only arrive of a value of $\sim 10^{28}$ cm for the radius of the universe but also we obtain a value of its mass of about $\sim 10^{23} M_\odot$, $M_\odot$ being the solar mass ($M_\odot = 2 \times 10^{33} g$). All these numbers seem to be matching very nicely with the corresponding results known from other theories. Our estimated result for the average mass density of the present universe comes out to be $\approx 10^{-29} g/cm^3$, which is supposed to be the case as per expectation.

1. The Quantum Theory of the Universe

If we visualize the present universe to be a system composed of some self-gravitating fictitious particles, each of mass m, interacting through pair-wise gravitational interactions, then the Hamiltonian of the system can be written as

$$H = -\sum_{i=1}^{N}(\frac{\hbar^2}{2m})\nabla_i^2 + \frac{1}{2}\sum_{i=1, i\neq j}^{N}\sum_{j=1}^{N} v(|\vec{X}_i - \vec{X}_j|), \qquad (2)$$

where $v(|\vec{X}_i - \vec{X}_j|) = -g^2/|\vec{X}_i - \vec{X}_j|$, having $g^2 = Gm^2$, G being Newton's universal gravitational constant. As we will discusse later, recently it is shown that the Newton-Cartan limit (that is $c \to \infty$ and the 3-space being flat though the spacetime is still curved) of Einstein' General Theory of Relativity can be exactly quantized[17] without any socalled time problem and it leads to the above Hamiltonian Eq.(2).

The effective particles which constitute the universe are treated as fermions. In our calculation, we use the zero temperature formalism which can be justified on the ground that the present cosmic background temperature, being 2.728K, is close to 0K. It is well known that the temperature in the early universe was extremely high. In order to simulate the condition in the early universe, using the present theory, we have to assume that the present total kinetic energy of the system was then available in the form of an equivalent temperature T by virtue of the relation $<KE> = (\frac{3}{2})Nk_BT$, the factor $(\frac{3}{2})Nk_B$ comes from all the three degrees of freedom of a particle.

In the Thomas-Fermi approximation[14], we evaluate the total kinetic energy of the system using the relation :

$$<KE> = \left(\frac{3\hbar^2}{10m}\right)(3\pi^2)^{2/3}\int d\vec{X}[\rho(\vec{X})]^{5/3}, \qquad (3)$$

where $\rho(\vec{X})$ is the single particle distribution function which the constituent particles obey within the universe, such that

$$\int d\vec{X}\rho(\vec{X}) = N, \qquad (4)$$

N being the total number of these particles in the universe. The total potential energy of the system in the Hartree-approximation is obtained as

$$<PE> = -(\frac{g^2}{2})\int d\vec{X}d\vec{X}'\frac{1}{|\vec{X}-\vec{X}'|}\rho(\vec{X})\rho(\vec{X}') \qquad (5)$$

Evaluation of the integrals shown in Eq.(3) and Eq.(5) is done using a single particle density distribution of the form:

$$\rho(\vec{x}) = \frac{Ae^{-x}}{x^3}, \qquad (6)$$

where $x = (r/\lambda)^{1/2}$, $r = |\vec{X}|$, λ being the variational parameter and A is the normalization constant given as $A = \frac{N}{16\pi\lambda^3}$. Though $\rho(\vec{X})$ is a trial density, still the physics of the behaviour of the density should be incorporated into it while using one. As one can see from Eq.(6), $\rho(\vec{X})$ is singular at the origin. This looks to be consistent with the concept behind the Big Bang theory of the universe. The early universe was not only known to be super hot, but also it was superdense. To account for the scenario at the time of the Big Bang, we have, therefore, imagined of a single-particle density $\rho(x)$ for the system which is singular at the origin ($r = 0$). This is only true at the microscopic level, which is not so meaningful looking at things macroscopically. Although $\rho(x)$ is singular, the number of particles N, in the system is finite. Since the present universe has a finite size, its present density which is nothing but an average value is finite. At the time of Big Bang ($t = 0$), since the scale factor (identified as the radius of the universe) is supposed to be zero, the average density of the system can assume an infinitely large value, implying its superdense state. Having thought of a singular form of single-particle density at the time of the Big Bang, we have tried with a number of singular form of single particle densities of the kind $\rho(\vec{r}) = B\frac{exp[-(\frac{r}{\lambda})^\nu]}{(\frac{r}{\lambda})^{3\nu}}$ where $\nu = 1, 2, 3, 4...or\ \frac{1}{2}, \frac{1}{3}, \frac{1}{4}, ...$ Though the normalization constant B here is a function of ν, for $\nu = \frac{1}{2}$, B=A. Integer values of ν are not permissible because they make the normalization constant infinite. Out of the fractional values, $\nu = \frac{1}{2}$ is found to be most appropriate, because, it has been shown in our earlier paper that it gives the expected upper limit for the critical mass of a neutron star[15] beyond which black hole formation takes place and other parameters of the universe[14] satisfactorily correct. Also because, if ν goes to zero (like $1/n$, $n \to \infty$), $\rho(r)$ would tend to the case of a constant density as found in an infinite many-fermion system. In view of the arguments put forth above, one will have to think that the very choice of our $\rho(r)$ is a kind of ansatz in our theory, which is equivalent to the choice of a trial wave function used in the quantum mechanical calculation for the binding energy of a physical system following variational techniques. As mentioned earlier, singularity at $r = 0$ in the single particle distribution has nothing to do with the average particle density in the system, which happens to be finite (because of the fact that N is finite and volume V of the visible universe is finite), and hence it is not going to affect the large scale spatial homogeneity of the observed universe. Having accepted the value $\nu = \frac{1}{2}$, the parameter λ associated with $\rho(r)$ is determined after minimizing $E(\lambda) = <H>$ with respect to λ. This is how, we are able to find the total energy of the system corresponding to its lowest energy state.

After evaluating the integrals given in Eq.(3) and Eq.(5), the expression for total energy of the universe, $E(\lambda)$, becomes

$$E(\lambda) = \left(\frac{12}{25\pi}\right)\left(\frac{\hbar^2}{m}\right)\left(\frac{3\pi N}{16}\right)^{5/3}\frac{1}{\lambda^2} - \left(\frac{g^2 N^2}{16}\right)\frac{1}{\lambda} \qquad (7)$$

Now, minimizing $E(\lambda)$ with respect the λ and then evaluating it at $\lambda = \lambda_0$, where λ_0 is the value of λ at which the minimum occurs, the total binding energy of the universe, corresponding to its lowest energy state, becomes

$$E_0 = -(0.015442)N^{7/3}\left(\frac{mg^4}{\hbar^2}\right) \qquad (8)$$

In view of our earlier works[15, 14] and as discussed in the previous chapter, here also we identify $2\lambda_0$ result with the radius R_0 of the universe. It must be noted that the size of any compact object (either an atom or a star) is not well defined in quantum theory. The justification regarding the identification of the radius R_0 of a star with $2\lambda_0$ follows from the consideration of the so called tunneling effects used in quantum mechanics. Classically, it is known that a particle has a turning point where the potential energy becomes equal to the total energy[18, 19]. Since the kinetic energy and therefore the velocity are equal to zero at such a point, the classical particle is expected to be turned around or reflected by the potential barrier. For example, considering the case of an electron in the hydrogen atom ground state such classical turning point occurs where the potential $V(r) = -e^2/r = E_{total} = -e^2/2a_0$; that is at $r = 2a_0$. Quantum mechanically, the probability distribution $r^2\rho(r)$ has a non-zero value for $r > 2a_0$; that is, the electron has access to the region $r > 2a_0$ which is forbidden by classical theory. Such penetration or tunneling into or through the potential energy barriers is typical of results of quantum theory. If the electron had a value of $r > 2a_0$, then its kinetic energy would have to be negative to satisfy the condition $E_{total} = T + V$,with $V > E_{total}$. Since negative kinetic energy is physically absurd, $r = 2a_0$ is to be identified as the classical radius. Using the above idea, from the present theory one can easily see that at $\lambda = 2\lambda_0$, the potential energy of the system becomes equal to the the total energy, there by proving that the radius of the system $R_0 = 2\lambda_0$.

$$R_0 = 2\lambda_0 = (\frac{\hbar^2}{mg^2}) \times (4.047528)/N^{1/3} \tag{9}$$

2. Mach's Principle

We now invoke Mach's principle[20] which states that inertial properties of matter are determined by the distribution of matter in the rest of the universe. Mach had the view[1] that the velocity and acceleration of a particle would be meaningless had the particle been alone in the universe. We have to talk of acceleration only with respect to other bodies, just like we talk of velocities with respect to other bodies. This means that the inertial mass of a particle is the result of the particle feeling the presence of other particles in the universe. If we denote the inertial mass of the particle by m_{inert}, it is to be determined by its response to accelerated motion. As far as the universe is concerned, the distance particles beyond the Hubble length, which we take as the radius of the visible universe R_0 are unobservable and therefore do not contribute to the determination of local inertial mass. If M denotes the gravitational mass of the observable universe, the gravitational energy of the particle is given by $E_{gr} = \frac{GMm_{grav}}{R_0}$, where m_{grav} is the gravitational mass of the particle, that is, the mass determined by its response to gravity. In accordance with the spirit of Mach's principle, one must have $E_{gr} = \frac{GMm_{grav}}{R_0} = m_{inert}c^2$, where $m_{inert}c^2$ is the intrinsic energy of the particle. Since m_{inert} and m_{grav} are taken to be equal both in Newtonian theory and in the General Theory of Relativity, we have Mach's principle[20] expressed through the relation as

$$\frac{GM}{R_0c^2} \approx 1 \tag{10}$$

and using the fact that the total mass of the universe $M = Nm$, we are able to obtain the total number of particles N constituting the universe.

3. Results

Using Eq.(10) in the expression $R_0 = 2\lambda_0$, Eq.(9) one arrives at

$$N = 2.8535954 \left(\frac{\hbar c}{Gm^2}\right)^{3/2} \tag{11}$$

Substituting the expression for N from Eq.(11) in Eq.(9), the expression for R_0 becomes

$$R_0 = 2.8535954 \left(\frac{\hbar}{mc}\right)\left(\frac{\hbar c}{Gm^2}\right)^{1/2} \tag{12}$$

Using Eq.(12), we have made an estimation for the the radius of the present universe, R_0, by varying m, the mass of the fictitious particle and using the measured value for the gravitational constant G, that is by taking $G = 6.67 \times 10^{-8}\ dyn\ cm^2\ g^{-2}$. We have chosen a set of four values for m and calculate R_0, from which the age of the Universe

τ_0 is estimated using the relation $R_0 \simeq c\tau_0$. All these are shown in Table-I. Accepting $\tau_0 = 15 \times 10^9$ yr, to be the most correct value for the age, we find that this corresponds to a value of $R_0 \simeq 1.422 \times 10^{28}$ cm. This is obtained from Eq.(12) by choosing $m = 1.23891 \times 10^{-35}$ g. For this m, we calculate N following Eq.(11), which gives $N \sim 10^{91}$. This gives rise to the total mass of the universe as $M \simeq Nm \sim 10^{23} M_\odot$, $M_\odot$ being the solar mass, whose value is 2×10^{33} g.

Table I.

$m \times 10^{-35} g$	$R \times 10^{28} cm$	$N \times 10^{91}$	$M \times 10^{56} gm$	$\tau_0 \times 10^9 yr$
1.07299	1.896	2.38429	2.55832	20
1.23891	1.422	1.54865	1.91875	15
1.51744	0.948	0.84297	1.27916	10
2.14598	0.474	0.29804	0.639588	5

Using the results for M and R of the present universe of age $\approx 15 \times 10^9$ yr from Table-I, nucleon number density, the ratio of number of nucleons to number of photons etc. are calculated which are given in Table-II.

We now calculate the variation of the gravitational constant G with respect to time, $\dot{G}(t)$. For that, we, use the expression for G, as found from Eq.(12), which is

$$G = K/R_0^2, \tag{13}$$

where

$$K = (8.1430067)\left(\frac{\hbar^3}{m^4 c}\right) \tag{14}$$

Further ,we assume that the above expression for G is also valid for anytime 't' provided we take the value of R_0 at that time. Using Eq.(13), we therefore, find

$$\dot{G} = \left(\frac{\partial G}{\partial R_0}\right) \times \left(\frac{\partial R_0}{\partial t}\right) \simeq c\left(\frac{\partial G}{\partial R_0}\right) = \frac{-2cK}{R_0^3} \tag{15}$$

Following this, we make an estimation of $(\frac{\dot{G}}{G})$, for $m = 1.24 \times 10^{-35}$ g which gives

$$\left(\frac{\dot{G}}{G}\right) = -1.3 \times 10^{-10} yr^{-1} \tag{16}$$

The above value is in extremely good agreement with the recent estimates[6]. This is roughly also the value reported by Van Flandern[22] following an analysis of the data relating to the effects of tidal friction on the elements and shape of the lunar orbit of the earth-moon system. However Shapiro et al[23], have arrived at a limiting value $\mid (\frac{\dot{G}}{G}) \mid \leq 4 \times 10^{-10} yr^{-1}$, based on a method which is used for monitoring the planets for a possible secular increase in their orbital periods by employing a radar reflection system between the Earth,Venus and Mercury.

Table II.

$M_n \times 10^{56}g$	$N_n \times 10^{78}$	$N_\nu/N_n \times 10^9$	$N_\nu \times 10^{88}$	N_ν/V	$M_\nu \times 10^{56}g$	$m_\nu \times 10^{-32}g$
0.03 (1.91875)	3.4406	3.344	1.15	950	1.8612	1.6176
0.03 (1.91875)	3.4406	3.125	1.08	890	1.8612	1.7309
0.03 (1.91875)	3.4406	1.740	0.60	500	1.8612	3.0900
0.03 (1.91875)	3.4406	1.000	0.34	285	1.8612	5.4090

4. Discussion of Results

In arriving at the various results of the present theory, we have accepted the value for the age of the universe τ_0 to be $\sim 15 \times 10^9 yr$. The mass of the constituent particles corresponding to this is found to be $m \sim 1.24 \times 10^{-35}g$. It is also to be noted that in the present theory we have treated the constituent particles as fermions. For a mass of $m \sim 1.24 \times 10^{-35}g$, the radius of the universe becomes $R_0 \sim 1.422 \times 10^{28}$ cm, which is considered to be an acceptable result, and total mass of the universe M_0 becomes $M_0 \sim (1 \times 10^{23}) M_\odot$, $M_\odot$ being the solar mass. If we assume the mass 3% of M to be solely due to the nucleons that are there in the today's universe, it would correspond to $\sim 10^{78}$ nucleons to be present. This is nothing but the famous Eddington result[21, 20]. Let us now look at our calculated value for the average mass density of the universe. From Table-I we see that, corresponding to $m \sim 1.23891 \times 10^{-35}g$ the average mass density of the universe is $\sim 10.6 \times 10^{-30} g/cm^3$. This, we find to be almost thirty times larger then the observed mass density, ρ_0, of the present universe, $\rho_0 \sim 3 \times 10^{-31} g/cm^3$ where the latter is being thought to be arising solely due to nucleons and other heavy elements etc, present. Let us now assume that out of the entire mass M_0 of the universe, some percentage of it has gone into the formation of the nucleons and other heavy elements present today and the remaining part has been lying in the form of Dark Matter in the present universe. This amounts to saying that the total number of fictitious particles responsible in the formation of the early universe have been subsequently converted to appear in the present universe as nucleons and other heavy nuclei and some other form of particles contributing to the Dark Matter of the universe. Let us accept the view that not more than three percent of the entire mass of the universe is contributed by the nucleons and other heavy nuclei[9], the contribution from the heavy nuclei being negligibly small compared to that of the nucleons. This would therefore mean that the remaining ninety-seven percent of its total mass constitutes what is known as the Dark Matter and Dark energy of the universe. Having accepted this picture, we have calculated the ratio of the number of neutrinos to nucleons, (N_ν/N_n), and the neutrino number per unit volume of the universe, (N_ν/V), etc, V being the volume of the universe.

From the above table-II we find that the calculation done by assuming that only three percent of the total mass of the universe constitutes the observed density of the universe, is the most appropriate one. Because, for this case, the average density of the nucleons (which constitute the observable matter of the universe) comes out to be $\rho_0 \sim 4.8 \times 10^{-31} g/cm^3$, which is considered close to the accepted value. This corresponds to an age of 15 billion years for the universe, and for this the total nucleon number of the universe becomes $\sim$

3.5×10^{78}. This being so, the remaining ninety seven percent of the total mass of the universe is considered to be solely due to the Dark Matter and Dark Energy present within it. Accepting the fact that this mass is generated by the neutrinos present in the universe, we have made an estimation of the ratio (N_ν/N_n) and (N_ν/V) by varying the mass of the neutrinos. From table-II, one can see that for a neutrino mass of $m_\nu \sim 3.09 \times 10^{-32}$ g, one arrives at $(N_\nu/N_n) \simeq 1.74 \times 10^9$ and $(\frac{N_\nu}{V}) \simeq 500$. A mass of $m_\nu = 3.09 \times 10^{-32} g$ corresponds to an energy of $14\ eV$, which is again found to be the right order of magnitude for the upper limit of the neutrino mass.

Calculating the value for the average mass density of the present universe (Table-II) which comes out to be $\rho_U \simeq 10.6 \times 10^{-30} g/cm^3$, one can clearly see that it is obviously much larger than the observed mass density ρ_0 ($\rho_0 \simeq 4.8 \times 10^{-31} g/cm^3$). But it is interesting to note that the total mass density including the cosmological constant contribution (which is equivalently included here through the fictitious particles) is of the correct order of magnitude and equal to the critical density $\rho_U = (\rho_c \sim 6.8 \times 10^{-30}\ g/cm^3)$. In the next chapter we will include the contribution due to cosmological constant explicitly.

To conclude, we have shown that the results of the present calculation are obtained by choosing a singular form of single particle density for the particles constituting the universe and a singular density is consistent with the idea relating to the Big Bang theory. Since the standard Big Bang theory so far has not succeeded to explain for the ratio of $\approx 10^9$ for (number of photons/number of nuclei) and here we do reproduce the above number correctly, the present work seems to be justifying the so called Big Bang theory of the universe.

5. The Quantum Theory of Gravity

Our expanding universe is made up of gravitationally interacting particles which are described by particles of luminous matter, dark matter and dark energy. Representing dark energy by a repulsive harmonic potential among the points in the flat 3-space, we derive a quantum mechanical relation connecting, temperature of the cosmic microwave background radiation, age, and cosmological constant of the universe. When the cosmological constant is zero, we get back Gamow's relation with a much better coefficient. Otherwise, our theory predicts a value of the cosmological constant 2.0×10^{-56} cm^{-2} when the present values of cosmic microwave background temperature of 2.728 K and age of the universe 14 billion years are taken as input. We study[16, 15, 14] the self-gravitating system such as the universe using a well-known many-particle Hamiltonian, which is known from the early days of quantum mechanics, from a condensed matter point of view by using a quantum mechanical variational approach. This can also be viewed as a novel way of looking at the self-gravitating systems and it not only reproduces the results known from Einstein's General Theory of Relativity but also goes beyond by predicting certain relations and specifically the value of the cosmological constant. Instead of looking at the systems through the space-time dynamics, this theory treats the energy of the system directly. The above Hamiltonian has only recently been derived[17] as representing the exactly soluble sector of quantum gravity. Infact from the quantum gravity point of view after quantizing the GTR in the $c \to \infty$ (Newton-Cartan theory with spacial vanishing curvature), we have explicitly the above known Hamiltonian. We then special relativize the problem by using

Mach's principle in case of the universe. Figure 1 represents the underling physical theories relating to full-blown quantum theory of gravity (denoted by FQG) which is still an elusive one. But ours is an effective theory of quantum gravity with vanishing spacial curvature ETQGK0 (denoted by a circle with squares inside) of this FQG with special-relativization done on the problem which is described by the quantum many-particle Hamiltonian obtained by quantizing the Newton-Cartan theory which we write as NQG (Fig. 1). Any general-relativistic exotic theory "theory of everything" (string theory, loop quantum gravity etc) would not be physically relevant if it does not reduce to Newton quantum gravity interacting with quantum fields in the Galilean-relativistic limit. Any valid theory of quantum gravity must reduce[17] to NQG (Fig. 1) in the "$c \to \infty$" limit, to GTR in the "$\hbar \to 0$" limit and to QTF in the "$G \to 0$" limit. Since in our theory, we relativise the NQG, the elusive full-blown quantum theory of gravity must be consistent with the results obtained by our approach. The directions of research have been to go from GTR to FQG or from QTF to FQG. But the third possibility of going from NQG/QNG to FQG by special relativising the Newton-Cartan quantum gravity (undoing the $c \to \infty$ limit) opens up an exciting yet unexplored direction in the research of quantum gravity.

An expanding system of self-gravitating particles can be described by the Hamiltonian which is given as

$$H = -\sum_{i=1}^{N} \frac{\hbar^2}{2m}\nabla_i^2 + \frac{1}{2}\sum_{i=1,\, i\neq j, j=1}^{N}\sum^{N} v(|\vec{X}_i - \vec{X}_j|) - \sum_{i=1}^{N}\Lambda c^2|\vec{X}_i|^2 \qquad (17)$$

where $v(|\vec{X}_i - \vec{X}_j|) = -g^2/|\vec{X}_i - \vec{X}_j|$, with $g^2 = Gm^2$, G being the universal gravitational constant and m the mass of the constituent particles. Incase of our universe, the last term in the above Hamiltonian describes the expansion due to the Dark Energy, which can be taken as the nonrelativistic limit of the Λ term known as Einstein's cosmological constant term in General Theory of Relativity.

6. Present Universe: Exactly Soluble Sector of Quantum Gravity

In this section we describe how the Hamiltonian given by Eq. 17 can be derived as a soluble sector of quantum gravity. In a seminal work[17], recently a generally covariant but Galilean-relativistic quantum theory of gravity has been constructed. It is shown that the Galilean-relativistic limit of as-yet unrealized full quantum theory of gravity with matter is exactly soluble and in the very classical ($c = \infty$) domain the problem of time and causality and other related problems do not exist. A space-time reformulation of the Newtonian theory of gravity known as Newton-Cartan theory of gravity is obtained when the nonrelaivistic limit of Einstein's general theory of relativity is considered. The Newton-Cartan[25] theory lies between that of special theory of relativity with specetime completely fixed and general relativity with no background structure. The NC theory has two fixed and degenerate metrics, one temporal and another spatial metric but with a dynamical connection field like GTR metric connection field. Newton-Cartan theory of gravity is in general the true Galilean-relativistic form of GTR and it is a local field theory. It has a mutable space-time unlike Newtonian theory of gravity of immutable spacetime and in Newtonian-Cartan

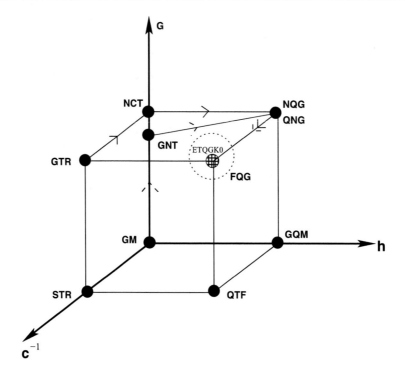

Figure 1. The great dimensional figure of physics indicating the fundamental roles played by G, h and c in various basic physical theories. The theories are GM=Galilean mechanics, STR=special theory of relativity, GTR=general theory of relativity, GQM=Galilean quantum mechanics, QTF=quantum theory of relativistic fields, NCT=Newtonian-Cartan theory, GNT=Galilean Newtonian theory, NQG=Newtonian quantum gravity, QNG=quantum Newtonian gravity. It turns out that when the NCT with (flat 3-space) is quantised the resulting quantum many-particle Hamiltonian is same as one would write the quantum many-particle Hamiltonian for a system of self-gravitating particles that is what we denote as QNG. Here FQG=The elusive full-blown quantum gravity represnted by the bigger circle with broken lines. If FQG turns out to require some additional fundamental constants-like the constant $\alpha' \equiv (2\pi T)^{-1}$ where T is the string tension of the string theory then instead of this cube of theories we will have a hypercube. But, in that case NQG would be a limiting theory of FQG with respect to total $\alpha' \to 0$ and $c \to \infty$. The position denoted by a circle with squares inside denotes our effective theory of quantum gravity (ETQGK0) in the limiting case when special relativisation is incorporated but the spacial curvature is zero, ETQGK0$\subset$ FQG. One can note two tracks to this limiting theory one shown by solid-arrow-heads and other by broken-arrow-heads. The cube has been adopted and modified from Ref.[17].

theory the instantaneous gravitational interactions between bodies propagate through the spacetime.

The connection field of Newtonian-Cartan theory depends on the distribution of matter. The quantum mechanically treated matter impels that the NC connection should be treated also quantum mechanically. So here a quantum theory of gravity is constructed at the

Galilean-relativistic level such that superposition principle holds for the states of matter and NC connection field. In the seminal paper[17] the following action functional in the inertial frame is constructed

$$\mathcal{I} = \int dt \int d\vec{x} \left[\frac{\Phi \Delta \Phi}{8\pi G} + \frac{\hbar^2}{2m} \delta^{ab} \partial_a \Psi \partial_b \Psi + \frac{i\hbar}{2}(\Psi \partial_t \bar{\Psi} - \bar{\Psi} \partial_t \Psi) - m\Psi\bar{\Psi}\Phi \right] \quad (18)$$

where $k = 0$ that is a flat space is considered and $\Lambda = 0$ is taken for simplicity.

When we extremize this functional by varying the scalar potential Φ, we get the Newton-Poisson equation

$$\Delta \Phi = \frac{4\pi G}{<\psi|\psi>} m\psi\bar{\psi}, \quad (19)$$

where $\psi := \sqrt{2\pi G}\Psi$. Also the extremization of the action with respect to the matter field Ψ gives the Schrodinger equation

$$i\hbar \frac{\partial}{\partial t} \psi = \left[-\frac{\hbar^2 \Delta}{2m} + m\Phi \right] \psi \quad (20)$$

These two equations describe a single Galilean-relativistic particle gravitationally interacting with its own Newtonian field. Solving the first equation gives

$$\Phi(\vec{x}) = -Gm \int d\vec{x}' \frac{\bar{\psi}(\vec{x}')\psi(\vec{x}')}{|\vec{x}' - \vec{x}|} \quad (21)$$

This when substituted in the second equation we obtain

$$i\hbar \frac{\partial}{\partial t} \psi(\vec{x}, t) = -\frac{\hbar^2 \Delta}{2m} \psi(\vec{x}, t) - Gm^2 \int d\vec{x}' \frac{\bar{\psi}(\vec{x}', t)\psi(\vec{x}', t)}{|\vec{x}' - \vec{x}|} \psi(\vec{x}, t) \quad (22)$$

But when ψ is written in second quantized form $\hat{\psi}$ as a annihilation operator satisfying $[\hat{\psi}(\vec{x}), \hat{\psi}^\dagger(\vec{x}')] = \delta(\vec{x} - \vec{x}')$, it describes a many-particle system in the Heisenberg picture. The system of particles is now described by a Hamiltonian

$$\hat{\mathcal{H}} = -\int d\vec{x} \hat{\psi}^\dagger(x) \frac{\hbar^2 \Delta}{2m} \hat{\psi}(\vec{x}) - \frac{1}{2} Gm^2 \int d\vec{x} d\vec{x}' \frac{\hat{\psi}^\dagger(\vec{x}')\hat{\psi}^\dagger(\vec{x})\hat{\psi}(\vec{x})\hat{\psi}(\vec{x}')}{|\vec{x}' - \vec{x}|} \psi(\vec{x}, t) \quad (23)$$

The Hamiltonian satisfies the Heisenberg equation of motion

$$i\hbar \frac{\partial}{\partial t} \hat{\psi}(\vec{x}, t) = [\hat{\psi}(\vec{x}, t), \hat{\mathcal{H}}] \quad (24)$$

It can be shown that when the Hamiltonian operator $\hat{\mathcal{H}}$ acts on a many-particle state gives

$$<\vec{x}_1...\vec{x}_N|\hat{\mathcal{H}}|\xi> = \left[-\frac{\hbar^2}{2m} \sum_{i=1}^{N} \nabla_i^2 - \frac{1}{2} Gm^2 \sum_{i,j,i\neq j} \frac{1}{|\vec{x}_i - \vec{x}_j|} \right] \times <\vec{x}_1...\vec{x}_N|\xi> \quad (25)$$

This is the quantum mechanical version of the classical manyparticle Hamiltonian with gravitational pair interactions. There is no self interaction since $\hat{\mathcal{H}}_I|\vec{x}> = 0$ and the number operator $\hat{N} = \int d\vec{x} \hat{\psi}^\dagger \hat{\psi}$ commutes with total Hamiltonian and it is conserved.

The Hamiltonian represents a system of N gravitating particles each of mass m interacting through a sum of pair-wise gravitational interaction. We special relativise the problem by using Mach's principle in case of the universe.

We use a variational approach[16, 15, 14] in our theory to study these systems. But the form of the trial densty chosen is so good that the calculated energy obeys the bound earlier known and results are very good. Our theory is provides a frame work to find a relation between time, temperature of the cosmic microwave background radiation and the cosmological constant of the universe. It determines the value of the Cosmological constant dynamically, where as in GTR the Cosmological constant is a free parameter.

7. Cosmological Constant Λ as the Dark Energy

The most important theory for the origin of the universe is the Big Bang Theory[2] according to which the present universe is considered to have started with a huge explosion from a superhot and a superdense stage. Theoretically one may visualize its starting from a mathematical singularity with infinite density. This also comes from the solutions of the type I and type II form of Einstein's field equations[1]. What follows from all these solutions is that the universe has originated from a point where the scale factor R (to be identified as the radius of the universe) is zero at time $t = 0$, and its derivative with time is taken to be infinite at this time. That is, it is thought that the initial explosion had happened with infinite velocity, although, it is impossible for us to picture the initial moment of the creation of the universe. The accelerated expansion of the universe has been conformed by studying the distances to supernovae of type Ia[26, 27]. For the universe, it is being said that the major constituent of the total mass of the present universe is made of the Dark Energy 70%, Dark Matter about 26% and luminous matter 4%. The Dark energy is responsible for the accelerated expansion of the universe since it has negative pressure and produces repulsive gravity. The cosmological constant[28, 20] of Einstein provides a repulsive force when its value is positive. The cosmological constant is also associated with the vacuum energy density[29] of the space-time. The vacuum has the lowest energy of any state, but there is no reason in principle for that ground state energy to be zero. There are many different contributions[29] to the ground state energy such as potential energy of scalar fields, vacuum fluctuations as well as of the cosmological constant. The individual contributions can be very large but current observation suggests that the various contributions, large in magnitude but different in sign delicately cancel to yield an extraordinarily small final result. The conventionally defined cosmological constant Λ is proportional to the vacuum energy density ρ_Λ as $\Lambda = (8\pi G/c^2)\rho_\Lambda$. Hence one can guess that $\rho_\Lambda = \Lambda c^2/8\pi G \approx \rho_{Pl} = c^5/G^2\hbar \sim 5 \times 10^{93}$ g cm^{-3}, where ρ_{Pl} is the Plank density. But the recent observations of the luminosities of high redshift supernovae gives the dimensionless density $\Omega_\Lambda = \rho_\Lambda/\rho_{cr} \equiv \Lambda c^2/3H_0^2 \approx 0.7$ where $\rho_{cr} = 3H_0^2/8\pi G \approx 1.9 \times 10^{-29}$ g cm^{-3}, which implies $\rho_\Lambda = \rho_{Pl} \times 10^{-123}$. This shows that the cosmological constant today is 123 orders of magnitude smaller.

This is known as the 'cosmological constant problem'. In the classical big-bang cosmology there is no dynamical theory[6] to relate the cosmological constant to any other physical variable of the universe. There have been some studies[30, 31, 32] regarding the universe to relate the space-time manifold to somekind of condensed matter systems.

Here by considering[16, 15, 14] the visible universe made up of self-gravitating particles representing luminous baryons and dark matter such as neutrinos (though only a small fraction) which are fermions and a repulsive potential describing the effect of Dark Energy responsible for the accelerated expansion of the universe, we in this chapter derive quantum mechanically a relation connecting temperature, age and cosmological constant of the universe. When the cosmological constant is zero, we get back Gamow's relation with a much better coefficient. Otherwise using as input the current values of $T = 2.728\ K$ and $t = 14 \times 10^9\ years$, we predict the value of cosmological constant as $2.0 \times 10^{-56}\ cm^{-2}$. Note that Λ is a completely free parameter in General Theory of Relativity. Also it is interesting to note that we obtain not only the value of the cosmological constant but also the sign of the parameter correct though it is a very small number.

8. Mathematical Formulation without Λ

We in this section derive a relation connecting temperature and age of the universe when cosmological constant is zero, by considering a Hamiltonian[16, 15, 14] used by us some time back for the study of a system of self-gravitating particles which is given as:

$$H = -\sum_{i=1}^{N}\left(\frac{\hbar^2}{2m}\right)\nabla_i^2 + \frac{1}{2}\sum_{i=1,\ i\neq j, j=1}^{N}\sum^{N} v(|\vec{X}_i - \vec{X}_j|) \quad (26)$$

where $v(|\vec{X}_i - \vec{X}_j|) = -g^2/|\vec{X}_i - \vec{X}_j|$, with $g^2 = Gm^2$, G being the universal gravitational constant and m the mass of the effective constituent particles describing the luminous matter and dark matter whose number is $N = \int \rho(\vec{X})d\vec{X}$.

As done in the earlier section, the total energy $E(\lambda)$ of the system which is given as

$$E(\lambda) = \frac{\hbar^2}{m}\frac{12}{25\pi}\left(\frac{3\pi N}{16}\right)^{5/3}\frac{1}{\lambda^2} - \frac{g^2 N^2}{16}\frac{1}{\lambda} \quad (27)$$

We minimize the total energy with respect to λ. Differentiating this with respect to λ and then equating it with zero, we obtain the value of λ at which the minimum occurs. This is found as:

$$\lambda_0 = \frac{72}{25}\frac{\hbar^2}{mg^2}\left(\frac{3\pi}{16}\right)^{2/3}\frac{1}{N^{1/3}} \quad (28)$$

Evaluating Eq.(27) at $\lambda = \lambda_0$, the total binding energy of the system is found as

$$E_0 \simeq -(0.015442)N^{7/3}\left(\frac{mg^4}{\hbar^2}\right) \quad (29)$$

Considering the case of the two-particle system (N=2), from Eq.(29), we find

$$E_0 = -(0.077823)\left(\frac{mg^4}{\hbar^2}\right) \quad (30)$$

This is seen to be quite high compared to the actual binding energy of the two-body system whose value is (-0.25) $\left(\frac{mg^4}{\hbar^2}\right)$. Comparing the two results, one should not consider Eq.(29) to be a drawback of the present theory, because it is supposed to be very accurate for very large N. Looking at Eq.(29), we find that E_0 varies as $N^{7/3}$ where N is the particle number. Such a dependence of the binding energy for the system on N was also found by Levy-Leblond[33] by assuming the particles to be fermions and looking at the distribution of N-points on a cubic lattice he was able to obtain both an upper and an lower bound for the binding energy of the system which for large N were given as

$$-(0.5)N^{7/3}\left(\frac{mg^4}{\hbar^2}\right) \leq E_0 \leq -(0.001055)N^{7/3}\left(\frac{mg^4}{\hbar^2}\right) \quad (31)$$

Anyway, comparing our result, as shown in Eq.(29), with Eq.(31), we find that it does not violate the inequalities established by Levy-Leblond[33].

Following the expression for $<KE>$ (as done in the earlier section and also according to virial theorem) evaluated at $\lambda = \lambda_0$, we write down the value of the equivalent temperature T of the system, using the relation

$$T = \frac{2}{3k_B}\left[\frac{<KE>}{N}\right] = \frac{2}{3k_B}(0.015442)N^{4/3}\left(\frac{mg^4}{\hbar^2}\right) \quad (32)$$

The expression for the radius R_0 of the universe, as found by us earlier[15, 14], is given as

$$R_0 = 2\lambda_0 = 4.047528\left(\frac{\hbar^2}{mg^2}\right)/N^{1/3} \quad (33)$$

Our identification of the radius R_0 with $2\lambda_0$ is based on the use of socalled quantum mechanical tunneling[18] effect. Classically, it is well known that a particle has a turning point where the potential energy becomes equal to the total energy. Since the kinetic energy and therefore the velocity are equal to zero at such a point, the classical particle is expected to be turned around or reflected by the potential barrier. From the present theory it is seen that the turning point occurs at a distance $R = 2\lambda_0$.

We now invoke Mach's principle[20] which states that inertial properties of matter are determined by the distribution of matter in the rest of the universe. Mach had the view[1] that the velocity and acceleration of a particle would be meaningless had the particle been alone in the universe. We have to talk of acceleration only with respect to other bodies, just like we talk of velocities with respect to other bodies. This means that the inertial mass of a particle is the result of the particle feeling the presence of other particles in the universe. If we denote the inertial mass of the particle by m_{inert}, it is to be determined by its response to accelerated motion. As far as the universe is concerned, the distance particles beyond the Hubble length which we take as the radius of the visible universe R_0 are unobservable and therefore do not contribute to the determination of local inertial mass. If M denotes the gravitational mass of the observable universe, the gravitational energy of the particle is given by $E_{gr} = \frac{GMm_{grav}}{R_0}$, where m_{grav} is the gravitational mass of the particle, that is, the mass determined by its response to gravity. In accordance with the spirit of Mach's principle, one must have $E_{gr} = \frac{GMm_{grav}}{R_0} = m_{inert}c^2$, where $m_{inert}c^2$ is the intrinsic energy of the particle. Since m_{inert} and m_{grav} are taken to be equal both in Newtonian theory and in the General Theory of Relativity, we have Mach's principle[20] expressed through the relation as $\left(\frac{GM}{R_0c^2}\right) = 1$, and using the fact that the total mass of the universe

$M = Nm$, we are able to obtain the total number of particles N constituting the universe, as

$$N = 2.8535954 \left(\frac{\hbar c}{Gm^2}\right)^{3/2} \tag{34}$$

Now, substituting Eq.(34) in Eq.(33), we arrive at the expression for R_0, as

$$R_0 = 2.8535954 \left(\frac{\hbar}{mc}\right)\left(\frac{\hbar c}{Gm^2}\right)^{1/2} \tag{35}$$

As one can see from above, R_0 is of a form which involves only the fundamental constants like $\hbar, c, G$ and the effective mass m which is ofcourse not fundamental. Now, eliminating N from Eq.(32), by virtue of Eq.(34),we have

$$T = \frac{2}{3}(0.0625019)\left(\frac{mc^2}{k_B}\right) \tag{36}$$

Since we are considering the visible universe, which is actually a patch with a horizon size determined by the speed of light and time that has passed since the Big Bang, we now assume that the radius R_0 of the visible universe is approximately given by the relation

$$R_0 \simeq ct \tag{37}$$

where t denotes the age of the universe at any instant of time.

Following Eq.(35) and Eq.(37), we write m as

$$m = \left(\frac{\hbar^3}{Gc^3}\right)^{1/4}(2.8535954)^{1/2}\frac{1}{\sqrt{t}} \tag{38}$$

It is interesting to see (as shown in Table. III) this variation of mass with time gives approximately the energy and hence the temperature scale of formation of elementary particles in different epochs of nucleosynthesis. We calculate temperatures in different epochs using our Eq.(40) to be derived shortly. This is in good agreement with the calculated values of temperature otherwise known from nucleosynthesis calculations[1, 6]. The period between $t = 7 \times 10^{-5}$ sec and 5 sec is called lepton era, while period before $t = 7 \times 10^{-5} sec$ is hadron era and the early era corresponding to the period $t < 10^{-43}$ sec is known as Planck era.

A substitution of m, from Eq.(38), in Eq.(36), enables us to write

$$\begin{aligned} T &= 0.070388\left(\frac{1}{k_B}\right)\left(\frac{c^5\hbar^3}{G}\right)^{1/4}t^{-1/2} \\ &= 0.06339\left[\frac{c^2}{G\,a_B}\right]^{1/4}t^{-1/2} \end{aligned} \tag{39}$$

This is exactly the Gamow's relation[6, 16] apart from the fact that Gamow's relation had the coefficient 0.41563 instead of 0.06339 as in our expression. Substituting the numerical value of a_B, which is equal to 7.56×10^{-15} $erg\,cm^{-3}K^{-4}$, and the present value for the universal gravitational constant G [$G = 6.67 \times 10^{-8}dyn.cm^2.gm^{-2}$], in Eq.(39),we obtain

$$T = (0.23172 \times 10^{10})t_{sec}^{-1/2}K \tag{40}$$

Table III.

Age of the universe (t) in sec.	Temperature (T) in K as calculated from Eq. (40)	Temperature (T) in K for the formation of elementary particles[1, 6]
5	$\approx 1 \times 10^9$	$\approx 6 \times 10^9 (e^+, e^-)$
1.2×10^{-4}	$\approx 2.1 \times 10^{11}$	$\approx 1.2 \times 10^{12} (\mu^+, \mu^-$ and their antiparticles)
7×10^{-5}	$\approx 2.8 \times 10^{11}$	$\approx 1.6 \times 10^{12} (\pi^0, \pi^+, \pi^-$ and their antiparticles)
1.5×10^{-6}	$\approx 1.9 \times 10^{12}$	$\approx 10^{13}$ (protons, neutron and their antiparticles)
10^{-43}	$\approx 0.73 \times 10^{31}$	$\approx 10^{32}$ (planck mass)

If we accept the age of the universe to be close to $14 \times 10^9 year$, which we have used here, with the help of Eq.(40), we arrive at a value for the Cosmic Microwave Background Temperature (CMBT) equal to $\approx 3.5K$. This is very close to the measured value of 2.728 K as reported from the most recent Cosmic Background Explorer (COBE) satellite measurements[5, 34]. However, if we use Gamow's relation, $t = 956$ billion years is required to obtain the exact value of 2.728 K for the cosmic background temperature from. Using our expression, Eq.(40), we would require an age of 22.832×10^9 year for the universe to get the exact value of 2.728 K. Long back a correction was made to Gamow's relation by multiplying it with a factor of $(\frac{2}{g_d})^{1/4}$ by taking into account the degeneracies of the particles, where $g_d = 9$. This correction effectively multiplies Gamow' relation with a factor of 0.68 and brings back the age of the universe to 425 billion years for the present CMBT. If we multiply our expression by the same factor to correct for the degeneracy of particles, we obtain a value of 2.4 K, which is less than the value of present CMBT. In the next section we see that by including the cosmic repulsion by the part given by cosmological constant we get back 2.728 K, This is physically correct since the cosmological term[20] has the meaning of negative pressure, it adds energy to the system by its tension when the universe expands, though the over all temperature decreases as the universe expands.

9. Entropy, Number of Photons and the Ratio $(\bar{N}_\gamma / N_n)$

In this section we estimate total entropy due to the CMBR, total number of photons and the ratio of number of photons to number of baryons. By virtue of the expression given in Eq. (39), we can rewrite T as

$$T = 0.070388 \left[\left(\frac{c^3}{G}\right) \left(\frac{\pi^2}{60\,\sigma}\right) \right]^{1/4} t^{-1/2} \tag{41}$$

where $\sigma = \frac{\pi^2 k_B^4}{60 \hbar^3 c^2}$ is the Stefan-Boltzman constant and its numerical value is equal to $5.669 \times 10^{-5} erg/cm^2.K^4.sec$, and we have

$$\sigma T^4 \simeq 2.4547 \times 10^{-5} \left(\frac{\pi^2 c^3}{60G}\right) \frac{1}{t^2} \tag{42}$$

The very form of the above equation suggests that the factor in its right hand side (rhs) can be identified as the energy density of the electromagnetic radiation at a time t. The radiation of this form is belived to follow the black- body law. The very agreement of our calculated result with the most accurate value for the temperature of the background radiation shows that age of the universe is very close to $\approx 14 \times 10^9 yr$. This also creates a kind of confidence in us regarding the correctness of our theory compared to others, inspite of its basic difference from the conventional approaches, relating to the evolution of the universe.

Having evaluated the expression in the rhs of Eq. (42), the energy of the electromagnetic radiation radiated per unit area per unit time is given as

$$u = 1.6345 \times 10^{33} \left(\frac{1}{t^2}\right) \tag{43}$$

where t is the age of the universe in sec at any instant of time. The entropy S associated with the microwave back-ground radiation is obtained as[35]

$$S = \frac{16Vu}{3cT} = 2.9058 \left(\frac{V}{T}\right) \times 10^{23} \left(\frac{1}{t^2}\right) \tag{44}$$

Assuming the present universe to be spherical, its volume V is given as $V = (\frac{4\pi}{3})R_0^3$, where R_0 denotes its radius. Taking $R_0 \simeq 1.325 \times 10^{28}$ cm, which corresponds to the age $t = 14 \times 10^9 yr$, since $(R_0 \approx ct)$, the photonic entropy of the present universe is calculated to give

$$S = 1.45 \times 10^{73} \left(\frac{1}{T}\right) erg/deg \tag{45}$$

For $T = 2.728K$, it becomes,

$$S = 0.5 \times 10^{73} \approx 10^{73} erg/deg \tag{46}$$

The equilibrium number of photons[35] associated with the microwave background radiation is given as

$$\overline{N}_\gamma = \frac{V 2\zeta(3)}{\pi^2 \hbar^3 c^3} k_\beta^3 T_0^3 \simeq (410.0)V \tag{47}$$

Following this, the photon density is found to be $(\frac{\overline{N}_\gamma}{V}) \simeq 410$, which is in very good agreement with the estimated value of 400 found[36] by doing a calculation of the total energy density carried by the cosmic microwave background radiation. Using Eq. (47), we have calculated the total number of photons in the present universe, which becomes

$$\overline{N}_\gamma = 0.4 \times 10^{88} \tag{48}$$

Considering the fact that the number of nucleons[37], N_n, in the present universe is $\approx 6.30 \times 10^{78}$, we obtain

$$\frac{\overline{N}_\gamma}{N_n} \simeq 0.063 \times 10^{10} \tag{49}$$

This agrees with the value $(0.14 \sim 0.33) \times 10^{10}$ as speculated by several earlier workers[38] following calculations on baryogenesis. So we find that the theory reproduces the temperature of the cosmic background radiation correctly. Besides, it also succeeds to reproduce the photon density associated with the background radiation, and the value of the ratio $(\bar{N}_\gamma/N_n)$, which nicely match with the results predicted by others.

10. The Generalized Gamow's Relation Connecting t, T and Λ

The cosmological constant term[28, 20] Λ associated with vacuum energy density was originally introduced by Einstein as a repulsive component in his field equation and when translated from the relativistic to Newtonian picture gives rise to a repulsive harmonic oscillator force per unit mass as $\sim (\Lambda c^2)\vec{r}$ between points in space when Λ is positive. The one-body operator corresponding to the potential can be written as $H_\Lambda = -\Lambda c^2 |\vec{X}|^2 \rho(\vec{X})$ where $\rho(X)$ here is measured in the unit of mass density and this term also contains a unit volume. Hence the energy corresponding to this repulsive potential can be written as:

$$<H_\Lambda> = -\int \Lambda c^2 |\vec{X}|^2 \rho^2(\vec{X}) \, d\vec{X} \qquad (50)$$

By including this contribution of H_Λ in the energy, we have the total energy

$$E(\lambda) = \frac{\hbar^2}{m} \frac{12}{25\pi} \left(\frac{3\pi N}{16}\right)^{5/3} \frac{1}{\lambda^2} - \frac{g_\Lambda^2 N^2}{16} \cdot \frac{1}{\lambda} \qquad (51)$$

where $g_\Lambda^2 = g^2 + \frac{3\Lambda c^2}{16\pi}$ and dimension of first term is same as the second term since the second term contains implicitely a product of unit mass and unit volume as can be easily seen by checking the single particle Hamiltonian H_Λ. Calculating as before, we have

$$N = 2.8535954 \left(\frac{1}{Gm^2}\right)^{3/4} \left(\frac{\hbar c}{g_\Lambda}\right)^{3/2} \qquad (52)$$

and

$$R_0 = 2.8535954 \left(\frac{\hbar}{mc}\right)^{1/2} \left(\frac{\hbar G^{1/4}}{g_\Lambda^{3/2}}\right) \qquad (53)$$

Now equating this R_0 with ct we have

$$Gm^{8/3} + \frac{3\Lambda c^2}{16\pi} m^{2/3} - Q = 0 \qquad (54)$$

where $Q = \frac{4.0475279 \hbar^2 G^{1/3}}{c^2} \times \frac{1}{t^{4/3}}$. Using $m' = m^{2/3}$, the above equation can be cast as a quartic equation in m'. We find[39] four analytic solutions for m' and hence for m. Three of the solutions are unphysical and the only solution which is physically correct is given as

$$m = \left(\frac{u^{1/2} + \sqrt{u - 4(u/2 - [(u/2)^2 + Q/G]^{1/2})}}{2}\right)^{3/2} \qquad (55)$$

where

$$u = [r + (q^3 + r^2)^{1/2}]^{1/3} + [r - (q^3 + r^2)^{1/2}]^{1/3} \qquad (56)$$

and $r = \frac{9\Lambda^2 c^4}{2(16\pi G)^2}$, $q = \frac{4Q}{3G}$. Now the Kinetic energy with the degeneracy factor as discussed in the previous section, is given as

$$T = \left(\frac{2}{g_d}\right)^{\frac{1}{4}} \frac{2}{3k_B} \left[\frac{<KE>}{N}\right] = \left(\frac{2}{g_d}\right)^{\frac{1}{4}} \frac{2}{3k_B} (0.015442) N^{4/3} \left(\frac{mg_\Lambda^4}{\hbar^2}\right) \qquad (57)$$

Using Eq.(52) and Eq.(55) in Eq.(57), we finally have the relation,

$$T = 0.0417\left(\frac{2}{g_d}\right)^{1/4} \frac{c^2}{k_B} \frac{[(\{u^{1/2} + \sqrt{4[(u/2)^2 + Q/G]^{1/2} - u}\}/2)^3 + \frac{3\Lambda c^2}{16\pi G}]}{(\{u^{1/2} + \sqrt{4[(u/2)^2 + Q/G]^{1/2} - u}\}/2)^{3/2}} \qquad (58)$$

This relation connects temperature T with time t and cosmological constant Λ since Q is a function of t and u is also a function of t and Λ. When $\Lambda=0$, we get back the relation Eq.(39) connecting T and t. Since we know the current values of $T = 2.728K$ and $t = 14 \times 10^9 year$, using Eq.(58), we solve for Λ. We do that in Fig.2 by plotting the left hand side and right hand side of Eq. (58) and finding the crossing point. This gives $\Lambda = 2.0 \times 10^{-56}$ cm^{-2} which is the value that has been derived dynamically here.

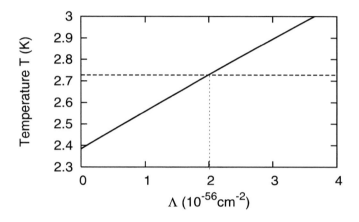

Figure 2. Determination of Λ by plotting the right hand side of Eq.(58) as a function of Λ (solid line) and left hand side as 2.728 K (thin broken line). The vertical dotted line indicates the value of $\Lambda = 2.0\ 10^{-56} cm^{-2}$

11. Conclusion

Considering the fact that the present universe might have been formed out of a system of fictitious self-gravitating particles, fermionic in nature, each of mass m, we are able to obtain a compact expression for the radius R_0 of the universe by using a model density distribution $\rho(r)$ for the particles which is singular at the origin. This singularity in $\rho(r)$ can be considered to be consistent with the so called Big Bang theory of the universe. By assuming that Mach's principle holds good in the evolution of the universe, we determine the number of particles, N, of the universe and its R_0, which are obtained in terms of the mass m of the constituent particles and the Universal Gravitational constant G only. It is seen that for a mass of the constituent particles $m \simeq 1.24 \times 10^{-35} g$, the age of the present universe, τ_0, becomes $\tau_0 \simeq 15 \times 10^9 yr$, or equivalently $R_0 \simeq 1.4 \times 10^{28} cm$. For this m, the total number of particles constituting the present universe is found to be $N \simeq 1.55 \times 10^{91}$ and its total mass ($M \simeq 1 \times 10^{23} M_\odot$), $M_\odot$ being the solar mass. All these numbers seem to be quantitatively agreeing with those evaluated from other theories.

Using the present theory, we have also made an estimation of the variation of the universal gravitational constant G with time which gives $(\frac{\dot{G}}{G}) = -1.3 \times 10^{-10}\ yr^{-1}$. This is again in extremely good agreement with the results of some of the most recent calculations. Lastly, a plausible explanation for the Dark Matter present in today's universe is given. Assuming neutrinos to be one of the most possible candidate for the Dark Matter, we have estimated the ratio of the number of neutrinos to nucleons and the number of neutrinos per unit volume of the universe, which respectively gives $(\frac{N_\nu}{N_n}) \sim 1.74 \times 10^9$, $(\frac{N_\nu}{V}) \sim 500$. Both these numbers seem to be in agreement with the findings of the recent observations. The present calculation gives a mass for the neutrino to be $m_\nu \sim 3.09 \times 10^{-32} g$ or equivalently, $14\ eV$, which is the right order of magnitude, as speculated by several workers If we consider that only 1% of the total mass of the universe is due to neutrinos, then mass of a neutrino can be estimated to be $(14/97)\ eV \approx 0.14\ eV$ which is in very good agreement with the limits set for its mass.

We study the universe as an expanding self-gravitating system using a many-particle Hamiltonian which is recently derived as representing the exactly soluble sector of quantum gravity and studied by us from a condensed matter point of view by using a quantum mechanical variational approach. This can also be viewed as a novel way of looking at the self-gravitating systems and it not only reproduces the results known from Einstein's General Theory of Relativity but also goes beyond by predicting certain relations and specifically the value of the cosmological constant. Instead of looking at the systems through the space-time dynamics, this theory treats the energy of the system directly. Infact from the quantum gravity point of view after quantising the GTR in the $c \to \infty$ (Newton-Cartan theory with spacial vanishing curvature), we have explicitly the above known Hamiltonian. We then special relativise the problem by using Mach's principle in case of the universe . The Hamiltonian describes the expanding universe as made up of gravitationally interacting particles such as particles of luminous matter and dark matter. The dark energy which is responsible for the accelerated expansion of the universe is taken as the a repulsive harmonic potential among the points in the flat 3-space as the nonrelativistic limit of the potential attributed to the cosmolgical constant. We derive a quantum mechanical relation connecting, temperature of the cosmic microwave background radiation, age, and cosmological constant of the universe. When the cosmological constant is zero, we get back Gamow's relation with a much better coefficient. Otherwise, our theory predicts a value of the cosmological constant $2.0 \times 10^{-56}\ cm^{-2}$ when the present values of cosmic microwave background temperature of 2.728 K and age of the universe 14 billion years are taken as input. It is interesting to note that in this flat universe, our method dynamically determines the value of the cosmological constant reasonably well compared to General Theory of Relativity where the cosmological is a free parameter.

References

[1] G. Contopoulos and D. Kotsakis, *Cosmology*, (Springer-Verlag, Heidelberg, 1987).

[2] Alan H. Guth, in *Bubbles, voids and bumps in time: the new cosmology* ed. James Cornell (Cambridge University Press, Cambridge, 1989).

[3] Utpal Sarkar, *Particle and Astroparticle Physics*, (Taylor & Francis, New York, 2007).

[4] Fang Li Zhi and Li Shu Xian, *Creation of the Universe*, (World Scientific, Singapore, 1989).

[5] A. A. Penzias and R. W. Wilson, *Astrophys. Jour.* **142**, 419 (1965).

[6] J. V. Narlikar, *Introduction To Cosmology*, (Cambridge University Press, London, 1993).

[7] G. S. Kutter, *The universe and life*, (Jones and Bartlett, USA, 1987).

[8] I. Novikov, *Black holes and the Universe*, (Cambridge University Press, Cambridge, 1990).

[9] F. Zwicky, *Helve. Phys. Act*, **6**, 10 (1933).

[10] James E.Gunn in *Bubbles,voids and bumps in time: the new cosmology* ed. James Cornel (Cambridge University Press, Cambridge, 1989).

[11] D. N. Schramm, *Phys. Today*, **36**, 27 (1983), G.Steigman, *Ann. Rev. Astron. Astrophys.*, **14**, 339 (1976).

[12] W. Buchmuller, P. Di Bari and M. Plumacher, *Phys. Lett.* **B 547**,128 (2002).

[13] R. Fardon, A. E. Nelson and N. Weiner, *JCAP*, **10**, 005 (2004).

[14] D. N. Tripathy and S. Mishra, *Int. J. Mod. Phys. D* **7**, 6, 917 (1998).

[15] D.N.Tripathy and S. Mishra *Int. J. Mod. Phys. D* **7**, 3, 431 (1998).

[16] S. Mishra, *Int. J. Theor. Phys.* **47**, 2655 (2008).

[17] J. Christian, *Phys. Rev. D*, 56, 4844 (1997).

[18] M. Karplus and R. N. Porter, *Atoms and Molecules*, (Reading Massachusetts: W. A. Benjamin, Inc. 1970).

[19] L.D.Landau and E.M.Lifshitz, *Quantum Mechanics*, (Pergamon Press, Oxford, 1965).

[20] E. R. Harrison, *Cosmology*, (Cambridge: Cambridge University Press, 1981).

[21] P. S. Wesson, *Cosmology and Geophysics*, (Adam Hilger Ltd, Bristol, 1978.)

[22] T. C. Van Flandern, *Mon. Not. R. Astron. Soc.*, **170**, 333 (1975).

[23] I. I. Shapiro, *BAAS*, **8**, 308 (1976).

[24] C. Brans and R. H. Dicke, *Phys. Rev*, **124**, 125 (1961).

[25] E. Cartan, *Ann. Ecole Norm. Sup.* **40**, 325 (1923)

[26] A. G. Riess *et al*[Supernova Search Team Collaboration], *Astron. J.* **116** 1009 (1998).

[27] S. Perlmutter *et al.* [Supernova Cosmology Project Collaboration], *Astrophys. J.* **517** 565 (1999).

[28] P. J. E. Peebles and B. Ratra, *Rev. Mod. Phys.* **75**, 559 (2003)

[29] S. M. Carroll, *Living Reviews in Relativity*, **4**, 1 (2001).

[30] G. E. Volovik, *The Universe in a Helium Droplet* (Clarendon, Oxford,2003).

[31] R. B. Laughlin, *Int. J. Mod. Phys. A*, **18**, 831 (2003).

[32] J. P. Hu and S. C. Zhang, *Phys. Rev. B.* **66**, 125301 (2002).

[33] J. M. Levy-Leblond, *J. Math. Phys.* **10**, 806, (1969)

[34] D. J. Fixen, E. S. Cheng, J. M. Gales, J. C. Mather, R. A. Shafer and E. L. Wright, *Astrophysics. J.* **473**, 576 (1996); A. R. Liddle, *Contemporary Physics*, **39**, no 2, 95,(1998).

[35] R. K. Pathria, *Statistical Mechanics*, (Pergamon Press, Oxford, 1972).

[36] A. M. Boesgaard and G. Steigman, *Ann. Rev. Astron.* **23**, 319 (1985)

[37] P. S. Wesson, *Cosmology and Geophysics*, (Adam Hilger Ltd, Bristol, 1978).

[38] I. Affleck and M. Dine, *Nucl. Phys.* **B249**, 361 (1985).

[39] M. Abramowitz and I. A. Stegun, *Handbook of Mathematical Functions*, (Dover, New York, 1965).

In: The Big Bang: Theory, Assumptions and Problems ISBN: 978-1-61324-577-4
Editors: J. R. O'Connell and A. L. Hale © 2012 Nova Science Publishers, Inc.

Chapter 8

ENTROPY GROWTH IN THE UNIVERSE

Marcelo Samuel Berman[*]
Instituto Albert Einstein/Latinamerica, Curitiba, PR, Brazil

Abstract

In a companion Chapter to this volume, Berman and Gomide proved that, under General Relativity theory, the total energy of a possibly rotating expanding Universe, is zero. Now, it will be shown that if the Universe is a zero-total energy entity and if, this energy is time-invariant, each type of energy contribution, to the total energy of the Universe, divided by Mc^2, yields a constant, during all times; for instance, if the present contribution of the cosmological "constant" Λ, drives the present Universe, it also must have driven alike, in all the lifespan of the Machian Universe (the relative contributions of energy densities of each kind, towards the total density, remain the same during all times). This fact is supported by the recent discovery that the Universe has been accelerating since a long time ago.

The immediate consequence of the above theory, is that all energy densities in the Universe are R^{-2} – dependent.

In consequence, we find that the absolute temperature, T, varies like $R^{-\frac{1}{2}}$; lambda, as R^{-2} so that the total entropy of the Universe grows with $R^{\frac{3}{2}}$, and, then, also as $M^{\frac{3}{2}}$.

These conclusions are supported by other calculations involving variants of General Relativity theory.

*This invited Chapter was accepted on 29 December 2010.

I. Introduction

According to a famous gravity researcher, the nature of gravitational entropy is an unresolved issue; many statements about it, in Physics textbooks are wrong when gravity is dominant; there is as yet no agreed definition of gravitational entropy that is generally applicable; and, until there is such an agreement, cosmological arguments relying on entropy concepts are ill-founded (Ellis, 2007).

---

[*]E-mail address: msberman@institutoalberteinstein.org; Address: Av. Candido Hartmann, 575 - # 17, 80730-440 - Curitiba - PR - Brazil

It is symptomatic that several well-known textbooks in General Relativity or Cosmology, omit completely the item *entropy* (some of the most recents are Padmanabhan, 2010; Ryder, 2009; Cooperstock, 2009; Hobson, Efstathiou and Lasenby, 2006). Nevertheless, we found a very authoritative account on relativistic thermodynamics in Plebański and Krasiński (2006). According to those authors, the determination of the absolute temperature and entropy, following the rules of classical phenomenological thermodynamics, depends on the kind of solution of the Einstein's equations employed in astrophysics and cosmology, which are "almost exclusively of high symmetry: they are spherically symmetric or stationary and axisymmetric, or homogeneous of Bianchi type".

It has been always problematic, according to standard cosmology, to explain why should the entropy be growing in the Universe. The purpose of this Chapter, is to show that, according to the zero-total energy of the Universe hypothesis, which was attributed to Machian universes (Berman, 2007a; c; d; 2008; a; b), we may find a time-varying entropy, depending on $R^{3/2}$, where R stands for the scale-factor (*read* the radius of the causally related Universe). The above result on zero-total energy, was proved in a companion Chapter in this volume, by Berman and Gomide (see also Berman and Gomide, 2010; 2011). The result for entropy, could not be obtained with the usual energy-momentum conservation equation, which implies that the entropy S is constant. We also shall review two papers by Berman (2009, 2009a), where we deal with variants of General Relativity theory, whereby growth in entropy is predicted.

In the beginning of the Universe, there was "nothing". We can say then, that the initial total energy of the Universe was zero-valued. By assuming the conservation of total energy, we may say that, up to now, the Universe is also zero-total-energy-valued (Feynman, 1962-3), (Berman, 2006; 2006a).

In the above cited references, it was shown that if the Machian Universe is depicted as having zero-total energy, not only it must obey several "generalized" Brans-Dicke equalities, but, the resulting cosmological model, has no initial infinite singularity. As a consequence, it was also shown that the so-called Pioneer anomaly (Anderson, 1999; Anderson et al, 2002), which appears as an unexplained deceleration to which several NASA spaceprobes were submitted, was indeed caused by a Machian related rotation of the Universe (Berman, 2007d). The fact that the zero-total energy theory explains such anomaly, and the similarity of the Brans-Dicke generalised conditions derived from it, with the known "lore", makes our entropy result very plausible. Barrow (1988) has pointed out that the Robertson-Walker's field equations for density and pressure, are the same as in Newtonian physics. That is the reason why Section II does not deal with General Relativistic tools.

II. Review of the Theory for Machian Universes

Berman has included in his Machian picture, the spin of the Universe, and replaced Brans-Dicke traditional relation, $\frac{GM}{c^2R} \sim 1$, with three different relations, which we call the Brans-Dicke relations for gravitation, for the cosmological "constant", and for the spin of the Universe (Berman, 2007a; 2007c; 2007d). We now review the theory, expanding the notion of Machian Universe, by including radiation (Berman, 2006b).

We shall consider a "large" sphere, with mass M, radius R, spin L, and endowed with a cosmological term Λ, which causes the existence of an energy density $\frac{\Lambda}{\kappa}$, where $\kappa = \frac{8\pi G}{c^2}$. We now calculate the total energy E of this distribution:

$$E = E_i + E_g + E_L + E_\Lambda + E_R, \tag{II.1}$$

where $E_i = Mc^2$, stands for the inertial (Special Relativistic) energy; $E_g \cong -\frac{GM^2}{R}$ (the Newtonian gravitational potential self-energy); $E_L \cong \frac{L^2}{MR^2}$ the Newtonian rotational energy; $E_\Lambda \cong \frac{\Lambda R^3}{6G}$ (the cosmological "constant" energy contained within the sphere), and $E_R = aT^4$, where a is a constant, and T stands for absolute temperature, while E_R represents radiational energy.

If we impose that the total energy is equal to zero, i.e., $E = 0$, and $\dot{E} = 0$ (the total energy is null, and time-invariant), we obtain from (II.1):

$$\frac{GM}{c^2 R} - \frac{L^2}{M^2 c^2 R^2} - \frac{\Lambda R^3}{6GMc^2} - \frac{E_R}{Mc^2} \cong 1. \tag{II.2}$$

As relation (II.2) above should be valid for the whole Universe, and not only for a specific instant of time, in the life of the Universe, and if this is not a coincidental relation, we can solve this equation by imposing that:

$$\frac{GM}{c^2 R} = \gamma_G, \tag{II.3}$$

$$\frac{L}{McR} = \gamma_L; \tag{II.4}$$

$$\frac{E_R}{Mc^2} = \frac{\rho_R R^3}{Mc^2} = \frac{aT^4 R^3}{Mc^2} = \gamma_R, \tag{II.4a}$$

and,

$$\frac{\Lambda R^3}{6GMc^2} = \gamma_\Lambda, \tag{II.5}$$

subject to the condition,

$$\gamma_G - \gamma_L^2 - \gamma_\Lambda - \gamma_R \cong 1, \tag{II.6}$$

where the $\gamma's$ are constants.

The above solution, of course, is unique, on condition that we expect R to be time-varying; other solutions to equation (II.2) could be found if it were not for the existence of the condition $\dot{R} \neq 0$.

It must be remarked, that our proposed law (II.3), is a radical departure from the original Brans-Dicke (Brans and Dicke, 1961) relation, which was an approximate one, while our present hypothesis implies that $R \propto M$. With the present hypothesis, one can show, that independently of the particular gravitational theory taken as valid, the energy density of the Universe obeys a R^{-2} dependence (see Berman, 2006; 2006a; Berman and Marinho, 2001).

More than that, we ensure, with the conditions (II.3)(II.4)(II.4a) and (II.5), that each different type of relative energy contribution, into the total energy, remains the same, during all periods of time, when compared with Mc^2.

We have now the following generalized Brans-Dicke relations, for gravitation, spin, radiation and cosmological "constant":

$$\frac{GM}{c^2 R} = \gamma_G , \tag{II.3}$$

$$\frac{GL}{c^3 R^2} = \gamma_G \cdot \gamma_L , \tag{II.7}$$

$$\frac{aGT^4 R^2}{c^4} = \gamma_R \cdot \gamma_G , \tag{II.7a}$$

and,

$$\frac{\Lambda R^2}{6c^4} = \gamma_\Lambda \cdot \gamma_G . \tag{II.8}$$

The reader should note that we have termed Λ as a "constant", but it is clear from the above, that in an expanding Universe, $\Lambda \propto R^{-2}$, so that Λ is a variable term. We also notice that $R \propto M$, and $L \propto R^2$, and $R \propto T^{-2}$, so that, we should have, $\rho_R \propto R^{-2}$. As relations (II.3), (II.7) and (II.8) have been reasonably well discussed by Berman elsewhere (see above cited references), it is our present purpose to check relation (II.7a).

The T^{-2} dependence on R, was dealt, earlier, for non-relativistic decoupled massive species by Kolb and Turner (1990).

It is clear from our previous hypotheses, that all the energy densities vary with R^{-2}. This can be checked one by one. For instance, from the definition of the inertial energy density,

$$\rho_i = \frac{M}{V} , \tag{II.9}$$

while,

$$V = \alpha R^3 , (\alpha = \text{constant}) \tag{II.10}$$

where ρ_i and V stand for the inertial (or, matter) energy density and tridimensional volume, we find:

$$\rho_i = \left[\frac{\gamma_G}{G\alpha}\right] R^{-2}. \tag{II.11}$$

We point out that, what we call the total energy density of the Universe, is the sum of all the positive energy densities for everything except the self-gravitational energy density, which is negative, and makes the effective total energy density of the Universe, zero-valued, and time-independent. This is how the Machian Universe escapes from the accusation of keeping the initial infinite singularity of the Universe: we are sure, because of the time-invariance of the null results for total energy and effective energy density, that,

$$\lim_{R \to 0} \rho = \lim_{R \to 0} E = 0 . \tag{II.11b}$$

We show now that the entropy of the Machian Universe, as described above, is not constant, but increases with time. On defining,

$$S = sR^3,$$

where, S and s represent total entropy, and entropy density, respectively for the Universe.

The guiding classical thermodynamical law is,

$$d(\rho V) + pdV = TdS.$$

The cosmic pressure definition is given by,

$$p = -\frac{dM}{dV}.$$

It can be seen from the above treatment, that $p \propto R^{-2}$.

According to the above formulae, we have:

$$S \propto sR^3 = \frac{\rho_R}{T} R^3 \propto T^3 R^3 \propto R^{\frac{3}{2}} \propto M^{\frac{3}{2}}, \tag{II.12a}$$

taken care of the relations given earlier, i.e., $R \propto T^{-2}$, and (II.3) above.

We have thus shown that S increases with time, likewise the scale factor to the power $\frac{3}{2}$. [This Machian property stands opposite to the usual assumptions in theoretical studies that, by considering $S =$ constant, found $R \propto T^{-1}$, which as we have shown, is incorrect in the Machian picture].

III. Entropy of the Universe in a Cosmological Newtonian Limit of GRT (Berman, 2009a)

There are many different ways in order to compare General Relativity with Newtonian Theory. In the next Section, we shall introduce a cosmological weak field limit. In such limit, we can not ignore the existence of cosmic pressure, and a cosmological "constant" term. The possibility of introducing cosmic pressure in this limit, was surligned by Peacock (1999). Earlier, Ray d'Inverno (1992) had worked with a cosmological repulsive acceleration in Newtonian Cosmology. I recall Peter Landsberg and Evans (1977) having done something similar.

Other limits are possible, like the Machian (Berman, 2008a). A linearized theory of the electromagnetic type, was also devised as some kind of Sciama's limit to the field equations (Sciama, 1953; Berman, 2008b).

Barrow (1988), has worked what we shall call the thermodynamical Newtonian "limit". He established the two energy conservation equations, namely, the Newtonian and the thermodynamical; the latter implies the following definition of pressure,

$$p = -\frac{dM}{dV},$$

where M and V represent the mass and volume of a given system. He showed that both equations led to Robertson-Walker's field equations of General Relativity, provided that the energy density term, instead of being related to mass energy, should be related to all forms of energy, like, for instance, radiation.

As to the Machian limits, Berman (2008a; 2007a) has shown that there are Newtonian-Machian and General Relativistic Machian limits, which, in fact, result in the same conditions for the relevant Physical quantities.

We must deny, that the strong but wrong impression in the air, surrounding Newtonian Cosmology as a pressureless model, is, in fact, true (Barrow, 1988).

Cosmological Newtonian Limit of Field Equations

Standard Cosmology (Weinberg, 2008), introduces constant entropy. We shall show in next Section, that the Universe, could bear a growing entropy, in the correct limit of Einstein's field equations.

It is well known that Einstein's field equations, in the so-called Newtonian limit, reduce to Poisson's equation,

$$\nabla^2 \Phi = 4\pi G \sigma , \qquad (\text{III}.1)$$

where Φ, G and σ stand for potential, gravitational constant and gravitational energy density. What probably never was told, is that σ represents not only the effective energy density, but also eventual pressure and cosmological "constant" terms. In fact, when applied in large scale, a pressure term may appear in the system, and in a cosmological scale, a lambda term is possible. In Whitrow's paper (Whitrow, 1946; Whitrow and Randall, 1951), he equated the inertial energy of the Universe (Mc^2) to the gravitational potential energy ($G\frac{M^2}{R}$), finding the approximate relation $G\frac{M}{R} \cong c^2$.

If we postulate _sphericity_ (the Universe resembles a "ball" of approximate spherical shape), _egocentrism_ (each observer sees the Universe from its center) and _democracy_ (each point in space is equivalent to any other one – all observers are equivalent), we may write, for each observer, the following Newtonian potential,

$$\Phi = -G\frac{M}{R} . \qquad (\text{III}.2)$$

In the needed interpretation, we shall see R as the radius of the causally related Universe.

Then, from (III.1) and (III.2), we find,

$$\nabla^2 \Phi = \frac{\partial^2 \Phi}{\partial R^2} = -2\pi G \frac{M}{R^3} = 4\pi G \left(\rho + 3\frac{p}{c^2} - \frac{\Lambda}{4\pi G} \right) . \qquad (\text{III}.3)$$

The obvious solution remains,

$$\rho = \rho_0 R^{-2} ,$$

$$p = p_0 R^{-2}, \tag{III.4}$$

$$\Lambda = \Lambda_0 R^{-2}.$$

In the above, ρ_0, p_0 and Λ_0 are constants. The resulting equation is, for baryonic matter,

$$M = \gamma R \equiv -2 \left[\rho_0 + 3\frac{p_0}{c^2} - \frac{\Lambda_0}{4\pi G} \right] R > 0. \tag{III.5}$$

We show that negative pressures are possible. As dark matter is represented by a positive cosmological term, for an accelerating Universe, we have $\Lambda_0 > 0$. On the other hand, the weak energy condition, stated as the positivity of energy density, should also apply and, thus, $\rho_0 > 0$. From (III.5), we find that $p_0 < \frac{\Lambda_0 c^2}{12\pi G} - \frac{1}{3}\rho_0 c^2$.

We can check that Whitrow's relation is retrieved in its generality.

Speaking of Thermodynamics

We saw above, that energy densities are R^{-2} – dependent. If all energy densities are as such, we can say that the radiation component ρ_{rad} has the same dependence, to wit,

$$\rho_{rad} = b\, R^{-2} = a\, T^4. \tag{III.6}$$

In the above, a and b are constants while T stands for absolute temperature, and the right hand side represents black-body radiation. We find the same relation that Kolb and Turner (1990) have already mentioned in another context, namely,

$$T^2 R = \text{constant.} \tag{III.7}$$

In this case, the entropy of such Universe is again given by (Sears and Salinger, 1975),

$$S = S_0 R^{3/2} \quad (S_0 = \text{constant}). \tag{III.8}$$

This entropy is growing, because $\dot{R} > 0$ for expanding Universes. Weinberg (1972), suggested that there would be dissipative processes in the Universe, that could be represented by viscosity terms.

Final Considerations

Some comments on the above Sections are necessary.

first) As a by-product of our presentation, we can check that there is no infinite singularity in the beginning of this Universe, i.e., according to (III.5),

$$\lim_{R \to 0} M \longrightarrow 0. \tag{III.9}$$

second) We have also found that pressure and density obey a perfect gas equation of state, for both are R^{-2} – dependent. As dark-energy is representable by the lambda-term, this has also the same dependence, which is consistent with Modern Cosmology (Weinberg, 2008).

third) The thermodynamical conclusion of ours, is also confirmed by prior work on the above subjects, which was done by Berman (2007; 2007a; 2007b; 2007c; 2008).

IV. Entropy of the Universe and Standard Cosmology (Berman, 2009)

The purpose of the present Section is to show that the entropy of the Universe grows with time, when we consider a Machian variation of Standard Cosmology. Even though the entropy is not constant, we shall find the correct Standard Cosmology result relating the absolute temperature with the age of the Universe.

In a previous paper (Berman, 2009a), according also to the above Section, we have shown that, in the correct Newtonian limit, Einstein's field equations yield a solution for energy density, cosmic pressure, and, possibly, a time-varying cosmological term, proportional to R^{-2}, where R stands for the radial distance in local Physics, or the scale-factor in cosmological situations. From it, we found a growing entropy in the Universe, scaled as $R^{\frac{3}{2}}$. Though Standard Cosmology (Weinberg, 2008), introduces constant entropy, we did show that the Universe had an absolute temperature T proportional to $R^{-\frac{1}{2}}$. This was contrary to the Standard Cosmology result, namely $TR = $ constant. Nevertheless, we shall show now, that the time-dependence of T, in our framework, yields the same $t^{-\frac{1}{2}}$ time-dependence as in Standard Cosmology, provided that the Universe is Machian, in the sense that the scale-factor R be linearly proportional to t. If not, we still show that other results are kept intact.

Time-Dependence of Temperature

Consider Robertson-Walker's Universe, with constant deceleration parameter (Berman, 1983; Berman and Gomide, 1988),

$$q \equiv m - 1 \equiv -\frac{\ddot{R}R}{\dot{R}^2} = \frac{\kappa}{6H^2} \sum \rho_i (1 + 3\alpha_i) . \tag{IV.1}$$

In the above, we are employing a multi-fluid model, with perfect gas equation of state, where q, ρ_i and α_i represent deceleration parameter, energy density of fluid number i ($i = 1, 2, 3...$) and $\alpha_i = $ constant.

Now, we define the density parameters,

$$\Omega_i \equiv \frac{\kappa}{3H^2} \rho_i(t) . \tag{IV.2}$$

Einstein's field equations, produce the following relation,

$$H^2 = \frac{\kappa}{3} \sum \rho_i . \tag{IV.3}$$

The multi-fluid cosmic pressures, are given by,

$$p_i = \alpha_i \rho_i \quad (\text{no summation}).\tag{IV.4}$$

From Berman's theory for $q = $ constant, we find:

$$H = \frac{1}{mt} = \frac{1}{(1+q)t} \quad (q \neq -1).\tag{IV.5}$$

The obvious solution for the above equations, with constant density parameters, is,

$$\rho_i = \rho_{i0} t^{-2} \quad (\rho_{i0} = \text{constant}).\tag{IV.6}$$

Notice that we may yet consider a vacuum constant energy density, but relation (IV.6) may, in many Cosmological theories, be of type (IV.6), i.e., as in Berman (2007a; 2007c; 2008), where, we have $\frac{\Lambda}{\kappa} \propto t^{-2}$.

If solution (IV.6) does apply to radiation, which is black-body's, we find,

$$\rho_{rad} = \rho_{rad0} t^{-2} = aT^4,\tag{IV.7}$$

where ρ_{rad0} and a are constants.

We find that,

$$T = \left(\frac{\rho_{rad0}}{a}\right)^{\frac{1}{4}} t^{-\frac{1}{2}}.\tag{IV.8}$$

Relation (IV.8) is pretty standard (Weinberg, 1972; 2008).

Entropy of the Universe - Machian Limit

Now recall our previous paper (Berman, 2009a), or the above Section, where the following law had been proposed,

$$\rho_{rad} \propto R^{-2},\tag{IV.9}$$

from which we would obtain,

$$T \propto R^{-\frac{1}{2}}.\tag{IV.10}$$

Relations (IV.9) and (IV.10), were based on the Newtonian limit of Einstein's field equations, and, in Section II above, we were talking of Standard Cosmology. The bridge between both formulations lies in the Machian relation,

$$R \approx ct.\tag{IV.11}$$

In the Machian approach, R stands for the causally related radius, but Berman has long ago stated that the scale-factor could be identified with this radius (Berman, 2009a).

We now find the following entropy dependence, through relation (IV.8) (Sears and Salinger, 1975):

$$S = S_0 t^{\frac{3}{2}} \quad (S_0 = \text{constant}). \tag{IV.12}$$

Of course, it matches our original result (Berman, 2008; 2009a). It is only when we make the Machian hypothesis above (cf relation IV.10), that with find the law (IV.12), because we "escape" from the framework of Standard Relativistic Cosmology, where $RT = $ constant, and, S is also constant.

Matching General Relativity

When we turn to the theory of constant deceleration parameters, we find, for each phase of the Universe, the scale-factor,

$$R = (mDt)^{\frac{1}{m}} \quad (D = \text{constant}). \tag{IV.13}$$

When Einstein's field equations are produced, and if we impose relation (IV.6), we find the correct solutions given by Standard Cosmology, i.e., the scale-factor depends on $t^{\frac{2}{3}}$ for matter, and $t^{\frac{1}{2}}$ for radiation. With these results, we find, as usual (Hobson et al., 2006),

$$\rho_{rad} \propto R^{-4}, \tag{IV.14}$$

and,

$$\rho_{matter} \propto R^{-3}. \tag{IV.15}$$

(We must caution that these energy densities are given now by non-Machian relations.) Relation (IV.6) is now validated, so now we reached a happy end.

Final Comment

We have found that the entropy of the Universe grows with time. This matches our previous Section result (Berman, 2009a). Our happy end resolves once more, one of the crucial problems of Standard Cosmology.

V. Concluding Remarks

Several remarks are necessary:

(A) allegations about the energy of the Universe, and, precisely, about its zero-value, can be traced to Feynman (Feynman, 1962-3), Rosen (Rosen, 1994-95), Cooperstock and Israelit (1995), Hawking (2001) and many others. Berman has derived this from Robertson-Walker's metric, so that it is a valid result in Relativistic Cosmology, for any tri-curvature value (Berman, 2006, 2006a). The existence of a "spectator", in order to measure the energy, is a philosophical question, rather than a scientific one;

(B) Machian properties have been proposed in different gravity theories, so there is no one single theory that owns such attributes (remember the origin of Brans-Dicke theory);

(C) the several generalized Brans-Dicke equalities, derived here from the energy equation, are just, the most simple set of solutions for the $E = \dot{E} = 0$ equation;

(D) the mentioned solutions, have very interesting properties: for instance, the relative contributions of each type of energy towards the total amount, is time-independent. This fact is coherent with the recently proclaimed and experimentally observed result that the Universe has been lambda-dominated since long ago;

(E) "Machian" conditions need not be "general relativistic" ones;

(F) you can not blame our Chapter for the fact that the angular momentum is high for the present Universe, because we have derived from this result, that the amount of angular velocity in the present Universe is small and it is undetectable with present technological tools;

(G) our framework is relativistic, in the low Newtonian limit, but this could be called, for instance, a Sciama gravitational theory (Sciama, 1953);

(H) by going back from the present time towards Planck's, no inconsistency with the energy densities would be found;

(I) because we can not suppose on the first stance, that the entropy is constant, we point out that the hypothesis that $RT = $ constant (and thus $S = $ constant), is inconsistent with $\dot{R} \neq 0$, and $\dot{E} = E = \dot{G} = \dot{c} = 0$. We do NOT have such constancy for RT, but, instead, we have $RT^2 = $ constant. In this case, the total entropy grows with $R^{3/2}$, while, according to standard Cosmology, it would be constant. It remains open, the possibility, to be featured in a new paper, of a static model of the Universe ($\dot{R} = 0$), with time-varying G and c;

(J) our Machian theory, as above, is pre-general relativistic, but both are not incompatible. The introduction of a Cosmological "constant", in Newtonian cosmology, is now a standard procedure. I think that it was introduced by Landsberg and Evans (1977). An available reference for it, is the book by D'Inverno (1992);

(K) we refer to the extremely important book by Sabbata and Sivaram (1994), where there are clues about the rotation of the Universe. The astronomical Pioneer anomaly, and the astrophysical laws, like Blackett formula, which relates spin and magnetic field, and Wesson's one, relating spin with mass, are discussed by us in another paper (Berman, 2007; Wesson, 1999; 2006).

(L) whether the present model is to be accepted or not, we point out that it does not contradict present experimental evidence. The inconvenient hypothesis that $S = $ constant, as in standard Cosmology, has now been substituted by an ever-increasing entropy, as long as $\dot{R} > 0$.

Acknowledgments

The author is indebted to Nova Science Publishers, specially its President Frank Columbus, for the kind invitation, and gratefully thanks his intellectual mentors, Fernando de Mello Gomide and the late M. M. Som, his colleagues Nelson Suga, Marcelo F. Guimarães,

Antonio F. da F. Teixeira, and Mauro Tonasse; I am also grateful for the encouragement by Albert, Paula, and messages by Dimi Chakalov. I dedicate this Chapter to M.M. Som, *in memoriam*.

References

Anderson, J.D. (1999) - *Planetary Report*, **19(3)**, 15.

Anderson, J.D. et al.(2002) - Study of the anomalous acceleration of Pioneer 10 and 11.- *Phys. Rev. D* **65**, 082004.

Barrow, J.D. (1988) - The Inflationary Universe, in *Interactions and Structures in Nuclei*, pp 135-150 (eds) R Blin Stoyle and W D Hamilton, Adam Hilger, Bristol. Paper presented at the University of Sussex, date 7-9th September, 1987, on *Interactions and Structures in Nuclei*.

Berman, M.S. (1983) - *Nuovo Cimento* **74B**, 182-186.

Berman, M.S. (1991) - *GRG* **23**, 465.

Berman, M.S. (1991a) - *Physical Review* **D43**, 1075.

Berman, M.S. (2006) - Energy of Black-Holes and Hawking's Universe in *Trends in Black-Hole Research*, Chapter 5. Edited by Paul Kreitler, Nova Science, New York.

Berman, M.S. (2006 a) - *Energy, Brief History of Black-Holes, and Hawking's Universe* in *New Developments in Black-Hole Research*, Chapter 5. Edited by Paul Kreitler, Nova Science, New York.

Berman, M.S. (2006b) - *On the Magnetic Field, and Entropy Increase, in a Machian Universe*. See Los Alamos Archives, http://arxiv.org/abs/physics/0611007 .

Berman, M.S. (2007) - The Pioneer Anomaly and a Machian Universe, *Astrophysics and Space Science*, **312,** 275. Posted in Los Alamos Archives, http://arxiv.org/abs/physics/0606117.

Berman, M.S. (2007a) - *Introduction to General Relativity and the Cosmological Constant Problem*, Nova Science, New York.

Berman, M.S. (2007b) - Is the Universe a White-Hole?, *Astrophysics and Space Science*, **311** , 359-361. See also, Los Alamos Archives, http://arxiv.org/abs/physics/0612007

Berman, M.S. (2007c) - *Introduction to General Relativistic and Scalar-Tensor Cosmologies*, Nova Science, New York.

Berman, M.S. (2008) - *A Primer in Black-Holes, Mach's Principle, and Gravitational Energy* , Nova Science, New York.

Berman, M.S. (2008a) - General Relativistic Machian Universe, *Astrophysics and Space Science*, **318**, 273-277. See also, Los Alamos Archives http://arxiv.org/abs/0803.0139.

Berman, M.S. (2008b) - On the Machian Origin of Inertia, *Astrophysics and Space Science*, **318**, 269-272. See also, Los Alamos Archives, http://arxiv.org/abs/physics/0609026.

Berman, M.S. (2009) - Entropy of the Universe and Standard Cosmology, *International Journal Theoret. Physics* **48**, 2286.

Berman, M.S. (2009a) - Entropy of the Universe – *International Journal Theoret. Phys.*, **48**, 1933. Los Alamos Archives http://arxiv.org/abs/0904.3135

Berman, M.S.; Gomide, F.M. (1988) - *GRG* **20**, 191-198.

Berman, M.S.; Gomide, F.M. (2010) - *General Relativistic Treatment of the Pioneers Anomaly*, pre-print Los Alamos Archives: http://lanl.arxiv.org/abs/1011.4627

Berman, M.S.; Gomide, F.M. (2011) - *On the Rotation of the Zero-Energy Expanding Universe*, Chapter accompanying this volume.

Berman, M.S.; Marinho, R.M. (2001) - *Astrophysics and Space Science*, **278**, 367.

Brans, C.; Dicke, R.H. (1961) - *Physical Review*, **124**, 925.

    Cooperstock, F.I. (2009) - *General Relativistic Dynamics*, World Scientific, Singapore.

Cooperstock, F.I.; Israelit, M. (1995) - *Foundations of Physics*, **25**, 631.

D'Inverno, R. (1992) - *Introducing Einstein's Relativity*, Clarendon Press, Oxford.

Ellis, G.F.R. (2007) - *Issues in the Philosophy of Cosmology*, in *Philosophy of Physics - Part B*, ed. J.Butterfield and J. Earman; included in the *Handbook of the Philosophy of Science*(eds D.M. Gabbay, P. Thagard and J. Woods). Elsevier, Amsterdam.

Feynman, R. (1962-3) - *Lectures on Gravitation*, Addison-Wesley, N.Y.

Hawking, S.W. (2001) - *The Universe in a Nutshell*, Bantam, N.Y.

Hobson, M.P.; Efstathiou, G.; Lasenby, A.N. (2006) - *General Relativity*, CUP, Cambridge.

Kolb, E.W.; Turner, M.S. (1990) - *The Early Universe*, Addison-Wesley, N.Y.

Landsberg, P.T.; Evans, D.A. (1977) - *Mathematical Cosmology*, OUP, Oxford.

Padmanabhan, T. (2010) - *Gravitation - Foundations and Frontiers*, CUP, Cambridge.

Peacock, J.A. (1999) - *Cosmological Physics*, Cambridge Universe Press, Cambridge, page 25, formula (1.85).

Plebański, J.; Krasiński, A. (2006) - *An Introduction to General Relativity and Cosmology*, CUP, Cambridge.

Raychaudhuri, A.K. (1979) - *Theoretical Cosmology*, OUP, Oxford.

Rosen, N. (1994) - *GRG* **26**, 319.

Ryder, L. (2009) - *Introduction to General Relativity*, CUP, Cambridge.

Sabbata, V.; Sivaram, C. (1994) - *Spin and Torsion in Gravitation*, World Scientific, Singapore.

Sabbata, V.; Gasperini,M. (1979) - *Lettere Nuovo Cimento* **25**, 489.

Sears, F.W.; Salinger,G.L. (1975) - *Thermodynamics, Kinetic theory, and Statistical Thermodynamics*, Addison-Wesley, New York.

Sciama, D.N. (1953) - *MNRAS* **113**, 34.

Weinberg, S. (1972) - *Gravitation and Cosmology*, Wiley, New York.

Weinberg, S. (2008) - *Cosmology*, Oxford University Press, Oxford.

Wesson, P.S. (1999) - *Space-Time-Matter*, World Scientific, Singapore.

Wesson, P.S. (2006) - *Five dimensional Physics*, World Scientific, Singapore.

Whitrow, G. (1946) - *Nature*, **158**, 165.

Whitrow, G.; Randall, D. (1951) - *MNRAS*, **111**, 455.

In: The Big Bang: Theory, Assumptions and Problems
Editors: J. R. O'Connell and A. L. Hale
ISBN: 978-1-61324-577-4
© 2012 Nova Science Publishers, Inc.

Chapter 9

DID THE BIG BANG REALLY TAKE PLACE?

Ernst Fischer[*]
Auf der Hoehe 82, D-52223 Stolberg, Germany

Abstract

The Big Bang model suffers from the occurrence of singularities and infinity conditions, where all known physical laws break down. Besides it is based on improvable new physics. The overwhelming part of the universe should consist of so called dark energy with very strange properties, unlike all we know. Besides not only the energy density should have been infinitely high, this state should have been present in an infinite space. The model takes its justification from the fact that it can explain the cosmological red shift, the microwave background radiation and (with restrictions) the chemical composition of matter.

But based on special and general relativity static spatially curved solutions of the field equations are possible, which allow a stable static homogeneous universe, if only potential energy is introduced as a part of the energy tensor and global Lorentz invariance exists.

But while a static homogeneous universe is stable as a whole, locally it is unstable, leading to the formation of structures, observed as galaxies and stars. There exists a continuous matter cycle, starting from a hot ubiquitous plasma, which locally cools down, first by gravitational loss of momentum, then by emission of electromagnetic radiation. This leads to fragmentation into clusters, galaxies and then into individual stars. The negative potential energy finally stops further collapse into a singularity. Instead this collapsed state may become unstable, leading to the emission of relativistic matter jets, delivering fresh hydrogen to the intergalactic plasma. The Big Bang is replaced by these small bangs.

This scenario is possible within the framework of the confirmed theory of relativity without singularities or new physics.

Keywords: Static universe, redshift, microwave background, element abundances, structure formation

---

[*]E-mail address: e.fischer.stolberg@t-online.de

1. Introduction

During the last century our view of the universe has dramatically changed. Based on Einstein's theories of special and general relativity a cosmological model has been developed, which assumes that the universe started from a singular state, the Big Bang, about 14 billion years ago and is expanding since then. The idea of expanding space replaced the static model, first proposed by Einstein, when the cosmological redshift was discovered, the shift of photon energies with distance of the source, which could be interpreted as a kind of Doppler shift.

The Big Bang model has remained the favoured description of the initial conditions of the universe, on which the development to its present state is based, though it starts from a singular state, where all our known laws of physics break down. It is not only the fact that at this state the energy density should have been infinite, in addition this infinite density should have been present also in an infinite space, as there is no expansion or contraction, which can transform infinite space to a finite volume. Besides there is no known mechanism, which could drive the expansion of space.

The main arguments for justification of the Big Bang model is that it is able to explain the observed cosmological redshift, the cosmic abundance of chemical elements and the existence of the microwave background radiation. But even the latter arguments remain questionable. The primordial element abundances may have been changed by several reactions during the expansion history and it is scarcely understandable that the microwave radiation background should have retained its black body spectrum during expansion by three orders of magnitude, when the equilibrium is not stabilised by interactions.

In addition, to explain the homogeneity of the observed universe, so called inflation had to be included in the expansion, a phase where the spatial scale changes exponentially by many orders of magnitude, to guarantee causal connection of the observed space. The cause of this inflation remains unknown or even unknowable. Besides that, in the last decades observations of distant supernova explosions have shown that the data can be fitted to the expansion history only, if there exists some form of gravitationally repulsive dark energy, which amounts to about 70% of the matter equivalent of the universe. But there exists no reasonable idea, what this dark energy might be and how it could be generated in the Big Bang.

It would lead too far to go into details of all the attempts to overcome the problems of the Big Bang model. Here we confine ourselves to demonstrate that there exist alternative explanations of all the phenomena, which have been used as corner stones in favour of the expanding world model: the cosmological red shift, the element abundances and the microwave radiation background. References to some of the innumerable scientific articles in the area will be included only, if direct use of their observational data or formulae is made.

In this chapter we will show that alternatives to the Big Bang theory are possible, which are based on the observationally confirmed laws of special and general relativity, but do neither require some singular state of matter nor new or unknown physics like inflation or dark energy. In the next section we will first demonstrate, how under the assumption of global homogeneity and isotropy of space a stable static universe is possible, similar to Einstein's original concept, and that redshift is a necessary consequence of curvature, when strict energy conservation is assumed together with a limitation of gravitational interaction to the speed of light.

In the subsequent sections first the formation of structures will be discussed and their role in a continuous matter cycle, which never ends, but leads back to its initial state of a homogeneous matter distribution in form of a hot diluted plasma. No matter is lost in singular black holes but recycled to the plasma in the form of matter jets, in which the primordial element composition is restored. Then we will show that continuous redshift will lead to the existence of a homogeneous background radiation field.

2. Back to Einstein

To see, where the alternative path to understanding the global properties of the universe has been missed, we must go back to the original propositions, which led Einstein to formulation of his theory of relativity and then to the description of gravitation by a purely geometrical model (see e.g. [2]).

Special relativity is based on the concept that physical laws must be independent of the state of motion, which can be defined only relatively between two systems, but not in an absolute way. The fact that the vacuum speed of light must be the same, when measured in two systems moving with respect to each other, led to the well known proposition of Lorentz invariance, which regards space and time as a unity, where the partition into temporal and spatial components of motion depends on the reference system.

The equality of inertial and gravitational mass, observed in numerous experiments, led Einstein to the description of gravitation by a purely geometrical model of gravitation, where motion under the influence of gravitation is replaced by geodesic motion in non-Euclidean space and where the deviations from Euclidean geometry are caused by the gravitating matter. By extension of the tensor formalism from special relativity to curved Riemannian space Lorentz invariance could be conserved at least locally. This does imply that the speed of gravitational interaction is limited to the speed of light, just like electromagnetic interaction.

Formally general relativity follows the path of differential geometry. The local path element ds is related to the spatial and temporal distance elements by an expression of the form

$$ds^2 = \sum g_{ij}dx^i dx^j, \qquad (i = 0, 1, 2, 3) \qquad (1)$$

where x_0 is the normalised time coordinate $x_0 = ct$ and x_1 to x_3 represent a triple of spatial coordinates. The components g_{ij} of the fundamental tensor contain the information on deviation from Euclidean geometry and thus on the local strength of gravity. The relation between the fundamental tensor and the matter distribution is given by the Einstein field equation, which relates the deviation from Euclidean space, the local curvature, to the matter distribution.

$$G_{ij} = R_{ij} - \frac{1}{2}Rg_{ij} = \kappa T_{ij} \qquad (2)$$

R_{ij} is the Ricci curvature tensor, $R = R_{ij}g^{ij}$ the curvature scalar. T_{ij} contains the matter and energy fields, which contribute to gravitation. The constant κ is related to the conventional units by $\kappa = 8\pi G/c^4$ (G is the gravitational constant). The second term on the left hand side of the field equation was introduced by Einstein to guarantee that the divergence of the left side is zero, since he considered as a fact that the divergence of the right hand side is zero, too, because of energy conservation [2]. The relation between the components of the Ricci-tensor and the metric is given by

$$R_{ij} = \frac{\partial \Gamma^k_{ij}}{\partial x_k} - \Gamma^k_{im}\Gamma^m_{kj} - \frac{\partial \Gamma^k_{ik}}{\partial x_j} + \Gamma^k_{ij}\Gamma^m_{km} \qquad (3)$$

and its relation to the Levi-Civit connection

$$\Gamma^k_{ij} = \frac{1}{2}g^{km}\left(\frac{\partial g_{im}}{\partial x_j} + \frac{\partial g_{jm}}{\partial x_i} - \frac{\partial g_{ij}}{\partial x_m}\right). \qquad (4)$$

This set of equations forms the basis of gravitational interaction and can be found in every textbook on gravitation (see e.g. [4], [5]).

But it is not only the nonlinear character of these equations, which makes their practical application difficult. Also to find a reasonable definition of the energy tensor in curved space may cause problems. In Euclidean geometry we are accustomed to define the density of matter in some volume by the ratio of its total mass to the size of the volume element $\varrho = dm/dV$. But in general relativity the size of volume elements depends on the geometry itself. Thus there is no longer a unique definition of ρ. Einstein has proposed a way out of this dilemma, using the density defined in a Euclidean tangent space as an ingredient of the energy tensor; that means, the density which would be measured, if all surrounding matter were removed to infinity.

But with this assumption another problem occurs. Integrating the density over a finite volume does not result in the total mass of a body. The resulting mass compared to the sum of the masses of the individual particles is enhanced just by the mass equivalent of the gravitational binding energy. The question is, which mass is responsible for the gravitational attraction of other outside masses. Or to put it the other way: Does the gravitational attraction of a body to outside matter increase, when it contracts under the influence of its own gravity? The answer is simple, if we regard the limiting case of a mass shrinking close to a singular volume (a black hole in the conventional picture). If binding energy were added to the gravitating mass, the overwhelming attraction of such collapsed systems would dominate the complete structure of the universe. As this is obviously not the case, we must conclude that, if we use the definition of matter density from Euclidean tangent space, we must subtract the binding energy from the mass integral to get the correct gravitational attraction to outside matter.

From these considerations it is immediately clear that gravitational binding energy must be regarded as a genuine part of the energy tensor. That Einstein's theory of general relativity was so successful in the description of motion in the solar system (the perihelion shift of Mercury and the aberration of starlight passing close to the sun), though he did not include binding energy in his calculations, is easily explained by the fact that the influence is negligibly small in these cases. The ratio of rest mass energy to potential energy near the surface of the sun is of the order 10^{-6}. Strong effects are expected only in extremely collapsed systems or, if tiny influences accumulate over very large distances like in cosmological observations.

The question remains, how to include potential energy into the tensorial framework of relativity. General relativity is a strictly local description of gravity. The influence of all surrounding matter at some point is encoded in the local curvature and in the metric tensor. Thus the potential energy part of the energy tensor can only contain the local matter density, the local curvature and the metric tensor. It must vanish, if the matter density is zero and if the curvature is zero. In the limit of Newtonian gravity it should reduce to the classical potential energy. A full mathematical discussion would be outside the scope of this article. To understand the implications on the general cosmological model we will confine ourselves to homogeneous solutions.

3. Homogeneous Cosmological Solutions

From many astronomical observations it is apparent that we are not living in a preferred position in space, but that on large scale space is homogeneous and isotropic. Thus its average local properties can be described by some constant matter density, some density of kinetic energy, expressed as an homogeneous pressure, and by some density of potential energy, expressed by the curvature and the matter density. Radiation fields can be neglected here.

The form of the potential energy term can be derived from the analogy to Newtonian gravity. The only possible ingredients are the matter density ϱ, the radius of curvature a, and the metric tensor g_{ij}. The dependence on a can be quantified by considering a homogeneously curved space filled with particles at rest, say the surface layer of a sphere with radius a. The gravitational force between individual particles scales with the reciprocal square of their distance. To expand the sphere to a larger radius work has to be done against the mutual gravitational attraction. This force scales as $1/a^2$. By integration we find that the gravitational potential scales with the radius of curvature as $1/a$. As the potential energy should vanish for vanishing curvature, the potential energy term must be of the form $\lambda \varrho/a \times g_{ij}$. λ is a constant which can be determined from the Newtonian limit.

Thus the energy tensor looks just like the one used by Einstein to describe his static cosmological model, if we identify the quantity $\Lambda = \lambda \varrho/a$ with Einstein's cosmological constant. But now Λ is not a true constant, but it scales with matter density and curvature. If we regard a as a dynamical quantity, as it is done in the expanding Big Bang model, the matter density scales as $1/a^3$ so that Λ scales as $1/a^4$. Thus unlike in Einstein's static model the universe is stable against changes of a. The variation of potential energy with a is steeper than the slope of ϱ and thus increasing a leads to negative da/dt, as can be verified from the field equation (2).

In the case of a homogeneous isotropic universe the path element must be the same for every point. The elements of the metric tensor can only depend on the temporal distance τ and the spatial distance σ between two points. The path element can be written in the form

$$ds^2 = -f(\sigma, \tau) d\tau^2 + h(\sigma, \tau) d\sigma^2, \tag{5}$$

where $d\sigma = \sqrt{dx_1^2 + dx_2^2 + dx_3^2}$ is the spatial distance element in some orthonormal basis. A special solution of the field equation (2) is obtained, when the dependence of f and h on σ is omitted. In this case by rescaling the time coordinate we can always set $f = 1$. This leads to expanding world models like the Big Bang model with its singularity and infinity problems and the need of introduction of new physics.

The occurrence of singularities can be avoided only, if h does not depend on τ. Also in this case we can reach $f = 1$ by a suitable gauge transformation, relating the time coordinate to some fixed observer by

$$t = \int_0^\tau \sqrt{-f(\sigma)}\, d\tau, \tag{6}$$

using the fact that in observations by radiation the relation $d\sigma = c\, d\tau$ must be valid due to the constancy of the vacuum speed of light. This transformation leads to a solution in the form of Einstein's static universe. But now we must consider the finite speed of interaction for every motion explicitly, and we must add the potential term to the energy tensor. While in the strictly local representation the space-time geometry is completely encoded in the curvature, extending the coordinate system to a finite range requires the introduction of the explicit potential energy term in the field equation to maintain global energy conservation.

Motion in a curved static universe is different from Newtonian physics. In curved space every free motion along a geodesic line must be regarded as an accelerated motion, the acceleration being perpendicular to the direction of motion. According to relativity this acceleration leads to a continuous change of the local time scale and thus to similar effects as the continuous local acceleration in an expanding space.

4. Redshift

The most prominent argument in favour of expanding space was the detection of cosmological redshift. But as has been indicated above in curved space relativistic effects lead to an energy loss of the same order without any expansion. We can demonstrate this as well with a locally fixed time coordinate according to eq.(6) as in the strictly local commoving system. While in the first case momentum loss appears as a consequence of retarded interaction with the gravitational potential, in a commoving system it is regarded as a consequence of the change of time scale along the path of a moving particle or energy quantum.

In the commoving system the functions $f(\sigma)$ and $h(\sigma)$ are easily determined. In a curved system of curvature radius a the relation between the distance in some linear direction x and the length measured along the curved line is given by

$$\sigma = a \arctan\left(\frac{x}{\sqrt{a^2 - x^2}}\right). \tag{7}$$

The derivative, needed to determine the curvature, is

$$d\sigma = \frac{a}{\sqrt{a^2 - x^2}}\, dx, \quad \text{leading to} \quad h(\sigma) = \frac{a^2}{a^2 - \sigma^2} \tag{8}$$

with the derivatives $h'(0) = 0$ (as must be expected from the condition of isotropy) and $h''(0) = 2/a^2$. Under the assumption that ordinary matter of density ϱ is the dominant source of curvature the field equation (2) can be solved for $f(\sigma)$ and the relation between the curvature radius a and the matter density:

$$G_{00} = R_{00} - \frac{f}{2} R = \frac{6}{a^2} = \kappa \varrho c^2 \tag{9}$$

$$G_{ii} = R_{ii} - \frac{h}{2}R = -\frac{f'^2}{2f^2} + \frac{f''}{f} - \frac{2}{a^2} = 0 \tag{10}$$

with the solution $f = e^{-2\sigma/a}$ and $1/a = 4/3\pi G\rho/c^2$. The time scale changes exponentially (or linear as a first approximation), equivalent to loss of momentum with the distance traveled by a moving particle. This general loss of momentum is unique to all kinds of particles, be it massive or massless like photons. From eq.(10) we can immediately see that $f = 1$ is not a valid solution, unless we replace the value zero on the right hand side by some negative pressure term $\kappa p = -2/a^2$, as it was introduced in Einstein's static solution, or we have to introduce it in form of a potential energy term, if we transform the field equations to a locally fixed time coordinate.

In the discussion above we have neglected minor contributions to the energy tensor like radiation and pressure or other kinetic energy fields. A more detailed description shows that the loss of momentum or kinetic energy is always compensated by a change of potential energy, just as we are accustomed in Newtonian gravity. Energy is globally conserved, unlike in the Big Bang model, where cooling with expansion leads to vanishing of energy from radiation fields and gain of energy by the so called dark energy field.

The magnitude of the red shift is of the same order as in an expanding universe. The only difference is that in our locally fixed coordinate system we have to add the potential energy term $-1/2\kappa\varrho g_{ij}$ to the energy tensor, leading to an equilibrium density of only 2/3 of the so called critical density $\varrho = 3H^2/(8\pi G)$, discussed in the Big Bang model with Euclidean space structure. Assuming a value of 70 km/s/Mpc for the Hubble constant H the total matter density thus would be about $6 \cdot 10^{-30}$g/cm^3.

From the equivalence of spatial curvature with a gravitational potential we see that in a homogeneous static universe gravitational attraction of matter is balanced by the repulsive action of its own potential energy. Globally this equilibrium is stable, but local perturbations lead to the formation of structures as we observe them as stars and galaxies.

5. The Matter Cycle

From observations we know that cosmic structures evolve in time. The Big Bang model ascribes this evolution to the occurrence of instabilities during the cooling period of a hot cosmic plasma. The final state will be either a infinitely dilute cold gas or matter collapsed into infinitely dense black holes. Apart from the fact that it is scarcely possible to explain, how the observed structures at high redshift could have been developed in the short time since the Big Bang, the origin and size of the primordial fluctuations remains unexplained.

If the universe is globally static, there are no such timing and singularity problems, but the continuous changes must be regarded as local fluctuations around a global equilibrium state. A matter cycle must exist which explains the formation of structures, but finally leads back to the initial state. The existence of such a cycle is possible only by the interplay of the positive pressure connected with motion, and the negative pressure connected with potential energy.

Many steps of structure formation can be described in a static universe similar as in the Big Bang model. The main difference is that no matter can vanish permanently from the cycle by formation of black holes. Thus we will first discuss the question, why this permanent loss of matter does not occur.

5.1. Black Holes

In the conventional theory of gravitational collapse there is no process which can prevent the matter from shrinking into a singularity. But as has been discussed above, a correct relativistic theory must include potential energy with its negative pressure term into the balance. It is the repulsive action of this negative pressure, which prevents collapse into a singularity. Instead contraction will proceed into a stable configuration, where the attraction of matter and kinetic pressure is balanced by the repulsion from potential energy. In a radially symmetric configuration this point is reached at the Schwarzschild radius $r_s = 2MG/c^2$.

In real cases such a collapsed system will not be radially symmetric, but the accretion of matter will bring with it angular momentum and magnetic fields. Though part of the rotation and the magnetic fields of individual matter clumps or stars, which fall into the 'black hole', will compensate, the central collapsed system will be rotating rapidly and contain strong magnetic fields. Thus there is no stable equilibrium configuration. In the direction of the axis of rotation there may be an overshot of repulsive potential energy, while in the plane of rotation the radial equilibrium is not reached. This leads to the ejection of matter in the direction of the axis of rotation.

The emission of matter jets from the collapsed cores of active galaxies is a well observed phenomenon. From the mechanism of formation we can understand the properties of the jets. The jets originate in an environment, where the potential energy of matter is comparable to the rest energy equivalent of matter itself. As at least the final steps of collapse must be regarded as nearly adiabatic, the kinetic energy must be of the same order, in the range of the rest energy of the proton. Under these conditions, at a temperature of 10^{13}K there exist no longer individual chemical elements, but a mixture of the basic building blocks of matter: protons, electrons and neutrons or eventually other unstable elementary particles.

The primary composition of the jets will reflect this composition: a relativistic plasma with a temperature of about 10^{13}K. The ejection of this plasma is strongly constricted to the axis of the magnetic field. Charged particles emitted perpendicular to the field lines would spiral back to the source. The conditions in the collimated jets are similar to those proposed in the Big Bang model at the time of formation of atoms. Thus we expect that on their way into free space part of the elementary particles combine into helium and other light elements, just as described in the Big Bang model. The jets with the so called primordial element composition partly react with the surrounding matter of the parent galaxy, making the jet structure visible over large distances, but most of the matter escapes into free space, forming the seed for new generations of galaxies. There is no Big Bang, but the collapsed galaxy cores with their jets form a continuous series of small bangs, which regenerate the primordial plasma, which is the material for the formation of new structures.

5.2. The Cosmic Plasma

Just like in the Big Bang model the extremely hot particles ejected into space must undergo a long cooling process, before they are able to condense into structures like galaxies or stars. In the Big Bang model this cooling is caused by the expansion of space. In a static

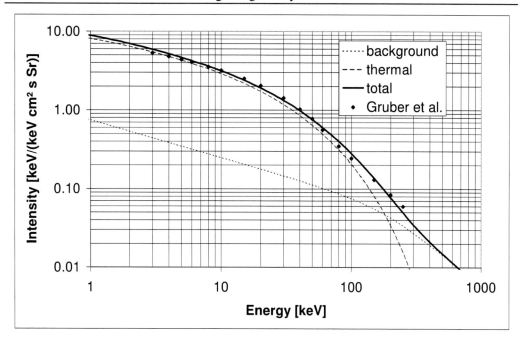

Figure 1. X-ray background radiation from the cosmic plasma. Measurements from Gruber et al. [3] and model calculations (see text).

cosmos momentum loss caused by movement in the gravitational potential, the gravitational radiation, is the primary cooling mechanism. Electromagnetic radiation will play a minor role. There will be emission of bremsstrahlung, but as long as the density of the plasma is low, electromagnetic interaction will only lead to thermalisation of the energy distribution. Cooling by bremsstrahlung will become important only, when overdense regions form by instability.

The loss of energy by gravitational radiation is slow compared to the time scale of changes in the life of galaxies. Thus we must expect that most of the matter content of the universe is present in the form of a hot plasma of thermal energy distribution with an high energy tail from the injection process. Though the emission of bremsstrahlung is very weak, it can be measured as a continuous x-ray background radiation. The spectrum and the intensity is well known, so that the temperature and density of the emitting plasma can be estimated. The accuracy is limited, however, by contributions from unresolved individual sources and by possible absorption or scattering by foreground matter concentrations.

Fig.1 shows data from Gruber et al.[3] together with a fitting curve with $\rho = 3.6 \cdot 10^{-30} g/cm^3$ and $kT = 110\,keV$. To account for the high energy tail from injection a contribution from $1 \cdot 10^{-30} g/cm^3$ high energy particles with an energy distribution $dn_e/dE \sim 1/E$ between 1 GeV and 100 keV has been added. Due to the decreasing collisional cross section the contribution of the high energy tail is considerably lower than that of the thermal plasma. Taking into account that there may be some attenuation by absorbing particles, these values are in good agreement with the total matter density of $6 \cdot 10^{-30} g/cm^3$ estimated from redshift.

In the Big Bang model there is no room for such a hot plasma. The X-ray continuum is ascribed to unresolved individual sources, though there are no known sources of radiation, which could explain the high energy part of the spectrum above 10 keV.

5.3. Structure Formation

Though the cosmic plasma constitutes the overwhelming part of the total matter content of the universe, there will always be residuals of former galaxies and some dust grains which have escaped from supernova explosions into free space. Thus there are always some density fluctuations which under suitable conditions can develop into new structures. The particle energy of fresh injected matter is so high that it scarcely reacts with existing matter. The mean free path is larger than the diameter of the universe. It remains homogeneously distributed until it has lost sufficient energy by gravitational radiation to get thermalised. In the hot thermal plasma the mean free path is considerably less. It reacts with the existing fluctuations, which can develop into large structures.

Gravitational instability is well known since the days of Jeans (1902). Perturbations in a homogeneous matter distribution lead to pressure waves, which can enhance the initial perturbation. There is a critical minimum mass involved in the initial perturbation, the Jeans mass, above which perturbations are not damped out. For a radially symmetric perturbation of the matter distribution with radius R_J and a mean density ϱ the Jeans mass is

$$M_J = \frac{4\pi}{3}\varrho R_J^3 = \frac{\pi}{6}\left(\frac{kT}{Gm}\right)^{3/2}\varrho^{-1/2}. \tag{11}$$

From this formula we find for the stability limit of the intergalactic plasma ($kT = 110\,\text{keV}$, $\varrho = 3.6 \cdot 10^{-30}\text{g/cm}^3$) $R_J = 107\,\text{Mpc}$, which corresponds to the largest observed structures in the universe. Primarily the underdense regions may become unstable. In overdense regions contraction leads to an increase of pressure, which prevents further collapse, underdense regions can lose their matter down to complete vacuum. Thus Jeans instability will result in a filamentary structure with enclosed vacuum bubbles, just the foamy structure which we see in astronomical observations.

In the densest regions of the filaments cooling by emission of electromagnetic bremsstrahlung becomes more important so that the pressure rise is counteracted by radiation loss. New instabilities arise inside the filaments. But now growth of overdense regions is preferred, as radiation loss increases with the square of the electron density. The filaments fragment into individual plasma clumps, which are the progenitors of galaxies and galaxy clusters. The individual clumps cool down and contract further with the internal pressure balancing gravitational attraction.

With the assumption that the individual regions cool down isobaric, they reach the recombination limit of about $kT = 10\,\text{eV}$ at a density of $4 \cdot 10^{-26}\text{g/cm}^3$. At this point by emission of recombination continua and atomic line radiation the cooling process is strongly accelerated so that a further fragmentation can be expected. This fragmentation corresponds to a stability limit of $R_J = 1.7\,\text{Mpc}$ and a mass of $2.5 \cdot 10^{43}\text{g}$, the typical mass of a galaxy.

If the gas cools isobarically down to about 10 K, the density would be of the order 10^{-22}g/cm^3. The critical mass in this case is $5 \cdot 10^{35}\text{g}$, corresponding to the mass of the largest stars. The distance between these stars would be about $2R_J = 6\,\text{pc}$, corresponding to the size of observed regions of star formation. Of course this extrapolation into the range of star formation should be looked at with caution. Here the dynamics is more complex, as the dust component of the medium is of considerable influence.

The very complex processes in galaxies cannot be discussed here in detail. Angular momentum, which individual matter clumps take up during fragmentation, dominates the evolution of spiral galaxies and prevents them from further collapse. Chemical reactions and nuclear fusion produce kinetic energy, which stabilises stars against gravitational forces. But on the Hubble time scale all motions lose kinetic energy on account of the gravitational potential. The motion of individual stars and the angular momentum of spiral galaxies will decrease. Stars cannot remain fixed on circular orbits, but will all spiral into the core of the galaxy. This is just what we observe. Galaxies are spirals, not circular objects like the ring structure of Saturn. Only if the rotation period of systems is short compared to the Hubble time, orbits look circular.

With the accretion of matter onto the central compact core of galaxies we come to the completion of the matter cycle, where perturbation of the balance between potential energy and rest energy of matter leads to the emission of hot matter jets and thus back to the starting point of the cycle. That all this occurs can be understood on the basis of the established laws of relativity, if only the finite speed of gravitational interaction and the role of potential energy in the conservation laws, expressed by the curvature of space, are taken into account correctly.

6. Microwave Background Radiation

The detection of the diffuse microwave radiation background was considered as one of the strongest arguments in favour of the Big Bang cosmological model. As the universe should have passed through a phase, when radiation was the dominant contribution to the energy budget, from this phase residual fields should have been left after formation of matter particles and decoupling of matter from the radiation field in the recombination phase. This radiation field would then be shifted into the microwave spectral range by the expansion of space.

But apart from the question, where the energy went, when the radiation cooled down, so that today the microwave background contributes only marginally to the total energy, it can be scarcely understood, that the radiation field has retained its black body spectrum all through the recombination era, the subsequent cooling phase and the re-ionisation of the major fraction of the universe. The state of a field, which is not coupled to a stabilising mechanism, will deviate more and more from initial equilibrium, when exposed to perturbations.

The observed homogeneity and the black body spectrum can be understood only, if a stabilising mechanism is at work still today. But from our basic physical knowledge it is evident that such a mechanism exists. It is the same mechanism, which brings the radiation field into equilibrium in every cavity and which is supposed to have established the equilibrium radiation field in the Big Bang model, before the first particles formed.

It is the process, which led Planck to the derivation of his well known radiation law. Based on the quantum nature of radiation, individual photons cannot be regarded as independent entities, when the spatial extension of the wave packages is in the order of their distance. Quantum interference leads to redistribution of the vibration modes of the electromagnetic field. Thus every radiation field in a closed volume tends to an equilibrium distribution, when the walls are kept on a fixed temperature, so that the field cannot take up or lose energy to the walls.

By principle this mechanism is valid also for the complete universe, but here the equilibrium is continuously disturbed by various emission or absorption processes. Additionally all the photons lose energy by the cosmological redshift mechanism. So most of the emitted photons will finally reach the wavelength range, where their wave packages overlap with other photons. Quantum interference begins and they are included into the thermal background field.

Of course, the photons of the background field lose energy, too. The equilibrium temperature will be fixed by the balance between this loss and the energy, which is supplied by the new photons from emission processes. The energy loss of the background field can easily be calculated, but the energy input is difficult to estimate, as our knowledge of the various emission processes is very limited. We have to take the background temperature of 2.7K as an observational fact, just like the Hubble constant.

We cannot expect that the background radiation will ever reach complete equilibrium. It results from a dynamical process. There is energy input, preferably at the high end of the energy distribution, the thermalisation by quantum interference and the loss by gravitational interaction. To confirm the process described here, one could try to detect possible deviations from the black body spectrum at the high energy end, but accurate observations in this spectral range are not possible at present due to the extreme foreground effects.

That the microwave background is not completely homogeneous, is well known from measurements by various satellites like COBE and PLANCK. The spatial fluctuations of the black body temperature have been measured with high accuracy. Such fluctuations must be expected from interactions between the radiation and the charged particles of the intergalactic plasma with its clumpy and filamentary structure. By inelastic scattering from the high energy electrons in the denser regions of the plasma and in large galaxy clusters (Sunyaev Zel'dovich effect) the microwave photons can take up energy from the plasma, shifting the background temperature to higher values.

The observed angular correlation length of the fluctuations shows a strong maximum at multipole moments around 240. This is just the length scale expected from the filamentary structure of the intergalactic plasma or the distribution of galaxy clusters resulting from this structure. The observed fluctuations of the radiation temperature are a consequence of the plasma structure. It is not, as discussed in Big Bang scenarios, the formation of structures as a consequence of temperature fluctuations of the radiation field at the time of recombination. The observed microwave background, its spectrum and also its fluctuations can be explained by well understood processes in a static universe and without invoking processes in a radiation dominated far past.

7. Supernovae

This chapter should not end without addressing the topic of type Ia supernovae. Though they are not directly related to the supposed Big Bang, they are used to constrain the expansion history of the universe. To obtain a reasonable fit of observational results to the model description, the introduction of so called dark energy with repulsive gravity was necessary, leading to an accelerated expansion of space.

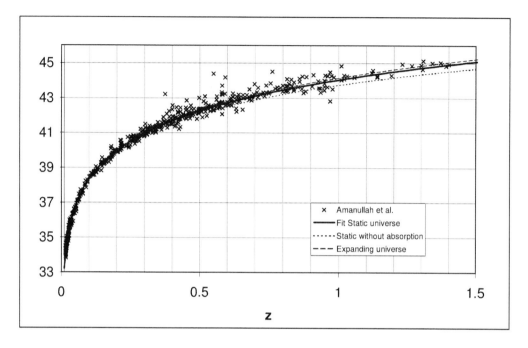

Figure 2. Distance module of SNIa supernovae. Observations (Union2 Compilation [1]) and fitting curves from expanding world model ($\Omega_m = 0.3, \Omega_\Lambda = 0.7$) and static universe.

In both cases, in static curved space and in expanding space, the observed luminosity of supernovae decreases with distance proportional to the size of the surface into which the radiation has expanded since its emission and additionally by cosmological energy loss and time dilatation. Besides in a static universe we must take into account possible attenuation by absorption from dust particles. Such dust particles may have been formed in the debris of former supernovae in former generations of galaxies. Due to their high velocities the largest particles will leave their parent galaxy into free space. Particles in the size range of μm will survive there for very long times and thus lead to absorption, which is nearly independent of wavelength all over the visible and near IR spectral range.

In the static case the observed intensity thus scales with the red shift parameter z as

$$I \sim \frac{1}{(1+z)^2} e^{-\kappa s} \frac{1}{F} \quad \text{with} \quad s = a\ln(1+z) \tag{12}$$

κ is the absorption coefficient, F the area into which the emitted radiation has expanded. In curved space of radius a the area at distance s is $F = 4\pi a^2 \sin^2(s/a)$.

In the Big Bang model absorption cannot be expected and space is Euclidean. But the distance traveled by the light depends on the expansion history. Denoting the actual expansion parameter by H_0 the commoving distance change is

$$\frac{ds}{dz} = \frac{c}{H_0} \frac{1}{\sqrt{\Omega_m(1+z)^3 + \Omega_\Lambda}}. \tag{13}$$

Ω_m and Ω_Λ are the relative fractions of matter and dark energy to the closing density ($\Omega_m + \Omega_\Lambda = 1$). A reasonable fit to observational data is possible only, if Ω_Λ is about 0.7. The overwhelming part of the universe must be this unknown repulsive energy. In Fig.2 observational data from the Union2 Compilation [1] are shown together with fitting curves for the Big Bang model and for a static universe. The data are presented by their distance module, a normalised logarithmic scale.

In the static case the best fit is obtained, if a slight absorption is included in the fit. Assuming dust grains of $r = 1\mu m$ radius with an absorption cross section $\sigma = \pi r^2$ and a density of $\varrho_d = 2 g/cm^3$, this corresponds to a fraction of 8×10^{-4} of the total matter density, considerably less than the concentration of heavy elements in galaxies and clusters.

This demonstrates that a reasonable fit to the supernova data is possible also in a static world model. But in this case we need no hypothetical dark energy with its strange properties. The data can be explained applying only well proved physics.

8. Conclusion

In this chapter we have demonstrated that the Big Bang model of an expanding universe is not the only possible solution to describe the global properties of the universe. On the contrary this model puts more unresolved questions than it answers. To stay in agreement with observations more and more new physics had to be introduced and the problem of unphysical singularities has not been solved.

A quasi-static description of the universe, in which all the observed changes and developments are fluctuations around a global equilibrium, is possible and can be obtained within the scope of well proved physical principles. On the basis of Einstein's theories of special and general relativity we can construct a model, where the basic phenomena, which have been used as convincing arguments in favour of the Big Bang model, find another natural explanation without introducing any new physics and without the occurrence of singularities.

Redshift is not the result of expansion, but in curved space it must even be regarded as a necessary consequence, if we propose that energy conservation and Lorentz invariance are valid not only locally, but also on global scale. The fact that globally gravitational action of matter is balanced by its own potential energy has the appealing consequence that the total energy of the universe can be regarded as zero. Such a universe can be created from nothing. Effects resulting from the limitation of gravitational interaction to the speed of light have been studied in various publications in the past. But astonishingly they have been left out of consideration in the global world models. If we take the concept of relativity serious, a global time scale, which can be used to describe the development of the entire universe, cannot exist.

At this time the Big Bang model is so firmly established that it appears as a great risk to question the basis of the expanding world model and thus completely collapsing the building, which has been constructed by generations of scientists. But the features of the static curved universe are so attractive that we should take the risk. The longer we stay on the wrong path and try to save it by introduction of new physics, the higher the hurdles to return to better alternatives.

References

[1] Amanullah et al. (The Supernova Cosmology Project), *ApJ* (2010).

[2] Einstein, A.: Vier Vorlesungen ueber Relativitaetstheorie (1920), *Reprint: Grundzuege der Relativitaetstheorie*, Vieweg, Braunschweig (Germany) (1973)

[3] Gruber, D.E. et al.: *ApJ*, **520**, 124 (1999)

[4] Hawking, S.W. and Ellis, G.F.R.: *The large scale structure of space-time*, Cambridge Univ. Press (UK) (1973)

[5] Wald, R.M.: *General Relativity*, Univ. Chicago Press (US) (1984)

In: The Big Bang: Theory, Assumptions and Problems
Editors: J. R. O'Connell and A. L. Hale

ISBN: 978-1-61324-577-4
© 2012 Nova Science Publishers, Inc.

Chapter 10

DE SITTER-FANTAPPIÈ UNIVERSE

E. Benedetto
Mathematics and Computer Science Department
University of Salerno, Italy

Abstract

In this work we analyze the development of de Sitter-Fantappiè Relativity by analyzing, particularly, its applications to cosmology. The cosmological applications seem to be interesting since they resolve many problems of the standard cosmology. In this scenario the space flatness is linked to the observer geometry and it is independent from the presence and distribution of matter-energy. This resolves the flatness problem without introducing inflationary hypotesis and the global spacetime structure is univocally individuated by the algebraic structure of the physical laws.

1 Introduction

Einstein's General Theory of Relativity revolutionized our thinking about the nature of space and time and it explained gravity in a different way from Newton's law. Gravity is a manifestation of the geometry of spacetime and the gravitational force becomes a metric force, resulting from the local curvature of spacetime. One of the very first applications of General Relativity concerned the Universe itself and the first attempts towards applying General Relativity to cosmology were made by Einstein himself in 1917. The current models of cosmology are based on the following Einstein's equations

$$R_{\mu\nu} - \frac{1}{2}g_{\mu\nu}R = (8\pi G/c^4)T_{\mu\nu} + \Lambda g_{\mu\nu} \qquad (1)$$

where $R_{\mu\nu}$ is the Ricci tensor, R is the Ricci scalar, $T_{\mu\nu}$ is the stress-energy tensor, and Λ is the cosmological constant. By assuming a standard perfect fluid matter we may write

$$T_{\mu\nu} = (p + \rho)u_\mu u_\nu - pg_{\mu\nu} \qquad (2)$$

where p is the pressure, ρ the energy density and u_μ the velocity. Therefore if we apply this equations to the whole Universe, we find the relativistic cosmology, in which the cosmological principle can be postulated and a model of constant spatial curvature obtained. In fact the Robertson-Walker metric describes a spacetime with homogeneous and isotropic spatial sections, so that the intrinsic spatial curvature is constant throughout the space and its general form is written as

$$ds^2 = c^2 dt^2 - a^2(t)[\frac{dr^2}{1-kr^2} + r^2(d\theta^2 + \sin^2\theta d\varphi^2)] \tag{3}$$

in which (r, θ, φ) are the comoving coordinates and $a(t)$ is the scale factor. The dynamical problem is completely set when matter evolution equations are given; they are the contracted Bianchi identities

$$T^\mu_{\nu;\mu} = 0 \tag{4}$$

A further equation has to be imposed in order to assign the thermodynamical state of matter. It, usually, is

$$p = \gamma c^2 \rho \tag{5}$$

where γ is a constant ($0 \leq \gamma \leq 1$ for standard perfect fluid matter). In such context we obtain the system

$$\frac{\ddot{a}}{a} = -\frac{4\pi G}{3c^2}(\rho + 3p) + \frac{\Lambda c^2}{3} \tag{6}$$

$$(\frac{\dot{a}}{a})^2 + \frac{kc^2}{a^2} = \frac{8\pi G}{3c^2}\rho + \frac{\Lambda c^2}{3} \tag{7}$$

$$\dot{\rho} + 3(\frac{\dot{a}}{a})(\rho + p) = 0 \tag{8}$$

Predictions of relativistic cosmology include the initial abundance of chemical elements formed in a period of primordial nucleosynthesis, the large-scale structure of the Universe and the existence and properties of a thermal echo from the early cosmos, the cosmic background radiation. Despite this, we have to pay close attention to General Relativity, where, inevitably, the application of Einstein's equations to cosmological problems requires an extreme extrapolation of their validity to very far regions of spacetime. The relativistic cosmology is unable to provide an explanation as to why the density of the Universe should be so close to the critical value. In fact we have

$$\Omega - 1 = \frac{k}{H^2 a^2} = \frac{k}{\dot{a}^2} \tag{9}$$

and as $\dot{a}^2$ decreases with time, $|\Omega - 1|$ must increase if k is non zero. This means that the Universe diverges from the flat case if $k \neq 0$ and the fact that it appears to be almost flat today means that Ω must have been very close to one in the early Universe. Besides at present, the cosmic microwave background is observed to be extremely homogeneous and isotropic on large scales, with temperature fluctuations of only $10^{-5}K$. This suggests that all regions of the sky were in causal contact at some time in the past, but is contradicted as follows. The horizon size is the distance light has travelled since the beginning of the Universe and is given by

$$d(t) = a(t) \int_{t_1}^{t_2} \frac{dt}{a(t)} \tag{10}$$

which remains finite as $a(t_1) \to 0$ if $\ddot{a} > 0$. When the microwave background was formed the region in causal contact would have been approximately $0.09 Mpc$. With the subsequent expansion this corresponds to a patch of the present microwave background subtending an angle of only 2 degrees. Finally magnetic monopoles should have been created in large numbers during phase transitions in the early universe. They are very massive, stable and survive annihilation, to quickly come to dominate the Universe, resulting in matter domination before the epoch of nucleosynthesis and an observable monopole density today. However, the light elements are observed in the abundances predicted by primordial nucleosynthesis in radiation dominated cosmology and no monopoles are observed, so a mechanism is needed to remove the monopoles prior to nucleosynthesis. The most principal problem is the singularity problem and, according to Hawking-Penrose theorems, the appearance of singularity in cosmological solutions of General Relativity is inevitable [1],[2]. Many physicists and cosmologists are inclined to believe that classical General Relativity must be revised in the case of extremely high energy densities, pressures and temperatures. The singularity must mean for cosmology that the classical Einsteinian theory is inapplicable in the beginning of cosmological expansion of the Universe. Due to these facts alternative theories have been considered as, for example, Extended Theories of Gravity which have become a sort of paradigm in the study of gravitational interaction based on the enlargement and the correction of the traditional Einstein scheme [3]. The paradigm consists in adding higher-order curvature invariants and scalar fields into dynamics which come out from quantum terms in the effective action of gravity. The minimal extension, discussed by Einstein himself to obtain a static Universe model, is the general relativity with a cosmological constant and in this case the action is the following

$$S = \frac{1}{16\pi G} \int \sqrt{-g}(R + 2\Lambda) d^4 x \tag{11}$$

Gravity lagrangians with terms of quadratic or higher order in the Ricci scalar have also been studied in cosmology. In a Riemannian spacetime, let be given the action of the fourth-order as[4]

$$S = \frac{1}{16\pi G} \int \sqrt{-g}(R + \alpha R^2 + 16\pi G_{matter}) d^4 x \tag{12}$$

In a Riemannian spacetime this action is minimized with respect any variation of the metric tensor if and only if we have that

$$(1 + 2\alpha R)(R_{\mu\nu} - \frac{1}{2}g_{\mu\nu}R) = 8\pi G T_{\mu\nu} - \frac{\alpha}{2}R^2 g_{\mu\nu} + 2\alpha(g_\mu^\lambda g_\nu^k - g_{\mu\nu}g^{\lambda k})R_{;\lambda k} \tag{13}$$

The previous relation is the fundamental field equation of the fourth-order gravity in a Riemannian spacetime. We can take into account the most general class of higher–order-theories in four dimensions derived from lagrangians that are functions not only of R but also $\Box R$ or $\Box^n R$, where $\Box$ is the d'Alembertian. They can be generated by the action

$$S = \frac{1}{16\pi G} \int \sqrt{-g}[F(R, \Box R, ..., \Box^n R) + 16\pi G_{matter}]d^4x \tag{14}$$

The field equations are obtained by varying action with respect to the metric getting

$$\Theta(R^{\mu\nu} - \frac{1}{2}g^{\mu\nu}R) = 8\pi GT^{\mu\nu} + \frac{1}{2}g^{\mu\nu}(F-\Theta)R + (g^{\mu\lambda}g^{\nu k} - g^{\mu\nu}g^{\lambda k})\Theta_{;\lambda k}$$

$$+\frac{1}{2}\sum_{i=1}^{n}\sum_{j=1}^{i}(g^{\mu\nu}g^{\lambda k} + g^{\mu\lambda}g^{\nu k})(\Box^{j-1}R)_{;k}(\Box^{i-j}\frac{\partial F}{\partial \Box^i R})_{;\lambda}$$

$$-g^{\mu\nu}g^{\lambda k}[(\Box^{j-1}R)_{;k}\Box^{i-j}\frac{\partial F}{\partial \Box^i R}]_{;\lambda} \tag{15}$$

with

$$\Theta \equiv \sum_{j=0}^{n}\Box^j(\frac{\partial F}{\partial \Box^j R}) \tag{16}$$

Other motivations come from the Mach's Principle, that played an important role in the development of General Relativity, as, for example, the Brans-Dicke theory in which the gravitational interaction is mediated by a scalar field as well as the tensor field of General Relativity and the gravitational constant G is not presumed to be constant but instead $1/G$ is replaced by a scalar field which can vary from place to place and with time [5]. Another theory in accord with Mach's Principle is de Sitter Relativity initially proposed by Fantappiè and subsequently developed by Arcidiacono and other authors [6-18]. At present, both Brans-Dicke theory and de Sitter Relativity are generally held to be in agreement with observation.

The paper is organized as follows: In Sect.2 we introduce de Sitter-Fantappiè group, while in Sect.3 we analyze the 5-dimensional formulation; Sect.4 is devoted to the Arcidiacono equations; in Sect.5 we give the conclusions.

2 De Sitter-Fantappiè Group

As reported in [16] Fantappiè's starting point was the study of classical and relativistic physics spacetime. Space-time of classical physics is a 4-dimensional manifold endowed with the following geometrical structures:
   1) Absolute time;
   2) Absolute space;
   3) Absolute spatial distances;
   4) Absolute temporal distances.

Space-time of special relativity, instead, is a 4-dimensional manifold endowed with the following geometrical structures:
1) Relative time;
2) Relative space;
3) Absolute space-time distances.

Fantappiè noted that general relativity follows an extraneous approach to the tradition of mathematical physics in that it does not follow the group structure of physics. This is different from classical mechanics and special relativity.

As it is well known let us remember that Galileo's group is the main group of classical physics and is formed by the composition of the following transformations:

a) Spatial Rotations - characterized by three parameters

$$x'_\mu = a_{\mu\nu} x_\nu \quad , \quad t' = t, \qquad (17)$$

where $[a_{\mu\nu}]$ è an orthogonal matrix whose determinant is +1.

b) Inertial Movements - characterized by the three components of velocity

$$x'_\mu = x_\mu + v_\mu t \quad , \quad t' = t. \qquad (18)$$

c) Spatial translations - characterized by three parameters

$$x'_\mu = x_\mu + a_\mu \quad , \quad t' = t. \qquad (19)$$

d) Temporal translations - characterized by only one parameter

$$x'_\mu = x_\mu \quad , \quad t' = t + t_0. \qquad (20)$$

Therefore Galileo's group has order 10 and expresses Galileo's well-known relativity principle. Moving on to relativistic physics, spatial rotations and inertial movements are blended in a unique operation, the rotations of an Euclidian space M_4, characterized by 6 parameters,

$$x'_i = a_{ik} x_k, \qquad (21)$$

where $|a_{ik}| = 1$, $x_1 = x$, $x_2 = y$, $x_3 = z$, $x_4 = ict$.

These transformations, called Lorentz's special transformations, form Lorentz's proper group and joining the reflections, form Lorentz's extended group. Then we need to add the translations of M_4

$$x'_i = x_i + a_i, \qquad (22)$$

characterized by 4 parameters, which comprise spatial and temporal translations. By composing the transformations of these two groups, we obtain Lorentz's general transformations which form Poincaré's group of 10 parameters

$$x'_i = a_{ik} x_k + a_i . \qquad (23)$$

Poincarè's group mathematically translates Einstein's relativity principle. When $c \to \infty$ so that $\frac{v}{c} \ll 1$, Minkowski's space–time reduces to that of Newton's and Poincarè's group reduces to Galileo's group.

Fantappiè went on this direction and tried to understand if Poincarè's group could be the limit of a more general group, in the same manner as Galileo's group is the limit of Poincarè's group. In [6] He wrote a new group of transformations which had as limit Poincarè's group and He was able also to demonstrate that his group was not able to be the limit of any continuous group of 10 parameters. That is, by limiting to groups of 10 parameters and to 4–dimensional spaces, what happened with Galileo's and Poincarè's groups cannot be repeated. For this reason this group is called the final group.

Fantappiè's group is characterized by two constants: speed of light c and a radius of space-time r. This group determines an universe endowed with a perfect symmetry: de Sitter's Universe. Let us remember that de Sitter's Universe is obtained from Einstein's equations with a positive cosmological constant and the solution of motion equation is the following

$$r(t) = r(0)e^{ct\sqrt{\Lambda/3}}.$$

This model is a 4-dimesional hyperboloid in real time and a 4-dimensional sphere in imaginary time.

Fantappiè, moving from a space–time with hyperbolic structure, showed that, through a flat projective representation, one could obtain a space–time which generalizes Minkowski's space–time.

Let us consider the homogeneous coordinates so defined

$$x_k = r\bar{x}_k/\bar{x}_5 , \qquad (24)$$

then these transformations form the group of 5–dimensional rotations and are three types (moving to the non–homogeneous coordinates):

a) *Time translations*: considering two observers still standing in the same place, but separated by a great distance in time, that is, the same observer in two different moments

$$\begin{cases} x'_\alpha = \frac{x\sqrt{1-\eta^2}}{1-\eta t/(r/c)}, \\ t' = \frac{t-T_0}{1-\eta t/(r/c)}, \end{cases} \qquad (25)$$

with $\eta = \frac{T_0}{r/c} = \frac{T_0}{t_0}$ where T_0 is the parameter of time translation. It follows that for $t = \pm r/c$, one has $x' = 0$.

These transformations for $r \to \infty$ are reduced to the classic time translations

$$x' = x \quad , \quad t' = t - T_0. \qquad (26)$$

b) *Spatial Translations*: considering two observers at the same time and still standing compared to each other, but separated by great distance in space (for example along the x

axis)

$$\begin{cases} x' = \frac{x-S}{1+\alpha x/r}, \\ y' = \frac{y\sqrt{1+\alpha^2}}{1+\alpha x/r}, \\ z' = \frac{z\sqrt{1+\alpha^2}}{1+\alpha x/r}, \\ t' = \frac{t\sqrt{1+\alpha^2}}{1+\alpha x/r}, \end{cases} \quad (27)$$

with $\alpha = \frac{S}{r}$ and where S is the parameter of translation along the x axis. At the relativistic limit, that is for $r \to \infty$, equations (27) reduce to

$$x' = x - S, \quad y' = y, \quad z' = z, \quad t' = t. \quad (28)$$

c) *Pullings*: considering two observers that initially coincide, and one moving rectilinearly and uniformly to the other, with velocity parallel to the x axis

$$\begin{cases} x' = \frac{x-Vt}{\sqrt{1-\beta^2}}, \\ y' = y, \\ z' = z, \\ t' = \frac{t-Vx/c^2}{\sqrt{1-\beta^2}}. \end{cases} \quad (29)$$

2.1 Minkowski's metrics and de Sitter-Fantappiè metrics

Let us recall that relativistic metric is given by

$$ds^2 = \eta_{\mu\nu} dx^\mu dx^\nu \quad (30)$$

with

$$\eta_{\mu\nu} = \begin{pmatrix} 1 & 0 & 0 & 0 \\ 0 & 1 & 0 & 0 \\ 0 & 0 & 1 & 0 \\ 0 & 0 & 0 & -1 \end{pmatrix} \quad (31)$$

By following Fantappié transformations, instead, the metric of spacetime can be written as

$$L^2 ds^2 = L(\sum_{i=1}^{4} dx_i dx_i) - (\sum_{i=1}^{4} \frac{x_i}{r} dx_i)^2 \quad (32)$$

with

$$L = \frac{x_1^2 + x_2^2 + x_3^2}{r^2} - \frac{c^2}{r^2}t^2 + 1 = \frac{x_1^2 + x_2^2 + x_3^2}{r^2} - \left(\frac{t}{t_b}\right)^2 + 1 = \frac{x_1^2 + x_2^2 + x_3^2}{r^2} - \eta^2 + 1. \quad (33)$$

At the relativistic limit, that is for $r \to \infty$, this metric is reduced to Minkowski's metrics in fact

$$\begin{cases} L \to 1 \\ \sum \frac{x_i}{r}dx_i \to 0 \end{cases} \quad (34)$$

In the Fantappié-Arcidiacono Universe, therefore, the metrics reads

$$ds^2 = f_{\mu\nu}dx_\mu dx_\nu \quad (35)$$

with

$$f_{\mu\nu} = \begin{pmatrix} \frac{1}{L} - \frac{x_1^2}{L^2 r^2} & -\frac{x_1 x_2}{L^2 r^2} & -\frac{x_1 x_3}{L^2 r^2} & -\frac{x_1 x_4}{L^2 r^2} \\ -\frac{x_1 x_2}{L^2 r^2} & \frac{1}{L} - \frac{x_2^2}{L^2 r^2} & -\frac{x_2 x_3}{L^2 r^2} & -\frac{x_2 x_4}{L^2 r^2} \\ -\frac{x_1 x_3}{L^2 r^2} & -\frac{x_2 x_3}{L^2 r^2} & \frac{1}{L} - \frac{x_3^2}{L^2 r^2} & -\frac{x_3 x_4}{L^2 r^2} \\ -\frac{x_1 x_4}{L^2 r^2} & -\frac{x_2 x_4}{L^2 r^2} & -\frac{x_3 x_4}{L^2 r^2} & \frac{1}{L} - \frac{x_4^2}{L^2 r^2} \end{pmatrix} \quad (36)$$

Therefore, the metric is function of spacetime also without gravitational field.

2.2 Relativistic proper time and de Sitter-Fantappiè proper time

In the special relativity, we have

$$ds^2 = dx^2 + dy^2 + dz^2 - c^2 dt^2 = -c^2 d\tau^2. \quad (37)$$

Therefore

$$c^2 d\tau^2 = c^2 dt^2 \left(1 - \frac{dx^2 + dy^2 + dz^2}{c^2 dt^2}\right) = c^2 dt^2 \left(1 - \frac{v^2}{c^2}\right) \quad (38)$$

and so

$$d\tau = dt\sqrt{1 - \beta^2}. \quad (39)$$

In the projective Universe instead

$$\begin{aligned} L^2 ds^2 &= L(dx^2 + dy^2 + dz^2 - c^2 dt^2) - \left(\frac{x}{r}dx + \frac{y}{r}dy + \frac{z}{r}dz - \frac{c^2 t}{r}dt\right)^2 = \\ &= L(dx^2 + dy^2 + dz^2 - c^2 dt^2) - \left(\frac{x}{r}dx + \frac{y}{r}dy + \frac{z}{r}dz - \eta c dt\right)^2 = \\ &= Lc^2 dt^2(\beta^2 - 1) - \left[cdt\left(\frac{xdx}{rcdt} + \frac{ydy}{rcdt} + \frac{zdz}{rcdt} - \eta\right)\right]^2 = \\ &= Lc^2 dt^2(\beta^2 - 1) - c^2 dt^2 \left(\frac{x}{r}\frac{v_x}{c} + \frac{y}{r}\frac{v_y}{c} + \frac{z}{r}\frac{v_z}{c} - \eta\right)^2 = \\ &= c^2 dt^2 \left[L(\beta^2 - 1) - \left(\frac{x}{r}\frac{v_x}{c} + \frac{y}{r}\frac{v_y}{c} + \frac{z}{r}\frac{v_z}{c} - \eta\right)^2\right]. \end{aligned}$$

Therefore we can write

$$ds^2 = \frac{c^2 dt^2 [L(\beta^2 - 1) - (\frac{x}{r}\frac{v_x}{c} + \frac{y}{r}\frac{v_y}{c} + \frac{z}{r}\frac{v_z}{c} - \eta)^2]}{L^2} = -c^2 d\tau^2.$$

Then we obtain

$$d\tau = \frac{dt\sqrt{L(1-\beta^2) + (\frac{x}{r}\frac{v_x}{c} + \frac{y}{r}\frac{v_y}{c} + \frac{z}{r}\frac{v_z}{c} - \eta)^2}}{L} = dt\frac{M}{L}. \tag{40}$$

2.3 Relativistic mass and de Sitter-Fantappiè mass

In relativistic spacetime, the mass is only function of velocity, in fact

$$m = m_0 \frac{dt}{d\tau} = \frac{m_0}{\sqrt{1-\beta^2}}. \tag{41}$$

In projective spacetime instead

$$m = m_0 \frac{dt}{d\tau} = m_0 \frac{L}{M}. \tag{42}$$

Therefore the mass is also function of spacetime coordinates. We study the following three interesting cases:

1) In the system of the observer for $t = 0$ we have

$$m = \frac{m_0}{\sqrt{1-\beta^2}}$$

in agreement with special relativity

2) For the masses that move following Hubble's law we obtain

$$m = m_0 \frac{\frac{x^2}{r^2} + 1}{\sqrt{(\frac{x^2}{r^2} + 1)(1 - \frac{H^2 x^2}{c^2}) + \frac{x^2}{r^2}\frac{H^2 x^2}{c^2}}} =$$

$$m_0 \frac{\frac{x^2}{r^2} + 1}{\sqrt{(\frac{x^2}{r^2} + 1)(1 - \frac{x^2}{r^2}) + \frac{x^4}{r^4}}} = m_0(\frac{x^2}{r^2} + 1).$$

The mass increases with the distance.

3) For $x = \beta = 0$ we have

$$m = m_0(1 - \eta^2).$$

At the time of Big Bang all the masses tend to zero.

3 5-Dimensional formulation

In [17] we considered the dynamical properties of projective spacetime and we shortly resume the gotten results. The projective 4-velocity is defined by the following relation

$$\begin{cases} U_\alpha = v_\alpha \frac{L}{M} \\ U_4 = c\frac{L}{M} \end{cases} \tag{43}$$

For $r \to \infty$ we have

$$\begin{cases} L \to 1 \\ M \to \sqrt{1-\beta^2} \end{cases} \tag{44}$$

then we obtain the relativistic 4-velocity. Fantappié's transformations represent 5-dimensional rotations, [8], and therefore we defined the following 5-vectors:

$$\begin{cases} \mathbf{U}_\mu = \frac{d\overline{x}_\mu}{d\tau} \\ \mathbf{P}_\mu = m_0 \mathbf{U}_\mu \end{cases} \quad (\mu = 1,2,3,4,5) \tag{45}$$

Remembering that

$$x_\mu = r\frac{\overline{x}_\mu}{\overline{x}_5}$$

we obtain

$$U_\mu = \frac{dx_\mu}{d\tau} = \frac{r(\mathbf{U}_\mu \overline{x}_5 - \mathbf{U}_5 \overline{x}_\mu)}{\overline{x}_5^2}. \tag{46}$$

By multiplying for m_0, it follows that

$$P_\mu = \frac{r(\mathbf{P}_\mu \overline{x}_5 - \mathbf{P}_5 \overline{x}_\mu)}{\overline{x}_5^2}. \tag{47}$$

These are the relations that link 4-dimensional and 5-dimensional vectors. The following equation

$$f = \frac{d}{dt}(m_0 v \frac{L}{M}) \tag{48}$$

represents the classical 3-dimensional force, where t is the projective time, and now we introduce 5-dimensional force defined as

$$\mathbf{F} d\tau = d\mathbf{P}. \tag{49}$$

By replacing the definition of the proper time we can write

$$\mathbf{F} dt \frac{M}{L} = d\mathbf{P} \tag{50}$$

For small variations of $\overline{x}_5$ it is possible to write

$$\begin{cases} \mathbf{P}_1 = P_1 \frac{\overline{x}_5}{r} \\ \mathbf{P}_2 = P_2 \frac{\overline{x}_5}{r} \\ \mathbf{P}_3 = P_3 \frac{\overline{x}_5}{r} \\ \mathbf{P}_4 = P_4 \frac{\overline{x}_5}{r} \\ \mathbf{P}_5 = 0 \end{cases} \qquad (51)$$

By dividing spatial component by temporal component we obtain

$$\begin{cases} \frac{\mathbf{P}_1}{\mathbf{P}_4} = \frac{P_1 \frac{\overline{x}_5}{r}}{P_4 \frac{\overline{x}_5}{r}} = \frac{m_0 L v_1}{M} \frac{M}{m_0 L c} = \frac{v_1}{c} \\ \frac{\mathbf{P}_2}{\mathbf{P}_4} = \frac{v_2}{c} \\ \frac{\mathbf{P}_3}{\mathbf{P}_4} = \frac{v_3}{c} \end{cases} \qquad (52)$$

It follows that

$$\frac{\widehat{P}}{\mathbf{P}_4} = \frac{v}{c} \Rightarrow v = \frac{\widehat{P}c}{\mathbf{P}_4}. \qquad (53)$$

Considering that, in the same approximation,

$$\mathbf{P}^2 = \left|\widehat{P}\right|^2 - \mathbf{P}_4^2 = -m_0^2 c^2 \qquad (54)$$

we obtain

$$\widehat{P} \cdot d\widehat{P} - \mathbf{P}_4 d\mathbf{P}_4 = 0. \qquad (55)$$

Consequently,

$$\frac{v}{c} \cdot d\widehat{P} = d\mathbf{P}_4 \Rightarrow f \cdot v dt \frac{\overline{x}_5}{r} = c d\mathbf{P}_4. \qquad (56)$$

Therefore the previous equation becomes

$$dE = \frac{r}{\overline{x}_5} c d\mathbf{P}_4. \qquad (57)$$

Then we have the relation

$$E = \mathbf{P}_4 c \frac{r}{\overline{x}_5} + A = m_0 c^2 \frac{L}{M} + A \qquad (58)$$

where A is an integration constant.

Now we consider a particle initially at rest, and then we apply a force on it. We can write

$$\frac{\overline{x}_5}{r} \int_0^t f \cdot v dt = [c\mathbf{P}_4]_0^t \tag{59}$$

From equations (44), (26) we have

$$L = E - E_0 = m_0 c^2 \frac{L}{M} - m_0 c^2 \frac{L}{\sqrt{\frac{x^2}{r^2} + 1}}. \tag{60}$$

For example let us write the following system

$$\begin{cases} \frac{d}{dt}(m_0 v \frac{L}{M}) = f \\ x(0) = 0 \\ v(0) = 0 \end{cases} \tag{61}$$

where f is a constant force. Then, we obtain

$$d(m_0 v \frac{L}{M}) = fdt \Rightarrow m_0 v = ft\frac{M}{L}. \tag{62}$$

In conclusion

$$v = at\frac{M}{L}. \tag{63}$$

In the relativistic limit this relation can be written

$$v = at\sqrt{1-\beta^2} = \frac{at}{\sqrt{1+(at/c)^2}}. \tag{64}$$

This is the expression of the uniformly accelerated motion in special relativity.

4 Arcidiacono Equations

Let us remember that, following the definition of Cartan, any Riemann manifold is associated with an infinite family of Euclidean spaces tangent to it in each of its P points. These infinity spaces are joined by a connection law and are individuated by a holonomy group. By introducing a local coordinates system y^i and a linear forms, ω^i, of differential dy^i we can write $ds^2 = \omega_s \omega^s$. If we consider, on the tangent space to a point P, four ortogonal vectors e_i we have[10]

$$\begin{cases} dP = \omega^i e_i \\ de_i = \omega_i^k e_k \\ e_i e_k = \delta_{ik} \end{cases} \tag{65}$$

where $\omega_k^i = \gamma_{ks}^i \omega^s$ and γ_{ks}^i are the Ricci rotation coefficients. If the point P and the associated reference frame describe a closed infinitesimal cycle on the tangent space, in general the vector e_i' doesn't coincide with e_i and the cycle is open. It can be closed through a translation Ω^i and a rotation Ω_k^i on the tangent space and we have

$$\begin{cases} \Omega^i = d\omega^i + \omega^i_s \wedge \omega^s \\ \Omega^i_k = d\omega^i_k + \omega^i_s \wedge \omega^s_k \end{cases} \qquad (66)$$

where Ω^i is the torsion and Ω^i_k is the curvature. To develop the projective general relativity we have to introduce a 5−dimensional Riemann manifold which allows as holonomy group the Fantappié one, isomorphic to the 5−dimensional rotations group, and the gravitation equations are

$$R_{AB} - \frac{1}{2}Rg_{AB} = \chi T_{AB} \qquad (A, B = 0, 1, 2, 3, 4) \qquad (67)$$

where g_{AB} are the coefficients of the five-dimensional metric. We immediately understand, from how much we have said above, that, in projective general relativity, is fundamental the geometry of projective connections and therefore let us remember the following concepts.

If we have a differentiable manifold M and it symmetric connection θ, the curvature tensor in local coordinate is

$$K^\alpha_{\beta\gamma\delta} = \frac{\partial \theta^\alpha_{\beta\delta}}{\partial x^\gamma} - \frac{\partial \theta^\alpha_{\beta\gamma}}{\partial x^\delta} + \sum_\sigma (\theta^\alpha_{\sigma\gamma}\theta^\sigma_{\beta\delta} - \theta^\alpha_{\sigma\delta}\theta^\sigma_{\beta\gamma}). \qquad (68)$$

The Ricci tensor of connection is

$$K_{\beta\delta} = \sum_\alpha K^\alpha_{\beta\alpha\delta}. \qquad (69)$$

By setting

$$A_{\beta\delta} = \frac{1}{2}(K_{\beta\delta} - K_{\delta\beta}) = \frac{1}{2}\sum_\alpha \left(\frac{\partial \theta^\alpha_{\alpha\delta}}{\partial x^\beta} - \frac{\partial \theta^\alpha_{\alpha\beta}}{\partial x^\delta}\right) \qquad (70)$$

the following tensor

$$W^\alpha_{\beta\gamma\delta} = K^\alpha_{\beta\gamma\delta} - \frac{2}{n+1}\delta^\alpha_\beta A_{\gamma\delta} - \frac{1}{n-1}(\delta^\alpha_\gamma K_{\beta\delta} - \delta^\alpha_\delta K_{\beta\gamma}) + \frac{2}{n^2-1}(\delta^\alpha_\gamma A_{\beta\delta} - \delta^\alpha_\delta A_{\beta\gamma}) \qquad (71)$$

is said projective curvature tensor. Two such connections are projectively equivalent if they define the same geodesics up to parametrization.

A n−dimensional differentiable manifold M with symmetric connection θ is said locally projectively flat if $\forall x \epsilon M$ there is a neighborhood U and a diffeomorphism from U to an open of $\mathbb{R}^n$ which transforms the images of geodesics of θ ,contained in U , into straight lines. Let us remember that M is locally projectively flat if and only if the projective curvature tensor is identically zero and, if the Ricci tensor is symmetric, the manifold M is locally projectively flat if and only if the curvature tensor can be written with the following relation

$$K^\alpha_{\beta\gamma\delta} = \frac{1}{n-1}(\delta^\alpha_\gamma K_{\beta\delta} - \delta^\alpha_\delta K_{\beta\gamma}). \qquad (72)$$

Given a Riemannian manifold M and u and v, two linearly independent tangent vectors at the same point x_0, we can define

$$K(u,v) = \left[\frac{\sum R_{\alpha\beta\gamma\delta}u^\alpha v^\beta u^\gamma v^\delta}{\sum(g_{\alpha\gamma}g_{\beta\delta} - g_{\alpha\delta}g_{\beta\gamma})u^\alpha v^\beta u^\gamma v^\delta}\right](x_0). \tag{73}$$

It can be shown that $K(u,v)$ depends only on the plane spanned by u and v and it is called sectional curvature. In a Riemannian manifold the relation (74) can be written

$$R^\alpha_{\beta\gamma\delta} = \frac{1}{n-1}(\delta^\alpha_\gamma R_{\beta\delta} - \delta^\alpha_\delta R_{\beta\gamma}) \tag{74}$$

and setting as usual

$$R_{\alpha\beta\gamma\delta} = \sum_\rho g_{\alpha\rho}R^\rho_{\beta\gamma\delta} \tag{75}$$

we can write

$$R_{\alpha\beta\gamma\delta} = \frac{1}{n-1}(g_{\alpha\gamma}R_{\beta\delta} - g_{\alpha\delta}R_{\beta\gamma}). \tag{76}$$

The condition

$$R_{\alpha\beta\gamma\delta} + R_{\beta\alpha\gamma\delta} = 0 \tag{77}$$

becomes

$$g_{\alpha\gamma}R_{\beta\delta} - g_{\alpha\delta}R_{\beta\gamma} + g_{\beta\gamma}R_{\alpha\delta} - g_{\beta\delta}R_{\alpha\gamma} = 0 \tag{78}$$

and therefore

$$\sum_{\alpha,\gamma}g^{\alpha\gamma}(g_{\alpha\gamma}R_{\beta\delta} - g_{\alpha\delta}R_{\beta\gamma} + g_{\beta\gamma}R_{\alpha\delta} - g_{\beta\delta}R_{\alpha\gamma}) = 0 \tag{79}$$

that is

$$nR_{\beta\delta} - R_{\beta\delta} + R_{\beta\delta} - Rg_{\beta\delta} = 0 \tag{80}$$

and we get

$$R_{\beta\delta} = \frac{R}{n}g_{\beta\delta} \tag{81}$$

where $R = \sum_{\alpha,\gamma}g^{\alpha\gamma}R_{\alpha\gamma}$ is the scalar curvature. By replacing equation (81) in equation (74) we obtain

$$R^\alpha_{\beta\gamma\delta} = \frac{R}{n(n-1)}(\delta^\alpha_\gamma g_{\beta\delta} - \delta^\alpha_\delta g_{\beta\gamma}). \tag{82}$$

We can conclude saying that a Riemannian manifold is locally projectively flat if and only if the sectional curvature is constant. Therefore while in classical general relativity the curvature tensor equal to zero means Minkowski spacetime, in de Sitter-Fantappiè general relativity curvature tensor equal to zero means de Sitter spacetime.

5 Conclusions

The idea of de Sitter-Fantappiè relativity is to require that the laws of physics are not fundamentally invariant under the Poincaré group of special relativity, but under the symmetry group of de Sitter space instead. With this assumption, empty space automatically has de Sitter symmetry, and what would normally be called the cosmological constant in General Relativity becomes a fundamental dimensional parameter describing the symmetry structure of space-time. The discovery of the accelerating expansion of the universe has led to a revival of interest in de Sitter invariant theories, in conjunction with other speculative proposals for new physics, like doubly special relativity.

Acknowledgements

The author wishes immensely to thank Prof. Ignazio Licata and Prof. Leonardo Chiatti. Besides He wants to thank Prof. Ettore Laserra for relevant suggestions during the years. Finally He wants to thank his dear friend Enzo Ricciardi for the stimulating discussions.

References

[1] R. Penrose, Structure of Space-Time. (Mir, Moscow, 1972).

[2] S. Hawking, G. Ellis, The Large Structure of Space-Time. (Nauka, Moscow, 1979).

[3] S. Capozziello, M. Funaro, Introduzione alla Relatività Generale, Liguori Editore (2005).

[4] S.Capozziello, M. Francaviglia, Extended Theories of Gravity and their Cosmological and Astrophysical Applications, Gen.Rel.Grav.40:357-420, (2008).

[5] C. H. Brans and R. H. Dicke, Phys. Rev. 124, 925 (1961).

[6] L. Fantappie', Su una nuova teoria di relatività finale, Rend. Lincei. Vol. 17, 5, (1954).

[7] G. Arcidiacono, Rend. Accad. Lincei XVIII, fasc. 4 (1955).

[8] G. Arcidiacono, Projective Relativity, Cosmology and Gravitation; Hadronic Press; Nonantum (1986)

[9] G. Arcidiacono, La teoria degli universi, vol. I. Di Renzo, Rome, (1997).

[10] G. Arcidiacono, La teoria degli universi, vol. II. Di Renzo, Rome, (2000).

[11] I. Licata, Universe Without Singularities A Group Approach to De Sitter Cosmology, EJTP 3, No. 10, 211-224, (2006).

[12] I. Licata, L. Chiatti, The Archaic Universe: Big Bang, Cosmological Term and the Quantum Origin of Time; Int. Jour. Theor. Phys. 48, 1003-1018 (2009).

[13] I. Licata, L. Chiatti, Archaic Universe and Cosmological Model: "Big-Bang" as Nucleation by Vacuum ; Int. Journ. Theor. Phys. 49 (10), 2379-2402 (2010).

[14] L. Chiatti, Fantappié-Arcidiacono Theory of Relativity Versus Recent Cosmological Evidences : A Preliminary Comparson, EJTP 4, No. 15, 17–36 (2007).

[15] L. Chiatti, The Fundamental Equations of Point, Fluid and Wave Dynamics in the De Sitter-Fantappié-Arcidiacono Projective Special Relativity; EJTP 7, 259-280 (2010).

[16] G. Iovane, E. Benedetto, El Naschie $\epsilon^{(\infty)}$ Cantorian spacetime, Fantappiè's group and applications in cosmology, Int.Jou.Nonlinear Scienze and Numerical Simulations 6(4) 357-370 (2005).

[17] E. Benedetto, Fantappiè-Arcidiacono Spacetime and Its Consequences in Quantum Cosmology, Int. Journ. Theor. Phys. 48 (6), 1603-1621 (2009).

[18] E. Benedetto, Quantum memory of the Universe, submetted EJTP (2010).

In: The Big Bang: Theory, Assumptions and Problems
Editors: J. R. O'Connell and A. L. Hale

ISBN: 978-1-60456-802-8
© 2012 Nova Science Publishers, Inc.

Chapter 11

CONSCIOUSNESS AND ENERGY

*Vikram H. Zaveri**
B-4/6, Avanti Apt., Harbanslal Marg, Sion, Mumbai, India

Abstract

This article presents further development of the periodic relativity theory (PR), which lead to the derivation of the quantum invariant resulting in a suggestion that the universe originated with a vibration in an ocean of unmanifested fundamental substance called energy. Here we propose that the ocean of unmanifest energy without any oscillation is ocean of infinite indivisible motionless pure consciousness beyond space and time. When this infinite indivisible consciousness becomes active, it oscillates by its own power and generates discrete quanta of consciousness which the physicists know as the discrete quanta of energy because of a very low degree of manifestation of the consciousness which makes the quanta appear almost like insentient matter. Thus the one becomes many and the laws of relativity becomes operative. Since the consciousness ever remains indivisible, all the individual consciousness always remain connected with the infinite motionless consciousness like the waves with the ocean. Therefore union of both is possible.

Keywords: Consciousness, Cosmology, Energy, Relativity.

1. Introduction

This article presents further development of the periodic relativity theory (PR) [9]. Many authors have discussed consciousness from quantum perspective and dealt with the individual consciousness and its role in the collapse of the wave function [1] during the quantum measurement experiments. Some authors have also mentioned the universal consciousness [2] and its connection with the individual consciousness. Choudhury et. al. [3] has discussed quantum consciousness parameter (QCP) and fractal consciousness in different life forms. According to Zizzi [4], consciousness might have a cosmic origin, with roots in the pre-consciousness ingrained directly from the Planck time and that the universe

---

*E-mail address: cons_eng1@yahoo.com

might have achieved consciousness at the end of inflation. D'Abramo [5] re-states the mind-body problem by using the brain duplication argument to suggest that the interdepedence of brain with the physical world gives rise to consciousness. Song [6] has reviewed the incompatibility between self-observing consciousness and the standard axioms of quantum theory. Rosenblum and Kuttner [7] mention that the encounter of physics with consciousness likely has no practical consequences for physics but something beyond physics which is metaphysics. As per the current understanding in the physical and life sciences, all these articles maintain strict distinction between consciousness and matter. The former is considered sentient and the later insentient. Cochran's article [8] is an exception where he proposes a rudimentary degree of consciousness in atoms and the fundamental particles.

Many people are of the opinion that the existence of consciousness in this universe is a reality and the big bang theory could not be considered complete till it can account for the presence of consciousness along with the other forms of insentient matter. In this article we will look at consciousness from quantum as well as astrophysical and cosmological perspective.

2. The Unmanifest

2.1. Relativistic Invariant

In our earlier work periodic relativity (PR) [9], we discussed the relativistic invariant s^2 presented by Minkowski, Lorentz and Einstein which relates two points in space-time by the expression,

$$s^2 = x^2 + y^2 + z^2 - c^2 t^2. \tag{1}$$

Here $x^2 + y^2 + z^2$ represents three dimensional space, c is the velocity of light and t the ordinary linear time as we generally know. Einstein's relativity theory is founded upon this simple equation and a hypothesis based on the Eötvös experiment which showed that the gravitational mass of a body is equal to its inertial mass. It is also well known that if we replace c with velocity $v < c$ for other massive particles, Eq. (1) no longer remain meaningful. One of the PR proposal is to recognize continuity between the electromagnetic wave spectrum and the massive prticle wave spectrum. Such argument can be supported if we can supplant Eq. (1) by an equation which is not only applicable to velocity of light but also to velocity of all the other particles which travel at speeds less than that of light.

2.2. Periodic Invariant

It is possible to propose one such equation which I call the periodic invariant. It can be written as

$$s^2 = \lambda^2 - V^2 T^2, \tag{2}$$

where $\lambda = h/p$ is the associated de Broglie wavelength[10, 11], $V = c^2/v$ the phase velocity, v the particle velocity, and T the period of the wave. One can see that Eq. (2)

does satisfy light particles as well as other massive particles. If we replace ordinary particle velocity with that of light, we get $v = c$, $V = c$ and

$$s^2 = \lambda^2 - c^2 T^2, \tag{3}$$

If we multiply Eq. (3) for light with a real number n^2 and set $(n\lambda)^2$ equal to the cartesian distance $x^2 + y^2 + z^2$ and $(nT)^2$ equal to the linear time t^2, Eq. (3) becomes equivalent to Eq. (1). Therefore Eq. (1) is a special case of Eq. (2). Eq. (3) implies that space-time is not only curved but also wavy and that time does not flow in one direction but is strictly a periodic or cyclic phenomenon. Eqs. (1) and (3) both behave identically in the absence of gravitational field, but in the presence of g field, for astronomical distances, Eq. (3) remains null and Eq. (1) yields time like geodesics. In PR, both light as well as massive particles always travel along null paths. This makes it difficult to solve mundane problems of macroscopic proportions. This is why it becomes necessary to introduce approximations in the form of linear time and linear euclidean distance in Eq. (2) which permits time like and space like geodesics for addressing the problems involving complex structures. However for certain fundamental measurements such as gravitational redshift and deflection of light, Eq. (3) should yield more accurate results than that given by Eq. (1) because the reality does not get compromised.

We can also say that Eqs. (1) and (3) both behave identically even in the presence of gravitational field when the space-time interval involves atomic and sub-atomic distances. Thus the validity of the algebraic structure (Clifford and Lie Algebra) associated with Eq. (1) and the related gauge and spinor groups of particle physics is maintained. It is only at astronomical distances that the difference between two equations become perceptible and can affect any local symmetry formalism based on diffeomorphism. Therefore the validity of Dirac equation [12] is maintained with respect to Eqs. (2) and (3). Same is true for the algebra of Lorentz transformation when the space-time interval involves atomic and sub-atomic distances.

We can write Eq. (2) as

$$s^2 = (ch/cp)^2 - (c^2/v\nu)^2, \tag{4}$$

where h is Planck's constant, p the particle momentum and $\nu = 1/T$ is the frequency of the associated de Broglie wave. This period-frequency relation is the only fundamental and basic equation that relates the concept of time to the physical world in an objectively real manner. The relativistic invariant relates the space and time continuum on a macrocosmic scale. The periodic invariant does the same on a microcosmic scale. If we introduce the energy-momentum invariant

$$E^2 = E_0^2 + (cp)^2. \tag{5}$$

in Eq. (4), we get,

$$s^2 = ((hc)^2/(E^2 - E_0^2)) - (c^2/v\nu)^2, \tag{6}$$

where E = total energy of the particle and $E_0 = m_0 c^2$ is the rest energy of particle. Relativistic mass is little used by modern physicists. Notwithstanding the modern usage we will use m for relativistic mass and m_0 for rest mass throughout the article.

2.3. Quantum Invariant

The invariant Eq. (6) has a general form applicable to all de Broglie particles. In relativity, the vanishing of the invariant s^2 given by Eq. (1) does not mean that the distance between two space-time points gets obliterated. It simply means that the two space-time points can be connected by a light signal in vacuum. The new invariant Eq. (6), however, can vanish in two different ways. First, in the characteristic relativistic sense implying that two points in space-time can be connected by a energy signal (which can be a light signal or a massive particle signal), and secondly in an absolute sense where both terms on the right also vanish individually like the Euclidean invariant. In the first case, we get the relation,

$$(E^2 - E_0^2)/\nu^2 = (h^2 v^2)/c^2. \tag{7}$$

Substituting the photon parameters $E_0 = 0$ and $v = c$ into Eq. (7) gives the quantum hypothesis of Max Planck, $E = h\nu$. This provides sufficient reason to declare that Eq. (7) is a general form of Max Planck's quantum hypothesis applicable to both massless as well massive particles.

Essentially there is no difference between the relativistic invariant Eq. (1) and the invariant Eq. (6), other than the fact that the former defines the space-time continuum and the latter defines the energy-vibration continuum. The equivalence of both these continuums will become clear when we define the quantum invariant with the assumptions that, given sufficient energy, all particles having rest masses can disintegrate into particles with zero rest masses; and that all particles having zero rest masses will have a constant velocity in space regardless of the inertial frames of reference and equal to the velocity of light. These two assumptions would allow us to adopt the hypothesis that the creation begins with a vibration in the primal energy. We can introduce the photon parameters $E_0 = 0$ and $v = c$ in Eq. (6) to simulate the initial state of the universe. This gives,

$$s^2 = (hc/E)^2 - (c/\nu)^2. \tag{8}$$

And since the path of a massless particle is a null geodesic, for $s^2 = 0$, Eq. (8) can be further simplified to a form which is independent of the law of propagation of light. We shall call this form the Quantum Invariant.

$$s^2 = (h/E)^2 - (1/\nu)^2. \tag{9}$$

The quantum invariant can vanish in an absolute sense when $E \to \infty$ and $\nu \to \infty$. In this case the space-time continuum connecting two points gets completely obliterated and the resulting sub-quantic medium resembles a black hole singularity. Such a singularity suggests an equilibrated state of primal energy devoid of ripples which we shall call the unmanifest energy. This however is not a very accurate description of the unmanifest. The unmanifest could not be described as energy because there are no oscillations in the unmanifest which is motionless, whereas the energy is always associated with the oscillations $E = h\nu$. Hence the better way to put this is to say that the unmanifest is something which gives birth to both the energy and the oscillations which are two faces of the same coin. When one face disappears, the other automatically disappers with it. Since the unmanifest is not the energy, it does not gravitate. Similarly, the vibrating energy and the spacetime are two faces of the same coin. When the vibrating energy disappears, the spacetime superimposed on it disappears automatically. So is the unmanifest a perfect vacuum? Again the answer is no because the unmanifest is not the nothingness. So how do you describe the unmanifest? The unmanifest can only be described by negation. That is, you keep asking whether it is this or that and the answer is always no, because it is one of a kind in the whole universe and there is nothing else to compare it with.

So this repose of the equilibrated state of the unmanifest is disturbed when initial vibration sets off a chain reaction of creative processes. Following the first vibration in the unmanifest, several subtle and yet undetected forms of energies may have been created. Eventually certain gross form of vibrating energy of a very unified, fundamental and primal kind becomes manifest followed by what we call inflation [13, 14, 15, 16] in Lambda-CDM model [17, 18].

With the vibration comes the periodic phenomenon. Therefore time begins with the first vibration. Concept of proper time assumes linear time and distance scales whereas the true nature of reality is founded upon non-linear periods and wavelengths of the subatomic particles. Nevertheless to deal with a compound wave of a massive object such as a planet is not as simple as analyzing an individual particle. Thus the concept of proper time is useful in such cases as an approximation.

The vanishing of the quantum invariant leads one to conclude that energy and vibrations are not independent entities. Nowhere in the observable universe can one find any form of energy which is not in a state of vibration. The analogy of oneness of the waves and the ocean when former subsides will suffice to explain the vanishing of the quantum invariant. Another conclusion is that the space and energy are equivalent. There is nothing like empty space. All space is either filled with vibrating energy or with the unmanifest energy in equilibrium. One cannot conceive of space without associating it with some form of energy. In other words, space-time of Einstein's theory are mere imaginary artifacts superimposed on vibrating energy which is the only real substance.

In PR we concluded that the unmanifest could not be detected by any instrument and can only be described by negation. So the perfect description of the unmanifest is not possible. It is like trying to describe the taste of the sugar. Describing the taste is one thing and experiencing the taste an altogether different thing. At this point we simply say that eventhough it is not possible to describe or experimentally detect the unmanifest, it is however, possible to experience the unmanifest. In this article we will try our best to describe the unmanifest.

3. The Indivisible Consciousness

We begin with a proposition that the unmanifest is absolute, indivisible, infinite, motionless ocean of pure consciousness in which there are no vibrations. This indivisible consciousness is formless and has no boundaries and is the only thing that existed before the universe was created and therefore it is one without a second. This ocean of consciousness was all knowing because prior to the creation of the universe, there was nothing else for it to know other than itself. This ocean of consciousness was also blissfull because the unmanifest consciousness and bliss are not two separate things. This ocean of consciousness is called unmanifest because it is not manifest to the five human senses of the observer.

This does not mean that it does not exist. This unmanifest when viewed through the five human senses appear as the empty space. Hence the unmanifest can be described as the existence knowledge bliss absolute. We call this absolute because in the absence of second there cannot be relativity. Therefore it is beyond space and time and the laws of relativity fail at the level of the unmanifest. In general, the unmanifest does not interact with any form of manifest energies, hence it does not gravitate or collide with protons like neutrinos. Due to this aloofness, the hot big bang or supernova explosions does not bother the unmanifest, nor can it be detected by a tankfull of heavy water or Michelson-Morley interferometer [19]. This unmanifest does not go anywhere, because it has no place to go; nor does it come anywhere because it is ever present everywhere. Coming and going is for the waves, not for the ocean. Hence it is immovable and cannot be evaluated by action principles of Nambu-Goto [20] and Polyakov [21]. Since the unmanifest is free from any vibrations, it can not be represented by Euler beta function either. It is neither uni-dimensional nor multi-dimensional, because there are no directions in it. Directions appear with the appearance of wave (i.e. force). When electro-weak, strong and gravitational forces are united, the result will be another fundamental wave with finite energy and finite frequency which can be easily represented by the quantum invariant but this will not be a description of the unmanifest. The theory of zero point energy [22] deals only with manifest energies, it does not apply to the unmanifest. In accelerator experiments, one may have some doubts about the existence of strings or gravitons or Higgs boson, but no one doubts the existence of energy. The unmanifest is neither a micro nor a macro because it cannot be established in the absence of a second. The string cannot be a fundamental building block of the universe, because if discovered there is always a chance that it can be further divided, but this is not the case with the indivisible infinite consciousness. This unmanifest has a potential of becoming anything and everything that exists in the manifest universe.

3.1. Discrete Quanta of Consciousness

The creation begins when the indivisible ocean of consciousness oscillates by its own power and makes waves. There is no other cause that causes these first oscillations. Here we make the second proposition that this process of oscillation causes the quantization of the indivisible consciousness into seemingly discrete quanta of consciousness which is but an appearance presented to the five human senses, while the consciousness ever remains indivisible. This is where the wave-particle duality comes into picture. The particle nature presents apparent multiplicity but the wave nature indicates indivisibility. The degree of manifestation of consciousness in the discrete quanta is relatively low compared to the motionless unmanifest and thus the energy is born which is nothing but the consciousness in waves or consciousness in motion. Due to discreteness of the quanta, it presents an appearance of multiplicity. Thus the one becomes many and the laws of relativity becomes operative. According to PR, with the appearance of the wave the space time make their appearance as the wavelength and the period of the wave. Similarly the force and the directions make their first appearance simultaneously with the first oscillation. "Beginning" invariably refers to time and vibrations are invariably associated with what we call frequency. And frequency is the inverse of the period of the wave and the period measures time between two successive events. The mathematical relation between period and frequency $T = 1/\nu$

is understood only with a partial clarity of vision. It is not yet fully recognized that this is the only fundamental and basic expression which relates the otherwise arbitrary conception of time with the objective world. The concept of time as adopted by Einstein's relativity and Newton's classical mechanics assumes that the time is linear and flow in one direction from past to present and from present to future. This prevailing concept of time moving in one direction is a self-imposed illusion of the mind. Other authors have arrived at similar conclusion by analyzing the block universe concept [23]. In PR linear time is replaced by periods of the waves which are inverse of frequency of vibrations.

Following the first vibration in the unmanifest several subtle energies with high degree of manifestation of the consciousness and the entire subtle universes unknown to present day physics may have been created. Eventually certain gross form of vibrating energy of a very unified, fundamental and primal kind with a very low degree of consciousness becomes manifest such that each quanta appears as though it is insentient. This is followed by what we call inflation [13, 14] in the Lambda-CDM model [17, 18]. Following the inflation and creation of massive particles, a relativistic spin dependent wave equation of the kind discussed in Ref.[24] can become operative within the grand unified theories. This new wave equation has following advantages.

(1) It is relativistic and accounts for spin without the use of Dirac matrices.
(2) The squared radial momentum operator is Hermitian like the Schrödinger wave equation.
(3) In addition to providing the Dirac energy levels for Hydrogen, it provides simple and precise solution for the quarkonium spectra.

As quantization of consciousness progresses, it also undergoes condensation during phase transition in steps as recognised in particle physics. This makes the visible universe appear to five senses of the observer out of nothingness of the empty space. This is like one single expanse of water with blocks of ice formed here and there which presents apparent multiplicity but in reality it is just one single expanse of water. In the same way the consciousness remains undivided despite the appearance of multiplicity in the visible universe. The relation between consciousness and energy is comparable to that between water and ice.

3.2. Individual Consciousness

Since the advent of the cell theory (1830-1860), the scientific community has recognized the cell as the basic unit of life. When it comes to viruses, genes and self replicating molecules, such unanimity does not exist amongst scientists [25]. A simplest tobacco mosaic virus contains about 5,250,000 atoms. Each atom has its own mass and as per mass energy equivalence, virus is made out of energy which is consciousness in waves according to our second proposal, but even at the level of the virus the degree of manifestation of consciousness is so low that it becomes difficult to decide whether the virus has a life or not. This difficulty arises mainly because of the definition of the life adopted by the present day scientists more than anything else. Here we propose a more common sense definition of the life. In order to determine whether a man is dead or alive we do not say, let us see whether he can reproduce or not, or we do not say, let us see whether he can use his intelligence to make some decisions, but we simply check his heart to see whether it oscillates. If it does, he is alive. So the most fundamental, basic and general definition of life is primarily associated with oscillations. Hence we declare that whatever oscillates has a life. This immediately endows all the energies of the universe with life because energy is always associated with oscillations. From this point of view, articles of Cochran and Zizzi [4, 8] are reasonable and the test conducted by Woodward et. al. [26] unreasonable.

There is one similarity between individual consciousness and energy. Both are seen to have myriads of forms and names. Our experience with the individual consciousness is restricted to planet earth, but with the energy, it is universal. We say that the life on earth would not be possible without light, air and water. How can insentient something make so much difference for the life forms. We do recognize some differences in the consciousness of men, animals, birds, insects and plants which can be explained in terms of the degree of manifestation of the consciousness. The atoms and particles can fall on the extreme low end of this scale. Under these circumstances, the elementary consciousness of the particle itself can play some role in the collapse of the wave function; so a superposition of frequencies might be resolved by the conscious observation of one of them.

The vast difference between the consciousness of matter and the living beings is due to following reasons. Having created the matter, when it reaches a certain degree of sophistication, the unmanifest itself enters into matter as an Individual Soul and for a time gets bound within the limited bodies made of matter mistaking it for the Self and considering the rest of the universe as non-self. This division is possible due to the emergence of ego which is the I consciousness in every living being.

Here we have two manifest aspects of the Unmanifest. The first is the lower aspect what we call Nature and the second is the higher aspect which we call Individual Soul. All beings are born from the mixing of these two. The mind, intellect and egoism are part of the lower Nature. The individual soul does not perform any action but simply remains a witness to all actions performed by lower nature and experiences pairs of opposites such as happiness and misery, heat and cold etc. This is the observer of general relativity.

Hence we conclude that, to be alive and to be conscious are two different things. When we say that the matter (energy) is alive, it may not have much to do with the consciousness because the higher aspect of the unmanifest is missing. The same thing can be observed in case of an unconscious man whose heart may be beating. So the sentient and insentient can be defined as life with consciousness and life without consciousness respectively. With these definitions, the cell can be considered the smallest unit of life with consciousness.

Since our theory proposes one indivisible consciousness in the entire universe, all the observed individual consciousness behind various life forms must have some connection with this infinite indivisible consciousness and therefore it must be possible to experience such an indivisible state of consciousness. This must also be the cause behind common experience of different beings with respect to the observed phenomenon or objects in the universe. For example when one person sees a cow, the other people also see a cow in that animal and not a donkey.

4. Conclusion

When the infinite indivisible motionless consciousness becomes active (creating, preserving and destroying), it acts like the energy, when it remains inactive, it becomes unmanifest. It is possible to realise the oneness of individual consciousness and the infinite indivisible consciousness. The consciousness and the energy are like milk and its whiteness, the fire and its power to burn, the ocean and its waves, they cannot be separated. The entire manifest universe arises out of the infinite indivisible consciousness like condensed blocks of ice in an ocean, which presents apparent multiplicity in a single expanse of water. The observer's perception of the entire universe is due to mistaking the absolute for the relative. As long as the false perception persists, the relative universe appears real.

Acknowledgment

Author is grateful to Prof. Richard Mould for comments and suggestions.

References

[1] D. J. Bierman, *Mind and Matter*, Vol.1, 45-57 (2003).

[2] E. Manousakis, *Found. Phys.* **36**(6), 795, (2006).

[3] P. P. Choudhury, S. K. Dutta, Sk. S. Hassan, S. Sahoo, arXiv:0907.1394v1[physics.gen-ph].

[4] P. A. Zizzi, *NeuroQuantology*, Vol.3, 295-311, (2003).

[5] G. D'Abramo, *NeuroQuantology*, Vol. 5, Issue 4, 392-395, (2007).

[6] D. Song, *NeuroQuantology*, **6**, 272, (2008).

[7] B. Rosenblum, F. Kuttner, *Quantum enigma: physics encounters consciousness*, Oxford Univ. Press, New York, 2006.

[8] A. A. Cochran, *Found. Phys.* **1**(3), 235–250, (1971).

[9] V. H. Zaveri, *Periodic relativity: basic framework of the theory*. Gen. Relativ. Gravit. **42**, No.6, 1345–1374, (2010), doi: 10.1007/s10714-009-0908-5.

[10] L. Broglie, *Matter and light*, 165–179, New York: W. W. Norton, 1939.

[11] L. Broglie, *Non-linear wave mechanics*, 3–7, New York: Elsevier Pub. Co., 1960.

[12] P. A. M. Dirac, *The principles of quantum mechanics*, 4th ed.(revised), London: Oxford Univ. Press, 1974.

[13] A. H. Guth, *Phys. Rev. D* **23**, 347, (1981).

[14] A. Linde, *Phys. Lett. B* **108**, 389, (1982).

[15] A. H. Guth, *Inflation*, COA Series, Vol. 2, ed. W. I. Freedman, Cambridge: Cambridge Univ. Press, astro-ph/0404546.

[16] A. Linde, *Inflation and String Cosmology*, hep-th/0503195.

[17] M. Tegmark et al., *Phys. Rev. D* **69**, 103501, (2004), astro-ph/0310723.

[18] D. N. Spergel et al., *Ap. J. Suppl.* **148**, 175, (2003), astro-ph/0302209.

[19] H. Muller, S. Herrmann, C. Braxmaier, S. Schiller, A. Peters, *Phys. Rev. Lett.* **91**, 020401, (2003).

[20] B. Zwiebach, *A First Course in String Theory*, Cambridge Univ. Press, 2004, ISBN 0-521-83143-1.

[21] A. M. Polyakov, *Phys. Lett.* **B103**, 207–211, (1981).

[22] S. K. Lamoreaux, *Phys. Rev. Lett.* **78**, 5–8, (1997).

[23] G. F. R. Ellis, *Gen. Relativ. Gravit.* **38**, 1797–1824, (2006).

[24] V. H. Zaveri, *Quarkonium and hydrogen spectra with spin dependent relativistic wave equation*, Pramana - *J. Phys.*, **75**, No.4, 579–598, (2010), doi: 10.1007/s12043-010-0140-6.

[25] W. M. Stanley, E. G. Valens *Viruses and the nature of life*, 1st ed., 8–35, New York: E. P. Dutton & Co.Inc., 1961.

[26] J. F. Woodward, A. de Klerk, G. Kahler, K. Leber, P. Pompei, D. Schultz, S. Stern, *Found. Phys.* **2**(2/3), 241–244, (1972).

In: The Big Bang: Theory, Assumptions and Problems
Editors: J. R. O'Connell and A. L. Hale

ISBN: 978-1-60456-802-8
© 2012 Nova Science Publishers, Inc.

Chapter 12

ON THE ROTATION OF THE ZERO-ENERGY EXPANDING UNIVERSE

Marcelo Samuel Berman*and Fernando de Mello Gomide
Instituto Albert Einstein/Latinamerica - Av. Candido Hartmann, 575 - # 17
80730-440 - Curitiba - PR - Brazil

Abstract

The usual expanding metric due to Robertson and Walker, results in two general relativistic equations, which are called Friedman-Robertson-Walker Cosmological equations. The first one expresses conservation of energy, and involves the energy density of the Universe. The second one can be thought of as a definition of cosmic pressure, as the volume-derivative of energy, with negative sign. If rotation is present, both equations are altered. In Berman (2008a; b), we find a general relativistic treatment of rotation plus expansion. The results of our published research on the subject imply that not only such rotation is to be expected on General Relativity or even Newtonian theories, but also in other frameworks like Sciama's inertia model (Berman, 2008d; 2009e), and even in P.A.M. Dirac's large numbers hypothesis, for time-varying fine-structure along with other variable "constants" (Berman 2009a; 2010).

The constancy of the (zero) energy of Universe, was dealt elsewhere (Berman, 2009c). We now go a step further, and show that even when a metric temporal co-efficient is added to the usual Robertson-Walker's metric, and thus responds for the rotational state in addition to the expansion, by pseudo-tensor calculation, the zero-total-energy hypothesis is still valid. The Pioneers anomaly is given now a general relativistic full explanation. Another explanation, which is equivalent, is obtained by the theory of time-varying speed of light, which yields the same Pioneers anomalous deceleration.

Keywords: Cosmology; Universe; Energy; Pseudotensor; Hawking; Mach.

1. Introduction

Rotating metrics in General Relativity were first studied by Islam (1985), but Cosmology was not touched upon. However, it would be necessary an extreme perfect fine-tuning,

---
*E-mail addresses: msberman@institutoalberteinstein.org, marsambe@yahoo.com

in order to create the Universe without any angular-momentum. The primordial Quantum Universe, is characterized by dimensional combinations of the fundamental constants "c", "h" and "G" respectively the speed of light in vacuo, Planck's and Newton's gravitational constants. The natural angular momentum of Planck's Universe, as it is called, is, then, "h". It will be shown that the angular momentum grows with the expanding Universe, but the corresponding angular speed decreases with the scale-factor (or radius) of the Universe, such being the reason for the difficulty in detection of this speed with present technology. Notwithstanding, the so-called Pioneers' anomaly, which is a deceleration verified in the Pioneers space-probes launched by NASA more than thirty years ago, was attributed to a "Machian" ubiquitous field of centripetal accelerations, due to the rotation of the Universe. Berman's calculation rested on the assumption that the zero-total energy of the Universe was a valid result for the rotating case, but the proof was not supplied in that paper (Berman, 2007b). By "proof", one thinks on the pseudotensor energy calculations of General Relativity – the best gravitational theory ever published.

In his three best-sellers (Hawking, 1996; 2001; 2003), Hawking describes inflation (Guth, 1981; 1998), as an accelerated expansion of the Universe, immediately after the creation instant,while the Universe, as it expands,borrows energy from the gravitational field to create more matter. According to his description, the positive matter energy is exactly balanced by the negative gravitational energy, so that the total energy is zero,and that when the size of the Universe doubles, both the matter and gravitational energies also double, keeping the total energy zero (twice zero).Moreover, in the recent, next best-seller,Hawking and Mlodinow(2010) comment that if it were not for the gravity interaction, one could not validate a zero-energy Universe, and then, creation out of nothing would not have happened. On the other hand, Berman (2008a; b) has shown that Robertson-Walker's metric, is a particular, non-rotating case, of a general relativistic expanding and rotating metric first developed by Gomide and Uehara (1981). The peculiarity of the general metric is that instead of working with proper-time τ, one writes the field equations of General Relativity with a cosmic time t related by:

$$d\tau = (g_{00})^{1/2} dt \qquad (1)$$

where,

$$g_{00} = g_{00}(r, \theta, \phi, t) \qquad (2)$$

It will be seen that when one introduces a metric temporal coefficient g_{00} which is not constant, the new metric includes rotational effects. In a previous paper Berman (2009c) has calculated the energy of the Friedman-Robertson-Walker's Universe, by means of pseudo-tensors, and found a zero-total energy. *Our main task will be to show why the Universe is a zero-total-energy entity, by means of pseudo-tensors, even when one chooses a variable g_{00} such that the Universe also rotates, and then, to show how General Relativity predicts a universal angular speed, and a universal centripetal deceleration, numerically coincident with the observed deceleration of the Pioneers space-probes.* The first calculation of this kind, with the Gomide -Uehara generalization of RW's metric, was undertaken by Berman (1981), in his M.Sc. thesis, advised by the present second author, but where the rotation of the Universe was not the scope of the thesis.

The pioneer works of Nathan Rosen (Rosen, 1994), Cooperstock and Israelit, (1995), showing that the energy of the Universe is zero, by means of calculations involving pseudotensors, and Killing vectors, respectively, are here given a more simple approach. The energy of the (non-rotating) Robertson-Walker's Universe is zero, (Berman, 2007;2009c). Berman (1981) was the first author to work, in pseudotensor calculations for the energy of Robertson-Walker's Universe. He made the calculations on which the present paper rest, and, explicitly obtained the zero-total energy for a closed Universe, by means of LL-pseudotensor, when Robertson-Walker's metric was generalised by the introduction of a temporal-time-varying metric coefficient. However, the present authors, were unaware, in the year 1981, of the exact significance of their findings.

The zero-total-energy of the Roberston-Walker's Universe, and of any Machian ones, have been shown by many authors (Berman 2006; 2006a; 2007; 2007a; 2007b). It may be that the Universe might have originated from a vacuum quantum fluctuation. In support of this view, we shall show that the pseudotensor theory (Adler et al, 1975) points out to a null-energy for a rotating Robertson-Walker's Universe.Some prior work is mentioned,(Berman 2006; 2006a; 2007; 2007a; 2007b; Rosen, 1995; York Jr, 1980; Cooperstock, 1994; Cooperstock and Israelit, 1995; Garecki,1995; Johri et al.,1995; Feng and Duan,1996; Banerjee and Sen,1997; Radinschi,1999; Cooperstock and Faraoni,2003). See also Katz (2006, 1985); Katz and Ori (1990); and Katz et al (1997). Recent developments include torsion models (So and Vargas, 2006), and, a paper by Xulu(2000).

The reason for the failure of non-Cartesian curvilinear coordinate energy calculations through pseudotensors, resides in that curvilinear coordinates carry non-null Christoffel symbols, even in Minkowski spacetime, thus introducing inertial or fictitious fields that are interpreted falsely as gravitational energy-carrying (false) fields.

Carmeli et al.(1990) listed four arguments against the use of Einstein's pseudotensor: (1) the energy integral defines only an affine vector; (2) no angular-momentum is available; (3) as it depends only on the metric tensor and its first derivatives, it vanishes locally in a geodesic system; (4) due to the existence of a superpotential, which is related to the total conserved pseudo-quadrimomentum, by means of a divergence, then the values of the metric tensor, and its first derivatives, only matter, on a surface around the volume of the mass-system.

We shall argue below that, for the Universe, local and global Physics blend together. The pseudo-momentum, is to be taken like the linear momentum vector of Special Relativity, i.e., as an affine vector. In a previous paper (Berman, 2009c), we stated that "if the Universe has some kind of rotation, the energy-momentum calculation refers to a co-rotating observer". Such being the case, we now go ahead for the actual calculations, involving rotation. Birch (1982; 1983) cited inconclusive experimental data on a possible rotation of the Universe, which was followed by a paper written by Gomide, Berman and Garcia (1986).

2. Field Equations for Gomide-Uehara-R.W.-Metric

Consider first a temporal metric coefficient which depends only on t. The line element becomes:

$$ds^2 = -\frac{R^2(t)}{(1+kr^2/4)^2}\left[d\sigma^2\right] + g_{00}(t)\, dt^2 \tag{3}$$

The field equations, in General Relativity Theory (GRT) become:

$$3\dot{R}^2 = \kappa(\rho + \tfrac{\Lambda}{\kappa})g_{00}R^2 - 3kg_{00} \tag{4}$$

and,

$$6\ddot{R} = -g_{00}\kappa\left(\rho + 3p - 2\tfrac{\Lambda}{\kappa}\right)R - 3g_{00}\dot{R}\,\dot{g}^{00}. \tag{5}$$

Local inertial processes are observed through proper time, so that the four-force is given by:

$$F^\alpha = \tfrac{d}{d\tau}(mu^\alpha) = mg^{00}\ddot{x}^\alpha - \tfrac{1}{2}m\,\dot{x}^\alpha\left[\tfrac{\dot{g}_{00}}{g_{00}^2}\right]. \tag{6}$$

Of course, when $g_{00} = 1$, the above equations reproduce conventional Robertson-Walker's field equations.

We must mention that the idea behind Robertson-Walker's metric is the Gaussian coordinate system. Though the condition $g_{00} = 1$ is usually adopted, we must remember that, the resulting time-coordinate is meant as representing proper time. If we want to use another coordinate time, we still keep the Gaussian coordinate properties.

From the energy-momentum conservation equation, in the case of a uniform Universe, we must have,

$$\tfrac{\partial}{\partial x^i}(\rho) = \tfrac{\partial}{\partial x^i}(p) = \tfrac{\partial}{\partial x^i}(g_{00}) = 0 \qquad (i = 1,2,3). \tag{7}$$

The above is necessary in the determination of cosmic time, for a commoving observer. We can see that the hypothesis (3) – that g_{00} is only time-varying – is now validated.

In order to understand equation (6), it is convenient to relate the rest-mass m, to an inertial mass M_i, with:

$$M_i = \tfrac{m}{g_{00}}. \tag{8}$$

It can be seen that M_i represents the inertia of a particle, when observed along cosmic time, i.e., coordinate time. In this case, we observe that we have two acceleration terms, which we call,

$$a_1^\alpha = \ddot{x}^\alpha, \tag{9}$$

and,

$$a_2^\alpha = -\tfrac{1}{2g_{00}}(\dot{x}^\alpha \dot{g}_{00}). \tag{10}$$

The first acceleration is linear; the second, resembles rotational motion, and depends on g_{00} and its time-derivative.

If we consider a_2^α a centripetal acceleration, we conclude that the angular speed ω is given by,

$$\omega = \tfrac{1}{2} \left(\tfrac{\dot{g}_{00}}{g_{00}} \right). \tag{11}$$

By comparison between the usual $\tau-$metric, and the field equations in the $t-$metric, we are led to conclude that the conventional energy density ρ and cosmic pressure p are transformed into $\bar{\rho}$ and $\bar{p}$, where:

$$\bar{\rho} = g_{00} \left(\rho + \tfrac{\bar{\Lambda}}{\kappa} \right), \tag{12}$$

and,

$$\bar{p} = g_{00} \left(p - \tfrac{\bar{\Lambda}}{\kappa} \right). \tag{13}$$

We plug back into the field equations, and find,

$$\bar{\Lambda} = \Lambda - \tfrac{3}{2\kappa} \left(\tfrac{\dot{R}}{R} \right) \dot{g}^{00}. \tag{14}$$

For a time-varying angular speed, considering an arc ϕ, so that,

$$\omega(t) = \tfrac{d\phi}{dt} = \dot{\phi}, \tag{15}$$

we find, from (11),

$$g_{00} = Ce^{2\phi(t)}. \quad (\ C = \text{constant}) \tag{16}$$

Returning to (14), we find,

$$\bar{\Lambda} = \Lambda + \tfrac{3}{\kappa C} \left(\tfrac{\dot{R}}{R} \right) \omega e^{-2\phi(t)}. \tag{17}$$

This completes our solution.

The case where g_{00} depends also on r, θ and ϕ was considered also by Berman (2008b) and does not differ qualitatively from the present analysis, so that, we refer the reader to that paper.

3. Energy of the Rotating Robertson-Walker's Universe

Even in popular Science accounts(Hawking,1996;2001;2003 — and Moldinow,2010; Guth,1998), it has been generally accepted that the Universe has zero-total energy. The first such claim, seems to be due to Feynman(1962-3). Lately, Berman(2006, 2006 a) has proved this result by means of simple arguments involving Robertson-Walker's metric for any value of the tri-curvature ($0, -1, 1$).

The pseudotensor t_ν^μ, also called Einstein's pseudotensor, is such that, when summed with the energy-tensor of matter T_ν^μ, gives the following conservation law:

$$[\sqrt{-g}\,(T_\nu^\mu + t_\nu^\mu)]_{,\mu} = 0. \tag{18}$$

In such case, the quantity

$$P_\mu = \int \{\sqrt{-g}\, [T_\mu^0 + t_\mu^0]\}\, d^3x, \tag{19}$$

is called the general-relativistic generalization of the energy-momentum four-vector of special relativity (Adler et al., 1975).

It can be proved that P_μ is conserved when:

a) $T_\nu^\mu \neq 0$ only in a finite part of space;   and,
b) $g_{\mu\nu} \to \eta_{\mu\nu}$ when we approach infinity, where $\eta_{\mu\nu}$ is the Minkowski metric tensor.

However, there is no reason to doubt that, even if the above conditions were not fulfilled, we might eventually get a constant P_μ, because the above conditions are sufficient, but not strictly necessary. We hint on the plausibility of other conditions, instead of a) and b) above.

Such a case will occur, for instance, when we have the integral in (19) is equal to zero.

For our generalised metric, we get exactly this result, because, from Freud's (1939) formulae, there exists a super-potential, (Papapetrou, 1974):

$$_F U_\lambda^{\mu\nu} = \frac{g_{\lambda\alpha}}{2\sqrt{-g}} (\bar{g}^{\mu\alpha}\bar{g}^{\nu\beta} - \bar{g}^{\nu\alpha}\bar{g}^{\mu\beta})_{,\beta},$$

where the bars over the metric coefficients imply that they are multiplied by $\sqrt{-g}$, and such that,

$$\kappa\sqrt{-g}(T_\lambda^\rho + t_\lambda^\rho) = {}_F U_\lambda^{\rho\sigma}{}_{,\sigma},$$

thus finding, after a brief calculation, for the rotating Robertson-Walker's metric,

$$P_\lambda = 0.$$

The above result, with von Freud's superpotential, which yields Einstein's pseudotensorial results, points to a zero-total energy Universe, even when the metric is endowed with a varying metric temporal coefficient.

A similar result would be obtained from Landau-Lifshitz pseudotensor (Papapetrou, 1974), where we have:

$$P_{LL}^\nu = \int (-g)\, [T^{\nu 0} + t_L^{\nu 0}]\, d^3x, \tag{20}$$

where,

$$\kappa\sqrt{-g}(T^{\mu\rho} + \tilde{t}^{\mu\rho}) = \tilde{U}^{\mu\rho\sigma}{}_{,\sigma},$$

and,

$$\tilde{U}^{\mu\rho\sigma} = \bar{g}^{\lambda\mu}\, {}_F U_\lambda^{\rho\sigma},$$

A short calculation shows that, for the rotating metric, too, we keep valid the result,

$$P_{LL}^\nu = 0 \qquad (\nu = 0, 1, 2, 3). \tag{21}$$

Other superpotentials would also yield the same zero results. A useful source for the main superpotentials in the market, is the paper by Aguirregabiria et al. (1996).

The equivalence principle, says that at any location, spacetime is (locally) flat, and a geodesic coordinate system may be constructed, *where the Christoffel symbols are null. The pseudotensors are, then, at each point, null. But now remember that our old Cosmology requires a co-moving observer at each point. It is this co-motion that is associated with the geodesic system, and, as RW's metric is homogeneous and isotropic, for the co-moving observer, the zero-total energy density result, is repeated from point to point, all over space-time. Cartesian coordinates are needed, too, because curvilinear coordinates are associated with fictitious or inertial forces, which would introduce inexistent accelerations that can be mistaken additional gravitational fields (i.e.,that add to the real energy). Choosing Cartesian coordinates is not analogous to the use of center of mass frame in Newtonian theory, but the null results for the spatial components of the pseudo-quadrimomentum show compatibility.*

4. An Alternative Derivation

Though so many researchers have dealt with the energy of the Universe, our present original solution involves rotation. We may paraphrase a previous calculation, provided that we work with proper time τ instead of coordinate time t (Berman, 2009c). Then, the rotation of the Universe will be automatically included. We shall now consider, first, why the Minkowski metric represents a null energy Universe. Of course, it is empty. But, why it has zero-valued energy? We resort to the result of Schwarzschild's metric, (Adler et al., 1975), whose total energy is,

$$E = Mc^2 - \frac{GM^2}{2R}$$

If $M = 0$, the energy is zero, too. But when we write Schwarzschild's metric, and make the mass become zero, we obtain Minkowski metric, so that we got the zero-energy result. Any flat RW's metric, can be reparametrized as Minkowski's; or, for closed and open Universes, a superposition of such cases (Cooperstock and Faraoni, 2003; Berman, 2006; 2006a).

Now, the energy of the Universe, can be calculated at constant time coordinate τ. In particular, the result would be the same as when $\tau \to \infty$, or, even when $\tau \to 0$. Arguments for initial null energy come from Tryon(1973), and Albrow (1973). More recently, we recall the quantum fluctuations of Alan Guth's inflationary scenario (Guth,1981;1998). Berman (see for instance,2008c), gave the Machian picture of the Universe, as being that of a zero energy. Sciama's inertia theory results also in a zero-total energy Universe (Sciama, 1953; Berman, 2008d;2009e).

Consider the possible solution for the rotating case. We work with the τ-metric, so that we keep formally the RW's metric in an accelerating Universe. The scale-factor assumes a power-law, as in constant deceleration parameter models (Berman,1983;—and Gomide,1988),

$$R = (mD\tau)^{1/m}, \qquad (22)$$

where, m, $D =$ constants, and,

$$m = q + 1 > 0, \tag{23}$$

where q is the deceleration parameter.

For a perfect fluid energy tensor, and a perfect gas equation of state, cosmic pressure and energy density obey the following energy-momentum conservation law, (Berman, 2007, 2007a),

$$\dot{\rho} = -3H(\rho + p), \tag{24}$$

where, only in this Section, overdots stand for τ-derivatives. Let us have,

$$p = \alpha\rho \quad (\ \alpha = \text{ constant larger than } -1\ ). \tag{25}$$

On solving the differential equation, we find, for any $k = 0,\ 1,\ -1$, that,

$$\rho = \rho_0 \tau^{-\frac{3(1+\alpha)}{m}} \quad (\ \rho_0 = \text{ constant}). \tag{26}$$

When $\tau \to \infty$, from (26) we see that the energy density becomes zero, and we retrieve an "empty" Universe, or, say, again, the energy is zero. However, this energy density is for the matter portion, but nevertheless, as in this case, $R \to \infty$, all masses are infinitely far from each others, so that the gravitational inverse-square interaction is also null. The total energy density is null, and, so, the total energy. Notice that the energy-momentum conservation equation does not change even if we add a cosmological constant density, because we may subtract an equivalent amount in pressure, and equation (24) remains the same. The constancy of the energy, leads us to consider the zero result at infinite time, also valid at any other instant.

We refer to Berman (2006; 2006a) for another alternative proof of the zero-energy Universe. If we took τ instead of t, these references would provide the zero result also for the rotational case.

5. Pioneers' Anomaly Revisited

Einstein's field equations (4) and (5) above, can be obtained, when $g_{00} =$ constant, through the mere assumptions of conservation of energy (equation 4) and thermodynamical balance of energy (equation 5), as was pointed out by Barrow (1988). The latter is also to be regarded as a definition of cosmic pressure, as the volume derivative of energy with negative sign ($p = -\frac{d(\rho V)}{dV}$).

Now, let us consider a time-varying g_{00}. We may write the energy (in fact, the "energy-density")– equation, as follows:

$$\frac{3\dot{R}^2}{g_{00}} - \kappa(\rho + \frac{\Lambda}{\kappa})R^2 = -3k = \text{ constant}. \tag{27}$$

The r.h.s. stands for a constant. We can regard the l.h.s. as the a sum of constant terms, thus finding a possible solution of the field equations, such that each term in the l.h.s. of (27) remains constant. For example, let us consider,

$$\rho = \rho_0 R^{-2}, \tag{28}$$

$$\Lambda = \Lambda_0 R^{-2}, \tag{29}$$

$$g_{00} = 3\gamma^{-1}\dot{R}^2, \tag{30}$$

where, ρ_0, Λ_0 and γ are non-zero constants.

When we plug the above solution to the cosmic pressure equation (5), we find that it is automatically satisfied provided that the following conditions hold,

$$2\Lambda_0 = \kappa\rho_0(1+3\alpha), \tag{31}$$

$$p = \alpha\rho \qquad (\alpha = \text{constant}), \tag{32}$$

and,

$$\gamma = \kappa\rho_0 + \Lambda_0 - 3k \tag{32a}$$

As we found a general-relativistic solution, so far, we are entitled to the our previous general relativistic angular speed formula (11), to which we plug our solution (30), to wit,

$$\omega = \frac{\ddot{R}}{\dot{R}} = H + \frac{\dot{H}}{H}.$$

For the power-law solution of the last Section,

$$H = \frac{1}{mt},$$

so that,

$$\omega = -\frac{q}{mt} \approx t^{-1},$$

where we roughly estimated the present deceleration paramenter as $-1/2$, while, the centripetal acceleration,

$$a = -\omega^2 R \approx -t^{-2} R \simeq 8.10^{-8} \text{ cm.s}^{-2}.$$

Notice that the same result would follow from a scale-factor varying linearly with time. This is the sort of scale-factor associated with the Machian Universe. In fact, the field equations that we had (equations (4) and (5)), were not enough in order to determine the exact form of the scale-factor, because we had an extra-unknown term, the temporal metric coefficient. When we advance a given equation of state, the original RW's field equations, with constant g_{00}, may determine the scale-factor's formula. Just to remember, our solution is a particular one. Berman and Gomide(2011 submitted) have obtained a larger general class of solutions for the Pioneers annomaly problem, instead of the present particular one.

This is a general relativistic result. It matches Pioneers anomalous deceleration.

In an Appendix to this Section, we go ahead with the alternative calculation with a simple naive Special Relativistic - Machian analysis, as had been made in Berman(2007b).

Appendix to This Section

As we now have the pseudo-tensorial zero-total energy result, for rotation plus expansion, we might write in terms of elementary Physics, a possible energy of the Universe equation, composed of the inertial term of Special Relativity, Mc^2, the potential self-energy $-\frac{GM^2}{2R}$, and the cosmological "constant" energy, $\frac{\Lambda}{\kappa}(\frac{4}{3}\pi R^3)$, and not forgetting rotational energy, $\frac{1}{2}I\omega^2$, where I stands for the moment of inertia of a "sphere" of radius R and mass M. The energy equation is equated to zero, i.e.,

$$0 = Mc^2 - \frac{GM^2}{2R} + \frac{\Lambda}{\kappa}(\tfrac{4}{3}\pi R^3) + \tfrac{1}{2}I\omega^2 . \tag{33}$$

It must be remembered that R is a time-increasing function, while the total-zero energy result must be time-invariant, so that the principle of energy conservation be valid. A close analysis shows that the above conditions can be met by solutions (28) and (29), which were derived or induced from the general relativistic equations. When we plug the inertia moment,

$$I = \tfrac{2}{5}MR^2 , \tag{34}$$

we need also to consider the following Brans-Dicke generalised relations,

$$\frac{GM}{c^2 R} = \Gamma = \text{constant} , \tag{35}$$

and,

$$\omega = \tfrac{c}{R} . \tag{36}$$

If we calculate the centripetal acceleration corresponding to the above angular speed, we find, for the present Universe, with $R \approx 10^{28}$cm and $c \simeq 3.10^{10}$cm.s^{-2},

$$a_{cp} = -\omega^2 R \cong -8.10^{-8} cm/s^2 . \tag{37}$$

This value matches the observed experimentally deceleration of the NASA Pioneers' space-probes.

We observe that the Machian picture above is understood to be valid for any observer in the Universe, i.e., the center of the "ball" coincides with any observer; the "Machian" centripetal acceleration should be felt by any observed point in the Universe subject to observation from any other location.

We solve also other mistery concerning Pioneers anomaly.It has been verified experimentally, that those space-probes in closed (eliptical) orbits do not decelerate anomalously, but only those in hyperbolic flight.The solution of this other enigma is easy, according to our view.The eliptical orbiting trajectories are restricted to our local neighbourhood, and do not acquire cosmological features, which are necessary to qualify for our Machian analysis, which centers on cosmological ground.But hyperbolic motion is not bound by the Solar system, and in fact those orbits extend to infinity, thus qualifying themselves to suffer the cosmological Machian deceleration.

6. Dirac's Rotating Universe

Berman (2009a; 2010) has calculated Dirac's generalised Large Numbers Hypothesis, for a model involving variable fundamental constants of Nature, including the fine-structure "constant". It shall be seen that rotation of the Universe is part of the picture. Two different frameworks were devised by Berman, the first through a time-varying speed of light, and the second, through time-varying electromagnetic permittivities. We shall only present the first type (Berman, 2010), with the introduction of time-varying speed of light, which causes a time-varying fine-structure "constant", and a possible rotation of the Universe, either for the present time, or for inflationary periods. This paper is a sequel to a previous one dealing with the more or less equivalent consequences of a time-varying electric and magnetic permittivity (Berman, 2009a).

The rotation of the Universe (de Sabbata and Sivaram, 1994; de Sabbata and Gasperini, 1979) may have been detected experimentally by NASA scientists who tracked the Pioneer probes, finding an anomalous deceleration that affected the spaceships during the thirty years that they took to leave the Solar system. This acceleration can be explained through the rotation of the Machian Universe (Berman, 2007b). A universal spin has been considered by Berman (2008a; b).

A time-varying gravitational constant, as well as others, were conceived by P.A.M. Dirac (1938; 1974), Eddington (1933; 1935; 1939), Barrow (1990) through his Large Numbers Hypothesis. Later, Berman supplied the GLNH – Generalised Large Numbers Hypothesis (Berman, 1992; 1992a; 1994). This hypothesis arose from the fact that certain relationships among physical quantities, revealed extraordinary large numbers of the order 10^{40}. Such numbers, instead of being coincidental and far from usual values, were attributed to time-varying quantities, related to the growing number of nucleons in the Universe. In fact, such number N, for the present Universe, is estimated as $(10^{40})^2$. The number is "large" because the Universe is "old". At least, this was and still is the best explanation at our disposal.

The four relations below, represent respectively, the ratios among the scalar length of the causally related Universe, and the Classical electronic radius; the ratio between the electrostatic and gravitational forces between a proton and an electron; the mass of the Universe divided by the mass of a proton or a nucleon; and a relation involving the cosmological "constant" and the masses of neutron and electron.

If we call Hubble's "constant" H; electron's charge and mass e, m_e; proton's mass m_p, cosmological constant Λ, speed of light c, and Planck's constant h, we have:

$$\frac{cH^{-1}}{\left(\frac{e^2}{m_e c^2}\right)} \cong \sqrt{N} . \tag{38}$$

$$\frac{e^2}{G m_p m_e} \cong \sqrt{N} . \tag{39}$$

$$\frac{\rho(cH^{-1})^3}{m_p} \cong N . \tag{40}$$

$$ch(m_p m_e/\Lambda)^{1/2} \cong \sqrt{N} . \tag{41}$$

We may in general have time-varying speed of light $c = c(t)$; of $\Lambda = \Lambda(t)$; of $G = G(t)$; etc. We define the fine structure "constant" as,

$$\alpha \equiv \frac{e^2}{\hbar\, c(t)}, \tag{42}$$

and consider $\alpha = \alpha(t)$, because of the time-varying speed of light.

Power-law variations

One can ask whether the previous Section's constant-variations could be caused by a time-varying speed of light: $c = c(t)$. We refer to Berman (2007) for information on the experimental time variability of α. Gomide (1976) has studied $c(t)$ and α in such a case, which was later revived by Barrow (1998; 1998a; 1997); Barrow and Magueijo (1999); Albrecht and Magueijo (1998); Bekenstein (1992). This could explain also Supernovae observations. We refer to their papers for further information. Our framework now will be an estimate made through Berman's GLNH.

We express now Webb et al's (1999; 2001) experimental result as:

$$\left(\frac{\dot{\alpha}}{\alpha}\right)_{\exp} \simeq -1.1 \times 10^{-5}\, t^{-1}, \tag{43}$$

where t responds for the age of the Universe.

From (42) we find:

$$\frac{\dot{\alpha}}{\alpha} = -\frac{\dot{c}}{c}. \tag{44}$$

Again, we suppose that the speed of light varies with a power law of time:

$$c = At^n \quad (A = \text{constant}). \tag{45}$$

From the above experimental value we find:

$$n \approx 10^{-5}. \tag{46}$$

From (45) and (46) taken care of (44), we find:

$$\frac{\dot{\alpha}}{\alpha} = -\frac{\dot{c}}{c} = nt^{-1}. \tag{47}$$

From relations (38), (39), (40) and (41) we find:

$$N \propto t^{2+6n}. \tag{48}$$

$$G \propto t^{-1-3n}. \tag{49}$$

$$\Lambda \propto t^{-2-4n}. \tag{50}$$

$$\rho \propto t^{-1+3n}. \tag{51}$$

We see that the speed of light varies slowly with the age of the Universe. For the numerical value (46), we would obtain:

$$N \propto t^{2.0001}, \qquad (52)$$

and then:

$$G \propto t^{-1.00005}. \qquad (53)$$

$$\Lambda \propto t^{-2.0001}. \qquad (54)$$

$$\rho \propto t^{-0.99995}. \qquad (55)$$

This is our solution, based on Berman's GLNH, itself based on Dirac's work (Dirac, 1938; 1974). A pre-print with a preliminary but incomplete solution was already prepared by Berman and Trevisan (2001; 2001a; 2001b).

As a bonus we found possible laws of variation for N, G, ρ, and Λ. The Λ–term time variation is also very close and even, practically indistinguishable, from the law of variation $\Lambda \propto t^{-2}$.

It is clear that in this Section's model, the electric permittivity of the vacuum, along with its magnetic permeability, and also Planck's constant are really constant here. We point out again, that in the long run, it will be only when a Superunification theory becomes available, that the different models offered in the literature, could be discarded, (hopefully) but one.

Exponential Inflation

On remembering that relations (38) and (40) carry the radius of the causally related Universe, cH^{-1} , we substitute it by the exponential relation,

$$R = R_0 e^{Ht}. \qquad (56)$$

With the same arguments above, but, substituting, (45) by the following one,

$$c = c_0 e^{\gamma t}, \qquad (c_0, \ \gamma = \text{constants}) \qquad (57)$$

we would find:

$$N \propto e^{[H+2\gamma]\, t}, \qquad (58)$$

$$G \propto e^{-[\frac{H}{2}+\gamma]\, t}, \qquad (59)$$

$$\rho \propto e^{-2[H-\gamma]\, t}, \qquad (59a)$$

and,

$$\Lambda \propto e^{-H\, t}. \qquad (59b)$$

It seems reasonable that inflation decreases the energy density, and the cosmological term while N grows exponentially; of course, we take $H > \gamma$.

Rotation of the Universe

A closely related issue is the possibility of a Universal spin. Consider the Newtonian definition of angular momentum L,

$$L = RMv, \tag{60}$$

where, R and M stand for the scale-factor and mass of the Universe.

For Planck's Universe, the obvious dimensional combination of the constants $\bar{h}$, c, and G is,

$$L_{Pl} = \bar{h}. \tag{61}$$

From (60) and (61), we see that Planck's Universe spin takes a speed $v = c$. For any other time, we take, then, the spin of the Universe as given by

$$L = RMc. \tag{62}$$

In the first place, we take the known values of the present Universe:

$$R \approx 10^{28} cm,$$

and,

$$M \approx 10^{55} grams,$$

so that,

$$L = 10^{93} cm.gram.cm/s = 10^{120}\, \bar{h}. \tag{63}$$

We have thus, another large number,

$$\tfrac{L}{\bar{h}} \propto N^{3/2}. \tag{64}$$

For instance, for the power law, as in standard cosmology, we would have,

$$L \propto t^{3+9n} = t^{3(1+3n)}. \tag{65}$$

For exponential inflation,

$$L \propto e^{\frac{3}{2}[H+2\gamma]\, t}. \tag{66}$$

We now may guess a possible angular speed of the Universe, on the basis of Dirac's LNH. For Planck's Universe, the obvious angular speed would be:

$$\omega_{Pl} = \tfrac{c}{R_{Pl}} \approx 2 \times 10^{43} s^{-1}, \tag{67}$$

because Planck's Universe is composed of dimensional combinations of the fundamental constants. We recall a paper by Arbab (2004), that attaches a meaning to the above angular speeds, as yielding minimal accelerations in the Universe. The argument runs as follows. From manipulation with the constants that represent the Universe (c, h, G) we can construct, not only Planck's usual quantities, but also a dimensionally correct acceleration. With this acceleration, we would construct, if we call it a centripetal $a = -\omega^2 R$ term, the angular speed of our present calculation. But Arbab failed to interpret the existing Planck's constant as representing an angular rotation. However, he says that this centripetal acceleration is a consequence of the vacuum energy, and calculates correctly its present value.

In order to get a time-varying function for the angular speed, we recall Newtonian angular momentum formula,

$$L = R^2 M \omega. \tag{68}$$

In the case of power-law c – variation, we have found, from relation (60), that, $L \propto N^{3/2}$, but we also saw from (68) that $L \propto \rho R^5 \omega$, because $R = cH^{-1} \propto \sqrt{N}$ and $M \propto \rho R^3 \propto N$.

Then, we find that,

$$\omega = \omega_0 t^{-1+6n} = AR^{-(1-6n)} \qquad (\omega_0, A = \text{constants}). \tag{68a}$$

We are led to admit the following relation:

$$\omega \lesssim \tfrac{c}{R}. \tag{69}$$

For the present Universe, we shall find,

$$\omega \lesssim 3 \times 10^{-18} s^{-1}. \tag{70}$$

It can be seen that present angular speed is too small to be detected by present technology.

For the inflationary model, we carry a similar procedure:

$$\omega \propto \tfrac{N^{\tfrac{3}{2}}}{R^5 \rho} = e^{[-\tfrac{9}{2}H+\gamma]t}. \tag{71}$$

The condition for a decreasing angular speed in the inflationary period, is, then,

$$\gamma < \tfrac{9}{2}H. \tag{72}$$

7. On Sciama's Universe (Part I): Zero-Energy

The purpose of the present Section is to show that the Machian Universe, as described above (Berman, 2007; 2007a; 2007b; 2008; 2008a; 2008b), matches Sciama's linearized theory of "electrodynamical" gravitation; and the Sciama's Universe has zero-total energy.

Amidst several alternative theories of Gravity, which modify General Relativity, there stands an electrodynamical-type gravitational theory, put forward by Sciama (1953). The idea behind inertia, would be, according to Sciama, that the Universe obeys Mach's principle, i.e., at each point of space, the total force acting on a particle, is null, being composed by the second Newtonian law of force $m\vec{a}$, summed with a negative equal inertia force, which would be originated from the rest of the Universe and applied at that point. Thus, the total energy of the Universe would also be zero. The interaction of the Universe in a local point, would be made through linearized equations originated from Maxwellian-type fields (Arbab, 2004). Berman (2008d; 2009e), has detailed the cosmological consequences of Sciama's theory, and showed that the Universe has expansion plus rotation, while the angular speed is inversely proportional to the scale-factor (the "radial" coordinate), and the Universe obeys Brans-Dicke relation in the form of the equality,

$$G\frac{M}{c^2 R} = \gamma \sim 1, \qquad (73)$$

where $\gamma =$ constant.

The research on the origin of Inertia is a problem that involved passionately a number of physicists especially from the crisis of Classical Physics and from the birth of General Relativity. Einstein himself underlined the importance of his meditation on this argument in the development of his theory of gravitation. Moreover he formulated precisely his reflections on inertia origin in the Mach's Principle - that in its simpler form says that inertia properties of matter are determined in some manner by the other bodies of the Universe. Even though with the development of General Relativity, Einstein rejected explicitly his first considerations on Inertia, the Principle represented, in a compact form, a research project that guided many scientists in the development of gravitational theories alternative to General Relativity (for example Brans-Dike Theory) - with the following cosmological implications - and the reformulations of Classical Mechanics based on Mach's reflections on Inertia (see for example the model proposed by Shrödinger).

Berry (1989) has posed a Machian query. Consider a body of mass m, acted on by a large one M located at a distance r, while the large mass has an acceleration $\dot{\vec{a}}$ relative to the small one. In order to satisfy Mach's principle, the force exerted on the small mass by the larger, must contain a part proportional to $m\vec{a}$. By means of dimensional analysis, we find that the correct force should be proportional to $m\vec{a}$ and also to other terms: M, r, G and c, at some powers. According to Newton's third law, the power of M and m must be the same, i.e., they occur symmetrically in the force equation. The solution is,

$$\vec{F} = -GM\frac{m}{c^2 r}\vec{a}. \qquad (74)$$

This looks like the force whose acceleration measures mutually accelerated charges. For the gravitational case, Sciama has given a name to it: law of inertial induction. For one thing, we may understand from the analogy, that if electromagnetic radiation is possible, then we would also have gravitational radiation.

If law (74) is to be applied to the distant masses of the Universe, in the Machian picture, and if Newton's second law should be valid, we need the following Brans-Dicke relation to be valid:

$$\frac{GM}{c^2 r} = 1, \tag{75}$$

so that,

$$\vec{F} = -GM \frac{m}{c^2 r} \vec{a} = -m\vec{a}. \tag{76}$$

This section was a digression on a Berry's argument. It must be said that from formula (76), we have the same kind of zero-total force applied to each and all particles in the Universe, (inertial force plus gravitational force, equals zero) so that we retrieve a zero-total energy of the Universe.

8. On Sciama's Universe (Part II): Rotation

The purpose of the present Section is to show that Sciama's Universe is rotating, and that due to rotation, (gravitational) radiation yields the same constant power, when the whole Universe is considered; and that the entropy grows while keeping the radius proportional to the inverse square of absolute temperature.

The Machian postulates are, *sphericity* (the Universe resembles a "ball" of approximate spherical shape), *egocentrism* (each observer sees the Universe from its center) and *democracy* (each point in space is equivalent to any other one – all observers are equivalent)(Berman, 2009).

Consider the rotating and expanding Universe. The angular speed is given by,

$$\omega = \frac{c}{R}. \tag{77}$$

The reason for the above formula, is that, if we suppose that the Universe has constant zero-total energy, and is rotating, we would write the energy equation as:

$$E = 0 = Mc^2 - G\frac{M^2}{2R} + \frac{L^2}{MR^2}. \tag{78}$$

In the above, the inertial energy is represented by the first term in the r.h.s. of (78) followed by the potential energy and rotation terms, where L is the angular momentum of the Universe. We need such term in order to explain the Pioneer anomaly (Berman, 2007b), which was an anomalous constant deceleration which Berman says to be caused by the rotation of the Machian Universe, appearing as a centripetal one, and implying that the angular speed is given by relation (77) . The reason is that we find the Brans-Dicke "generalised" relations, which solve equation (78), namely,

$$G\frac{M}{c^2 R} = \gamma \sim 1, \tag{73}$$

$$\frac{L^2}{c^2 M R^2} = \gamma', \tag{79}$$

where $\gamma' =$ constant, is that there is no other solution with time-invariant zero-energy equation (see equation 6), that allows time-varying $R = R(t)$. In fact, we find from (79) and (73), that $L \propto R^2$. On the other hand, we know from Newtonian physics that $L \cong RM(\omega R)$. From the two last relations, we find necessarily a relation of type (77).

If we further calculate the power produced in the rotational motion of the "ball", we find,

$$P = \tau\omega = F_t R\omega = (M\alpha R) R\omega = MR^2\alpha\omega, \tag{80}$$

where τ is the torque of the tangential force F_t, which produces angular acceleration α.

From (77) we find,

$$\alpha = \frac{d\omega}{dt} = \frac{d}{dt}\left(\frac{c}{R}\right) = -cR^{-2}\dot{R} = -c^2 R^{-2}, \tag{81}$$

where we have made use of the Machian relation for the radius of the causally connected Universe,

$$R = ct. \tag{82}$$

At last, we obtain,

$$P = -c^2 M\omega = -\gamma\frac{c^5}{G}. \tag{83}$$

We would have obtained the same result, by introducing the equivalent of the electrodynamical Larmor power formula (Reitz et al, 1979), which yields the power radiated by an electric dipole, or an accelerated charge. The equivalent gravitational law is,

$$P_{Larmor} = \frac{2}{3}\left[\frac{GM^2}{c^3}\right] a^2, \tag{84}$$

where a is the acceleration. For the rotating case, $a = \omega^2 R$, and then, relation (84) becomes,

$$P_{Larmor} = \frac{2}{3}\left[\frac{GM^2}{c^3}\right] \omega^4 R^2 = \frac{2\gamma^2}{3}\left[\frac{c^5}{G}\right]. \tag{85}$$

Both formulae ((83) and (85)) are pretty similar. It sounds as if Machian Universe is akin with Sciama's linearized gravitation. It is expected that the power-loss will be radiated as gravitational waves, as we shall discuss in the next Section.

Quadrupole and Dipole Radiation

Einstein's quadrupole radiation formula (Weinberg, 1972), is of the type,

$$P_{einst} \approx \tfrac{G}{c^5}\omega^6 Q^2 \approx \tfrac{G}{c^5}\omega^6 \left[M^2 R^4\right]. \tag{86}$$

For our Machian Universe, the above will yield,

$$P_{einst} \approx \tfrac{c^5}{G}\gamma. \tag{87}$$

The formula for electrodynamical radiation depended on the dipole term's second time derivative, while the gravitational Einstein's radiation depends on the second time derivative of the quadrupole term. Nevertheless, for the Universe, we obtain the same result. Observe that we are in face of a constant radiating power, as was some time ago expressed by Berman (2008c). The fact, that for the Machian Universe, the dipole and the quadrupole power formulae coincide, does not mean that we may just take one instead of the other in local situations.

Temperature of the Universe. Entropy.

The power of a black-body radiator is given by (Halliday, Resnick, Walker, 2008):

$$P_{bb} = \sigma A T^4, \tag{88}$$

where A represents the radiating surface, at temperature T and σ is a constant.
For the Universe, we would find, on equating (88) with (83),

$$4\pi R^2 \sigma T^4 = \tfrac{c^5}{G}. \tag{89}$$

This results in the dependence of R with T^{-2}. This relation was found by Berman in several papers and books (see for instance, Berman, 2007, 2007a, 2008b). Now, let us calculate the entropy,

$$dS = \rho \left(4\pi R^2 dR\right) T^{-1}, \tag{90}$$

where ρ stands for the energy density of radiation, i.e.,

$$\rho = a T^4, \tag{91}$$

so that,

$$S = \tfrac{4\pi}{3} a T^3 R^3 \propto R^{\tfrac{3}{2}}. \tag{92}$$

We have found that the entropy of the Universe grows with $R^{\tfrac{3}{2}}$. This is a result of Sciama's theory, albeit Mach's theory. Berman has arrived to this formula in other cases (Berman, 2007, 2007a, 2008b, 2009).

Sciama's linear theory (Sciama, 1953), has been expanded, through the analysis of radiating processes (Berman, 2008d; 2009e).

Larmor's power formula, in the gravitational version, leads to the correct constant power relation for the Machian Universe. However, we must remember that in local Physics, General Relativity deals with quadrupole radiation, while Larmor is a dipole formula; for the Machian Universe the resultant constant power is basically the same, either for our Machian analysis or for the Larmor and general relativistic formulae.

9. Final Comments and Conclusions

We have obtained a zero-total energy proof for a rotating expanding Universe. The zero result for the spatial components of the energy-momentum-pseudotensor calculation, are equivalent to the choice of a center of Mass reference system in Newtonian theory, likewise the use of comoving observers in Cosmology. It is with this idea in mind, that we are led to the energy calculation, yielding zero total energy, for the Universe, as an acceptable result: we are assured that we chose the correct reference system; this is a response to the criticism made by some scientists which argue that pseudotensor calculations depend on the reference system, and thus, those calculations are devoid of physical meaning.

Related conclusions by Berman should be consulted (see all Berman's references at the end of this Chapter). As a bonus, we can assure that there was not an initial infinite energy density singularity, because attached to the zero-total energy conjecture, there is a zero-total energy-density result, as was pointed by Berman elsewhere (Berman, 2008).The so-called total energy density of the Universe, which appears in some textbooks, corresponds only to the non-gravitational portion, and the zero-total energy density results when we subtract from the former, the opposite potential energy density.

As Berman(2009d; f) shows, we may say that the Universe is *singularity-free*, and was created *ab-nihilo*, nor there is zero-time infinite energy-density singularity.

Paraphrasing Dicke (1964; 1964a), it has been shown the many faces of Dirac's LNH, as many as there are about Mach's Principle. In face of modern Cosmology, the naif theory of Dirac is a foil for theoretical discussion on the foundations of this branch of Physical theory. The angular speed found by us, matches results by Gödel (see Adler et al., 1975), Sabbata and Gasperini (1979), and Berman (2007b, 2008b, c).

There is a *constant-rotation* condition, for $n = \frac{1}{6}$, in the power-law solution; likewise, with $\gamma = \frac{9}{2}H$, this is the *constant-rotation* condition of the inflationary angular speed formula. However, these cases are foreign to the idea of a weak time-varying formula for the fine-structure "constant". Of course, the *constant-rotation* includes zero-rotation as a particular case. Rotation of the Universe and zero-total energy were verified for Sciama's linear theory, which has been expanded, through the analysis of radiating processes, thus, extending a previous paper of the present author (Berman, 2008d).

Larmor's power formula, in the gravitational version, leads to the correct constant power relation for the Machian Universe. However, we must remember that in local Physics, General Relativity deals with quadrupole radiation, while Larmor is a dipole formula; for the Machian Universe the resultant constant power is basically the same, either for our Machian analysis or for the Larmor and general relativistic formulae.

Referring to rotation, it could be argued that cosmic microwave background radiation should show evidence of quadrupole asymmetry and it does not, but one could argue that the angular speed of the present Universe is too small to be detected; also, we must remark that CMBR deals with null geodesics, while Pioneers' anomaly, for instance, deals with time-like geodesics. In favor of evidence on rotation, we remark neutrinos' spin, parity violations, the asymmetry between matter and anti-matter, left-handed DNA-helices, the fact that humans and animals alike have not symmetric bodies, the same happening to molluscs.

We predict that chaotic phenomena and fractals, rotations in galaxies and clusters, may provide clues on possible left handed preference through the Universe.

Berman and Trevisan (2010) have remarked that creation out-of-nothing seems to be supported by the zero-total energy calculations. Rotation was now included in the derivation of the zero result. We could think that the Universes are created in pairs, the first one (ours), has negative spin and positive matter; the second member of the pair, would have negative matter and positive spin: for the ensemble of the two Universes, the total mass would always be zero; the total spin, too. The total energy (twice zeros) is also zero.

We may interpret the metric (3) as representative of a variable speed-of-light, by means of equating $g_{00}(t) \equiv c^2(t)$.

Acknowledgments

The authors thank Marcelo Fermann Guimarães, Nelson Suga, Mauro Tonasse, Antonio F. da F. Teixeira, and for the encouragement by Albert, Paula and Luiza Mitiko Gomide.

A special recognition is made to Frank Columbus, for his kind invitation to submit the present Chapter, as well as for several other invitations.

References

Adler, R.J.; Bazin, M.; Schiffer, M. (1975) - *Introduction to General Relativity,* 2nd Edition, McGraw-Hill, New York.

Aguirregabiria, J.M. et al. (1996) - *GRG* **28**, 1393.

Albrow, M.G.(1973) - *Nature,* **241**,56.

Albrecht, A.; Magueijo, J.(1998) *A time varying speed of light as a solution to cosmological puzzles*-preprint.

Arbab, A. I. (2004) - Quantum Universe and the Solution to the Cosmological Problems, *GRG* **36** , 3565.

Bahcall, J.N. ; Schmidt, M. (1967) - *Physical Review Letters* **19**, 1294.

Banerjee, N.; Sen, S. (1997) - *Pramana J.Phys.,* **49**, 609.

Barrow, J.D. (1988) - *The Inflationary Universe*, in *Interactions and Structures in Nuclei,* pp. 135-150, ed. by R. Blin Stoyle and W.D. Hamilton, Adam Hilger, Bristol.

Barrow, J.D. (1990) - in *Modern Cosmology in Retrospect*, ed. by B. Bertotti, R.Balbinot, S.Bergia and A.Messina. CUP, Cambridge.

Barrow, J.D. (1997) - *Varying G and Other Constants*, Los Alamos Archives http://arxiv.org/abs/gr-qc/9711084 v1 27/nov/1997.

Barrow, J.D. (1998) - in *Particle Cosmology,* Proceedings RESCEU Symposium on Particle Cosmology, Tokyo, Nov 10-13, 1997, ed. K. Sato, T. Yanagida, and T. Shiromizu, Universal Academic Press, Tokyo, pp. 221-236.

Barrow, J.D. (1998a) - *Cosmologies with Varying Light Speed*, Los Alamos Archives http://arxiv.org/abs/Astro-Ph/9811022 v1 .

Barrow, J.D. ; Magueijo, J. (1999) - Solving the Flatness and Quasi-Flatness Problems in Brans-Dicke Cosmologies with varying Light Speed. *Class. Q. Gravity*, **16**, 1435-54.

Bekenstein, J.D. (1982) - *Physical Review* **D25**, 1527.

Berman, M. S. (1981, unpublished) - M.Sc. thesis, Instituto Tecnológico de Aeronáutica, São José dos Campos, Brazil.Available online, through the federal government site www.sophia.bibl.ita.br/biblioteca/index.html (supply author's surname and keyword may be "pseudotensor"or "Einstein").

Berman,M.S. (1983)-Special Law of Variation for Hubble's Parameter,*Nuovo Cimento* **74B,**182-186.

Berman, M.S. (1992) - Large Number Hypothesis - *International Journal of Theoretical Physics*, **31**, 1447.

Berman, M.S. (1992a) - A Generalized Large Number Hypothesis - *International Journal of Theoretical Physics*, **31**, 1217-19.

Berman, M. S. (1994) - A Generalized Large Number Hypothesis - Astrophys. *Space Science*, **215**, 135-136.

Berman, M.S. (1996) - Superinflation in G.R. and B.D. Theories: An eternal Universe - *International Journal of Theoretical Physics* **35**, 1789.

Berman, M. S. (2006) - *Energy of Black-Holes and Hawking's Universe*, in Chapter 5 of *Trends in Black Hole Research*, ed by Paul V. Kreitler, Nova Science, New York.

Berman, M. S. (2006a) - *Energy, Brief History of Black-Holes,and Hawking's Universe*, in Chapter 5 of: *New Developments in Black Hole Research*, ed by Paul V. Kreitler, Nova Science, New York.

Berman, M. S. (2007) - *Introduction to General Relativity and the Cosmological Constant Problem,* Nova Science, New York.

Berman, M. S. (2007a) - *Introduction to General Relativistic and Scalar Tensor Cosmologies* , Nova Science, New York.

Berman, M.S. (2007b) - The Pioneer Anomaly and a Machian Universe - *Astrophysics and Space Science,* **312**, 275. Los Alamos Archives, http://arxiv.org/abs/physics/0606117.

Berman, M. S. (2007c) - Gravitomagnetism and Angular Momenta of Black-Holes - *RevMexAA*, **43**, 297-301.

Berman, M. S. (2007d) - Is the Universe a White-hole? - *Astrophysics and Space Science*, **311**, 359.

Berman, M. S. (2008a) - A General Relativistic Rotating Evolutionary Universe, *Astrophysics and Space Science*, **314**, 319-321.

Berman, M. S. (2008b) - A General Relativistic Rotating Evolutionary Universe - Part II, *Astrophysics and Space Science*, **315**, 367-369.

Berman, M.S. (2008c) - *A Primer in Black Holes, Mach's Principle and Gravitational Energy*, Nova Science, New York.

Berman, M.S. (2008d) - On the Machian Origin of Inertia, *Astrophysics Space Science*, **318**, 269-272. Los Alamos Archives, http://arxiv.org/abs/physics/0609026

Berman, M.S. (2008e) - *General Relativistic Machian Universe*, Astrophysics and Space Science, **318**, 273-277.

Berman, M.S. (2008f) - Shear and Vorticity in a Combined Einstein-Cartan-Brans-Dicke Inflationary Lambda Universe, *Astrophysics and Space Science*, **314**, 79-82. For a preliminary report, see Los Alamos Archives, http://arxiv.org/abs/physics/0607005

Berman, M.S. (2009g) - Gravitons, Dark Matter, *and Classical Gravitation, AIP Conference Proceedings* **1168**, 1068-1071. Preliminary version, see Los Alamos Archives http://arxiv.org/abs/0806.1766 .

Berman, M.S. (2009) - General Relativistic Singularity-free Cosmological Model, *Astrophysics and Space Science*, **321**, 157-160.

Berman, M.S.(2009a) - Simple Model with time-varying fine-structure "constant", *RevMexAA* **45**, 139-142.

Berman, M.S. (2009b) - Entropy of the Universe, *International Journal of Theoretical Physics*, **48**, 1933. DOI 10.1007/s10773-009-9966-4 . Los Alamos Archives http://arxiv.org/abs/0904.3135.

Berman, M.S. (2009c) - On the zero-energy Universe, *International Journal of Theoretical Physics*, **48**, 3278.

Berman, M.S. (2009d) - Why the initial infinite singularity of the Universe is not there?, *International Journal of Theoretical Physics*, **48**, 2253.

Berman, M.S. (2009e) - On Sciama's Machian Cosmology, *International Journal of Theoretical Physics*, **48**, 3257.

Berman, M.S. (2009f) - General Relativistic Singularity-Free Cosmological Model, *Astrophysics and Space Science*, **321**, 157. Los Alamos Archives http://arxiv.org/abs/0904.3141

Berman, M.S. (2010) - *Simple Model with time-varying fine-structure "constant" - Part II, RevMexAA* **46,** 23-28.

Berman, M.S.; Gomide, F.M.(1988) - Cosmological Models with Constant Deceleration Parameter - *GRG* **20,**191-198.

Berman, M.S.; Gomide, F.M.(2010) - Los Alamos Archives arxiv:1011.4627. *General Relativistic Treatment of the Pioneers Anomaly* - submitted.

Berman, M.S.; Gomide, F.M.(2011) - *General Relativistic Treatment of the Pioneers Anomaly Revisited* - submitted.

Berman, M.S.; Trevisan, L.A. (2001) - Los Alamos Archives http://arxiv.org/abs/gr-qc/0112011

Berman, M.S.; Trevisan, L.A. (2001a) - Los Alamos Archives http://arxiv.org/abs/gr-qc/0111102

Berman, M.S.; Trevisan, L.A. (2001b) - Los Alamos Archives http://arxiv.org/abs/gr-qc/0111101

Berman, M.S.; Trevisan, L.A. (2010) - Creation of the Universe out of Nothing, *International Journal of Modern Physics,* **D19,** 1309-1313. For a preliminary version, see Los Alamos Archives http://arxiv.org/abs/gr-qc/0104060

Berry, M.V. (1989) - *Principles of Cosmology and Gravitation,* Adam Hilger, Bristol.

Birch, P. (1982) - *Nature*, **298,** 451.

Birch, P. (1983) - *Nature*, **301,** 736.

Carmeli,M.; Leibowitz, E. ; Nissani, N. (1990) - *Gravitation:SL(2,C) Gauge Theory and Conservation Laws,* World Scientific, Singapore.

Cooperstock, F.I. (1994) - *GRG* **26,** 323.

Cooperstock, F.I.; Israelit, M. (1995) - *Foundations of Physics*, **25,** 631.

Cooperstock, F.I.; Faraoni, V.(2003) - Ap.J. 587,483.

de Sabbata, V.; Gasperini, M. (1979) - *Lettere al Nuovo Cimento,* **25,** 489.

de Sabbata, V.; Sivaram, C. (1994) - *Spin and Torsion in Gravitation,* World Scientific, Singapore.

Dicke, R.H. (1964) - *The many faces of Mach,* in *Gravitation and Relativity,* ed. by Chiu,H.-Y. and Hoffmann,W.F., Benjamin, New York.

Dicke, R.H. (1964a) - *The significance for the solar system of time-varying gravitation,* in *Gravitation and Relativity,* ed. by Chiu,H.-Y. and Hoffmann, W.F., Benjamin, New York.

Dirac, P.A.M. (1938) - *Proceedings of the Royal Society* **165** A, 199.

Dirac, P.A.M. (1974) - *Proceedings of the Royal Society* **A338,** 439.

Eddington, A.S. (1933) - *Expanding Universe,* CUP, Cambridge.

Eddington, A.S. (1935) - *New Pathways in Science.* Cambridge University Press, Cambridge.

Eddington, A.S. (1939) - *Sci. Progress, London,* **34**, 225.

Feng, S.; Duan, Y. (1996) - *Chin. Phys. Letters*, **13**, 409.

Feynman, R. P. (1962-3) - *Lectures on Gravitation* , Addison-Wesley, Reading.

Freud, P.H.(1939) - *Ann. Math,* **40**, 417.

Garecki, J. (1995) - *GRG,* **27**, 55.

Gomide, F.M. (1976) - *Lett. Nuovo Cimento* **15**, 515.

Gomide, F.M.; Berman, M.S.; Garcia, R.L. (1986) - *RevMexAA*, **12**, 46.

Gomide, F.M.; Uehara, M. (1981) - *Astronomy and Astrophysics,* **95**, 362.

Guth, A. (1981) - *Phys. Rev.* **D23**, 347 .

Guth, A. (1998) - *The Inflationary Universe,* Vintage, New York, page 12.

Halliday, D.; Resnik, R.; Walker, J. (2008) - *Fundamentals of Physics,* Wiley. 8th Edition. New York.

Hawking, S. (1996) - *The Illustrated A Brief History of Time,* Bantam Books, New York, pages 166-167.

Hawking, S. (2001) - *The Universe in a Nutshell,* Bantam Books, New York, pages 90-91.

Hawking, S. (2003) - *The Illustrated Theory of Everything,* Phoenix Books, Beverly Hills, page 74.

Islam, J.N. (1985) - *Rotating Fields in General Relativity,* CUP, Cambridge.

Johri, V.B.; et al. (1995) - *GRG,* **27**, 313.

Katz, J. (1985) - *Classical and Quantum Gravity* **2**, 423.

Katz, J. (2006) - Private communication.

Katz, J.; Ori, A. (1990) - *Classical and Quantum Gravity* **7**, 787.

Katz, J.; Bicak, J.; Lynden-Bell, D. (1997) - *Physical Review* **D55**, 5957.

Landau, L.; Lifshitz, E. (1975) - *The Classical Theory of Fields*, 4th. Revised ed.; Pergamon, Oxford.

Moffat, J.W. (1993) - *International Journal of Modern Physics,* **D2**, 351.

Papapetrou, A. (1974) - *Lectures on General Relativity*, Reidel, Boston.

Radinschi, I. (1999) - *Acta Phys. Slov.,* **49**, 789. Los Alamos Archives, gr-qc/0008034.

Reitz, J.R.; Milford, F.J.; Christy, R.W. (1979) - *Foundations of Electromagnetic Theory*, Addison-Wesley. Reading.

Rosen, N.(1994) - *Gen. Rel. and Grav.* **26**, 319.

Rosen, N.(1995) - *GRG,* **27**, 313.

Sabbata, de V.; Gasperini, M. (1979) - *Lettere al Nuovo Cimento,* **25**, 489.

Schmidt, B.P. (1998) - *Ap.J.,* **507**, 46-63.

Schwarzschild, B. (2001)- *Physics Today,* **54** (7), 16.

Sciama, D.W. (1953) - *M.N.R.A.S.,* **113**, 34.

So, L.L.; Vargas, T. (2006) - Los Alamos Archives, gr-qc/0611012.

Tryon, E.P.(1973) - *Nature,* **246**, 396.

Webb, J.K; et al. (1999) –*Phys. Rev. Lett.* **82**, 884.

Webb, J.K; et al. (2001) – *Phys Rev. Lett.* **87**, 091301.

Weinberg, S. (1972) - *Gravitation and Cosmology,* Wiley. New York.

Will, C. (1987) - in *300 Years of Gravitation*, ed. by W.Israel and S.Hawking, CUP, Cambridge.

Will, C. (1995) - in *General Relativity - Proceedings of the 46th Scotish Universities Summer School in Physics,* ed. by G.Hall and J.R.Pulhan, IOP/SUSSP Bristol.

Xulu, S. (2000) - *International Jounal of Theoretical Physics,* **39**, 1153. Los Alamos Archives, gr-qc/9910015.

York Jr, J.W. (1980) - *Energy and Momentum of the Gravitational Field*, in *A Festschrift for Abraham Taub*, ed. by F.J. Tipler, Academic Press, N.Y.

INDEX

A

additives, 122
age, x, 78, 80, 205, 206, 208, 211, 212, 213, 214, 219, 221, 222, 223, 225, 226, 236, 296, 297
algebraic geometry, 174
alters, 63
amplitude, viii, 103, 105, 111, 124, 125, 128, 129, 135, 137, 145, 153, 155, 174, 182
annihilation, 217, 261
ASI, 166
astrophysical S-factors, vii, 1, 2, 3, 9, 20, 44
asymmetry, 165, 304, 305
asymptotics, 10, 21, 40
atmosphere, 66, 73, 74, 84, 88, 89, 93
atomic distances, 277
atomic nuclei, vii, 1, 3, 6, 8, 17, 57, 60, 66, 70, 91, 104
atomic nucleus, 2, 3
atoms, 70, 73, 74, 75, 76, 81, 84, 86, 89, 92, 93, 94, 96, 250, 276, 281, 282
autonomy, 83
Azerbaijan, 57

B

background radiation, x, xi, 180, 205, 206, 218, 223, 226, 243, 244, 245, 251, 254, 304
barriers, 210
baryonic matter, viii, 103, 104, 105, 106, 107, 111, 112, 113, 114, 115, 116, 117, 118, 119, 120, 125, 126, 127, 128, 129, 158, 165, 207
baryonic matter's density perturbations, viii, 103, 104, 105, 119, 120, 125, 128, 129
baryons, 165, 207, 219, 222
base, 46, 62, 67, 125
batteries, 75
beams, 83

Belgium, 182
bending, 94
Big Bang, vii, viii, x, xi, 61, 62, 64, 78, 79, 80, 81, 97, 98, 100, 103, 104, 197, 198, 199, 200, 201, 205, 206, 207, 208, 209, 211, 213, 214, 215, 217, 218, 219, 221, 223, 225, 227, 229, 243, 244, 245, 247, 248, 249, 250, 251, 253, 254, 255, 256, 257,259, 267, 273, 275, 285
Big-Bang hypothesis, x, 197
binding energies, 208
binding energy, 208, 210, 219, 220, 246
birds, 282
black hole, viii, ix, 61, 64, 73, 81, 86, 94, 96, 97, 98, 100, 173, 181, 245, 246, 249, 250, 278
blame, 239
Boltzman constant, 222
Boltzmann constant, 111
bonding, 74, 85, 86
boson, 96
brain, 74, 89, 276
Brazil, 229, 285, 306
Brownian motion, 94
building blocks, 250

C

calculus, 98, 100
calibration, 155, 156
cancer, 74
candidates, 200, 207, 208
carbon dioxide, 74
catalyst, 64
causality, 215
celestial bodies, 149
chaos, 66, 77, 79, 137
chemical, xi, 74, 243, 244, 250, 260

Chicago, 66, 92, 94, 96, 165, 166, 167, 257
classical mechanics, 263, 281
classification, 17, 23, 25, 39, 44, 45, 46, 47, 50, 51, 52, 53, 56
closed string, 138
cluster model, vii, 1, 2, 3, 4, 8, 9, 10, 15, 17, 20, 23, 25, 34, 44, 50, 55, 56, 57, 60
cluster-cluster, vii, 1, 2
clustering, 74, 120, 158
clusters, vii, xi, 1, 3, 5, 6, 57, 74, 79, 80, 82, 83, 86, 100, 106, 111, 112, 158, 243, 252, 254, 256, 305
coal, 76
cocoon, 78
coefficients of penetrability, 193
coherence, 65, 83
collisions, 66, 77, 94
coma, 207
compatibility, 70, 72, 73, 74, 76, 82, 83, 89, 91, 95, 96, 291
competition, 104
complement, 62
complementarity, 66, 67, 69, 70, 72, 73, 75, 77, 78, 79, 82, 83, 85, 91, 96
complexity, 65
composition, xi, 176, 177, 182, 243, 245, 250, 263
comprehension, 198
compression, 79, 92
computation, 63, 64, 98, 99, 182
computer, 9, 21, 32, 55
conception, 68, 281
condensation, 281
conduction, 74
conductor, 70, 74, 75, 76, 86, 87
conductors, 76
conference, 57
configuration, 44, 45, 51, 65, 66, 67, 69, 83, 96, 250
conformity, 27
consciousness, xii, 275, 276, 279, 280, 281, 282, 283
conservation, xii, 64, 65, 66, 67, 68, 69, 70, 71, 72, 73, 76, 77, 78, 79, 82, 86, 88, 90, 91, 92, 94, 97, 98, 114, 131, 139, 230, 233, 245, 246, 248, 253, 256, 285, 288, 289, 292, 294
constituents, 86, 208
construction, 6, 15, 39, 46, 57, 76
contradiction, 65, 90
convention, 69
cooling, 249, 250, 251, 252, 253
cooling process, 250, 252
copper, 75

cosmic rays, 96, 100
cosmic waves, viii, 61, 67, 69, 73, 74, 82, 84, 89, 90, 94, 95, 96, 97
cosmological time, 104
cosmos, 251, 260
covering, 175, 176, 178, 179, 180, 181, 182
cracks, 93
creationism, 201
creative process, 279
creep, 62
critical density, 110, 206, 207, 214, 249
critical value, 260
criticism, 304
crust, 84
cycles, 67, 68, 69, 86, 87, 89, 90, 91

D

dark energy, viii, x, xi, 65, 103, 105, 106, 113, 114, 118, 120, 123, 129, 158, 200, 205, 214, 226, 244, 249, 255, 256
dark matter, viii, x, 61, 65, 66, 67, 69, 74, 75, 77, 78, 80, 84, 85, 86, 87, 88, 91, 93, 96, 97, 106, 200, 202, 205, 207, 208, 214, 219, 226, 235
decay, 189, 193
decomposition, ix, 173, 175, 178, 179
decoupling, 253
defects, 149, 169
deficiency, 72
democracy, 234, 301
density fluctuations, 119, 252
depression, 66, 70, 74, 77, 78, 79, 97
depth, 21, 33, 43, 44, 57, 79, 94
derivatives, 91, 117, 162, 192, 248, 287, 292
destiny, viii, 61, 80, 81, 85, 86, 97, 98
destruction, 89, 93, 97
detectable, 94
detection, 83, 248, 253, 286
deuteron, 10, 48
dichotomy, 95
differential equations, 134
diffusion, 135
Dirac equation, 277
direct measure, 80
direct observation, 81
disaster, 93
discreteness, 280
disorder, 74
dispersion, 136
displacement, 142, 152, 154
distribution, x, xi, 98, 107, 109, 110, 113, 135, 136, 137, 138, 158, 205, 208, 209, 210, 211, 216, 220, 225, 231, 245, 251, 252, 254, 259

distribution function, 208, 209
divergence, 165, 186, 246, 287
DNA, 305
DOI, 307
dominance, 106, 111, 113
double helix, 76
duality, ix, 76, 83, 85, 90, 97, 173, 280
dynamical properties, 268

E

early universe, 92, 99, 185, 196, 208, 209, 261
egocentrism, 234, 301
egoism, 282
electric current, 70, 86, 87
electrical resistance, 86
electrodes, 75
electrolyte, 75
electromagnetic, xi, 48, 49, 55, 56, 66, 76, 199, 200, 207, 223, 233, 243, 251, 252, 276, 295, 301
electromagnetic waves, 66
electromagnetism, 68, 69, 74, 75, 76, 96
electron, 66, 69, 70, 71, 73, 74, 89, 90, 177, 210, 252, 295
electrons, 72, 73, 74, 75, 77, 86, 89, 90, 93, 96, 250, 254
elementary particle, 63, 92, 96, 179, 221, 222, 250
emission, xi, 243, 250, 251, 252, 253, 254, 255
encoding, 65
energy density, xi, xii, 69, 87, 104, 114, 122, 187, 199, 218, 223, 224, 230, 231, 232, 234, 235, 236, 237, 244, 259, 285, 289, 292, 298, 303, 304
energy efficiency, 8
energy input, 254
entropy, xi, 175, 176, 178, 179, 222, 223, 229, 230, 232, 233, 234, 235, 236, 238, 239, 301, 303
environment, 89, 250
EPR, ix, 173, 175, 183
equality, 245, 300
equilibrium, 223, 244, 249, 250, 253, 254, 256
Euclidean space, 245, 249, 270
Eurasia, 57
evolution, viii, x, 2, 17, 35, 61, 65, 77, 79, 86, 95, 97, 99, 103, 104, 105, 107, 109, 111, 113, 119, 120, 129, 130, 137, 138, 149, 158, 164, 165, 186, 188, 195, 196, 199, 205, 223, 225, 249, 260
excitation, 26, 35, 45, 46
expulsion, 93

F

federal government, 306
fermions, 209, 213, 219, 220
Feynman diagrams, 186
field theory, 174, 215
finite speed, 248, 253
fission, 72, 91, 97
flatness, 259
flight, 86, 87, 90, 91, 93, 96, 294
fluctuations, 104, 106, 112, 113, 119, 137, 218, 249, 252, 254, 256, 260
fluid, 119, 125, 127, 236, 237, 259, 260, 292
football, 92
force, ix, 67, 70, 71, 73, 75, 76, 78, 79, 82, 83, 85, 87, 88, 91, 96, 97, 104, 105, 106, 107, 112, 134, 138, 144, 158, 159, 160, 161, 164, 165, 186, 189, 190, 218, 224, 247, 259, 268, 269, 270, 280, 288, 300, 301, 302
formation, vii, viii, x, xi, 3, 44, 79, 81, 84, 91, 92, 103, 104, 105, 112, 130, 187, 194, 196, 197, 198, 199, 210, 213, 221, 222, 243, 245, 249, 250, 253, 254
formula, 3, 4, 12, 39, 145, 191, 239, 241, 252, 293, 299, 301, 302, 303, 304
foundations, ix, 62, 63, 93, 173, 174, 182, 304
fractal structure, 66, 68
fragments, 89
freedom, 209
Freud, 290, 309
friction, 93, 119, 128, 129, 138, 212
fusion, 8, 92, 253

G

galactic scales, 96
galaxies, viii, ix, xi, 78, 79, 80, 81, 82, 85, 87, 95, 96, 100, 103, 104, 105, 106, 111, 112, 113, 118, 119, 130, 137, 158, 159, 162, 164, 165, 200, 206, 207, 243, 249, 250, 251, 252, 253, 255, 256, 305
Galaxy, 100, 158
galaxy formation, 104
Galileo, 84, 263, 264
gamma rays, 93
General Relativity, xi, xii, 169, 200, 230, 233, 234, 238, 240, 241, 242, 257, 259, 260, 261, 262, 285, 286, 288, 300, 304, 305, 306, 309, 310
genes, 74, 89, 281
genetic alteration, 74
genetic disease, 74

geology, 99
geometry, xi, 72, 174, 245, 246, 248, 259, 271
Germany, 59, 94, 182, 243, 257
global scale, 256
God, 201
grand unified theory (GUT), vii, 61, 62
gravitation, ix, 83, 89, 104, 105, 107, 118, 158, 160, 164, 165, 200, 230, 232, 245, 246, 300, 302, 308
gravitational collapse, 250
gravitational constant, x, 114, 208, 211, 212, 219, 221, 226, 234, 245, 262, 295
gravitational field, 86, 132, 145, 147, 149, 150, 155, 266, 277, 291
gravitational force, 107, 158, 159, 165, 208, 247, 253, 259, 280, 295, 301
gravitational potential energy, 234
gravitational pull, 83
gravity, viii, ix, x, 61, 62, 64, 66, 69, 71, 74, 75, 78, 79, 82, 83, 84, 85, 88, 94, 96, 97, 98, 99, 120, 139, 165, 173, 174, 185, 186, 211, 214, 215, 216, 218, 220, 226, 229, 239, 245, 246, 247, 255, 259, 261, 262, 286
growth, viii, 103, 105, 106, 111, 112, 118, 119, 125, 128, 129, 230, 252

H

hadrons, 104
Hamiltonian, x, 157, 186, 205, 208, 209, 214, 215, 216, 217, 218, 219, 224, 226
Hawking radiation, 181, 183
height, 194, 195
Heisenberg picture, 217
helium, 92, 100, 250
Higgs boson, 96, 280
Hilbert space, 182
history, 83, 84, 174, 197, 199, 200, 201, 202, 244, 254, 255
homogeneity, 120, 123, 130, 199, 206, 210, 244, 253
human, 198, 201, 279, 280
humidity, 94
hydrogen, xi, 8, 35, 72, 92, 100, 111, 210, 243, 284
hydrogen atoms, 92
hydrogen bomb, 92
hypercube, 216
hypothesis, x, xii, 197, 198, 199, 200, 201, 230, 231, 238, 239, 276, 278, 285, 288, 295

I

illusion, 66, 88, 281
image, 75, 89, 104, 177, 271
imagination, 62, 75
incompatibility, 202, 276
independence, 191
India, 61, 99, 183, 205, 275
individuals, 180, 182
induction, 300
inequality, 63, 121
inertia, xii, 285, 288, 291, 294, 300
inflation, 104, 113, 120, 185, 196, 200, 244, 276, 279, 281, 286, 298
ingredients, 247
initial state, 199, 245, 249, 278
institutions, 182
insulation, 93
integration, 5, 107, 109, 154, 160, 191, 247, 269
intellect, 282
intelligence, 281
interface, 67, 93
interference, 192, 193, 195, 254
intergalactic space, 200
international relations, 67
interstellar dust, 84
invariants, 261
ionization, 74, 75
ions, 56, 89
IR spectra, 255
iron, 73, 75, 76, 91
Islam, 285, 309
isospin, 17, 18, 20, 21, 23, 45, 46, 48, 51
isotope, 70, 72, 73, 92
Israel, 310
issues, 86, 91, 174, 177
Italy, 182, 259

J

Japan, 183

K

Kazakhstan, 1, 60, 103, 167, 170
kinks, 138

L

laws, vii, viii, x, xi, xii, 61, 62, 63, 64, 65, 66, 67, 78, 82, 92, 97, 197, 199, 239, 244, 253, 273, 275, 280, 297
lead, viii, xii, 3, 13, 17, 34, 40, 41, 43, 53, 75, 103, 105, 113, 118, 120, 130, 138, 144, 149, 163, 244, 245, 248, 249, 251, 252, 255, 275
learning, 62
lepton, 104, 221
life sciences, 276
light, vii, viii, ix, xii, 1, 3, 6, 10, 50, 57, 60, 61, 62, 65, 66, 67, 69, 70, 71, 72, 73, 78, 79, 80, 81, 84, 85, 86, 88, 91, 92, 94, 95, 96, 100, 138, 149, 174, 176, 177, 181, 200, 202, 203, 207, 250, 255, 256, 261, 276, 277, 278, 282, 283, 305
linear dependence, 145
liquids, 94
lithium, 75, 100
Lorentz invariance exists, xi
luminosity, 207, 255
luminous matter, x, 200, 205, 214, 218, 219, 226

M

magnet, 68, 69, 72, 73, 75, 76, 77, 87
magnetic field, 72, 73, 239, 250
magnetic moment, 4, 48
magnetism, 76
magnets, 74
magnitude, 83, 87, 90, 109, 110, 112, 115, 117, 120, 136, 164, 165, 208, 214, 218, 226, 244, 249
malleable materials, viii, 61, 74, 93
manipulation, 299
mantle, 84
mapping, ix, 173, 179
Mars, 82
massive particles, 207, 276, 277, 278, 281
materials, viii, 61, 74, 93
matrix, 26, 48, 50, 54, 263
matter-anti-matter interaction, viii, 61, 96
measurements, vii, 1, 2, 3, 7, 13, 15, 16, 24, 25, 26, 34, 40, 48, 49, 50, 57, 81, 198, 207, 222, 254, 277
melting, 182
melts, 93
membranes, 70, 74, 89
mental disorder, 74
Mercury, 82, 212, 246
metal fatigue, viii, 61, 94

methodology, vii, 61, 63, 64
microwave radiation, 244, 253
Milky Way, 78, 80, 82, 84, 85, 87, 89
minisuperspace, ix, 185, 187
Minkowski spacetime, 272, 287
mitosis, 79
mixing, 18, 45, 192, 194, 195, 208, 282
modelling, vii, 61, 63, 64, 69
models, 2, 13, 60, 99, 118, 165, 186, 200, 208, 248, 256, 259, 287, 291, 297
modulus, 189, 190, 191, 195
molecules, 65, 66, 70, 84, 88, 89, 93, 94, 281
momentum, xi, 4, 5, 33, 40, 45, 48, 79, 81, 85, 86, 87, 88, 89, 94, 95, 130, 131, 148, 158, 162, 164, 165, 230, 239, 243, 248, 249, 250, 251, 253, 277, 281, 286, 287, 288, 290, 292, 298, 299, 301, 304
Moon, 85
Moscow, 59, 60, 98, 166, 167, 170, 273
multidimensional, 186
multiplication, 188
muscular dystrophy, 74

N

NAS, 167, 170
NATO, 166
natural laws, 62, 64, 66, 78, 82, 97
natural science, 62, 197, 198, 199, 202
natural sciences, 62, 197, 198, 199, 202
neglect, 106, 126
neon, 66, 88, 90
neutral, 70, 72, 74, 86, 89
neutrinos, x, 205, 206, 207, 208, 213, 214, 219, 226, 280, 305
neutrons, 72, 73, 91, 206, 250
Newtonian gravity, 165, 216, 247, 249
Newtonian physics, 248, 302
Newtonian theories, xii
Newtonian theory, 111, 215, 216, 304
nodes, vii, 1, 3, 57
normalization constant, 210
nuclei, vii, 1, 3, 6, 8, 10, 17, 34, 57, 60, 66, 70, 73, 91, 92, 104, 206, 213, 214
nucleons, vii, x, 1, 40, 71, 72, 92, 212, 213, 223, 226, 295
nucleus, 2, 3, 6, 9, 10, 15, 18, 20, 21, 23, 31, 32, 33, 34, 39, 40, 41, 43, 44, 45, 46, 47, 48, 50, 51, 53, 54, 55, 56, 71, 72, 73, 85, 91, 92, 95, 96
nuclides, 206
null, 133, 192, 231, 232, 277, 278, 287, 291, 292, 300, 305

O

oceans, 84
oil, 74
old age, 78
opportunities, 9, 182
orbit, 72, 73, 82, 83, 212
orbital states, vii, 1, 3, 17, 23, 39, 50, 53, 56
oscillation, xii, 69, 72, 83, 86, 87, 88, 90, 91, 138, 153, 280
overlap, 89, 95, 254
oxygen, 73

P

Pacific, 66
parallel, x, 71, 74, 75, 76, 89, 90, 197, 201, 265
parity, 305
particle physics, 182, 277, 281
partition, 245
path integrals, 186
Pauli principle, vii, 1
PCM, 2, 3, 16, 39, 43, 44, 50, 56
penetrability, x, 185, 187, 193, 194, 195, 196
periodic relativity theory (PR),, xii, 275
periodicity, 65
permeability, 297
permit, 201, 208
permittivity, 295, 297
phase shifts, vii, 1, 3, 6, 7, 15, 16, 17, 18, 20, 25, 26, 27, 29, 30, 31, 33, 34, 35, 38, 44, 45, 46, 47, 51, 52, 56, 57
phase transitions, 261
Philadelphia, 100
photons, 61, 64, 69, 73, 74, 86, 87, 89, 90, 91, 97, 104, 177, 206, 212, 214, 222, 223, 244, 249, 253, 254, 278
physical laws, xi, 243, 245, 259
physical mechanisms, ix, 104, 105
physical properties, 119, 120
physical theories, 215, 216
physics, xi, 8, 58, 62, 63, 64, 73, 74, 87, 90, 92, 94, 97, 98, 99, 167, 174, 182, 185, 189, 197, 198, 200, 202, 209, 216, 230, 240, 241, 243, 244, 248, 256, 262, 263, 273, 276, 281, 283, 306, 307
plane waves, 107
planets, 77, 78, 82, 83, 84, 207, 212
plants, 282
plausibility, 290
Poisson equation, 217
polar, 88, 89, 95
polarity, 68, 69, 72, 74, 75, 76, 77, 89
positron, 89, 92
potential two-cluster model, vii, 1
power generation, 70, 74, 77, 208
present value, x, 205, 214, 221, 226
President, 239
primum, viii, 61, 67, 68, 69, 70, 71, 72, 85, 86, 87, 88, 89, 90, 91, 95, 97
principles, 63, 66, 70, 92, 97, 256, 280, 283
probability, ix, 2, 3, 31, 57, 73, 118, 185, 187, 195, 210
probability distribution, 210
probe, viii, 104, 105, 138, 141, 144, 147
project, 14, 15, 300
propagation, 66, 189, 278
proposition, 245, 279, 280
protons, 57, 71, 72, 73, 89, 91, 92, 96, 200, 206, 222, 250, 280
pumps, ix, 104, 105, 158

Q

qualitative concept, 56
quanta, xii, 4, 9, 275, 280, 281
quantization, ix, 65, 73, 75, 77, 185, 186, 188, 195, 280, 281
quantum cosmology, ix, 185, 196, 202
quantum field theory, 174
quantum fields, 215
quantum fluctuations, 113, 291
quantum gravity, ix, x, 62, 71, 74, 96, 98, 99, 173, 174, 185, 186, 205, 214, 215, 216, 226
quantum mechanical theory, x, 205
quantum mechanics, x, 193, 197, 198, 201, 210, 214, 283
quantum state, 186
quantum theory, vii, 185, 186, 198, 210, 215, 216
quarks, 70, 71, 72, 73, 89, 96
quasars, 79, 100, 104
query, 300

R

radar, 83, 212
radial distance, 236
radiation, ix, x, xi, 73, 91, 92, 104, 138, 145, 146, 148, 149, 180, 181, 183, 185, 187, 188, 190, 193, 194, 195, 196, 206, 214, 223, 230, 232, 234, 235, 237, 238, 243, 244, 247, 248, 249, 251, 252, 253, 254, 255, 260, 261, 301, 303, 304

radius, x, 5, 9, 10, 11, 15, 20, 21, 31, 32, 40, 41, 43, 47, 48, 51, 53, 54, 55, 63, 80, 96, 107, 112, 164, 205, 208, 209, 210, 211, 213, 218, 220, 221, 223, 225, 230, 234, 237, 247, 248, 250, 252, 255, 256, 264, 286, 294, 295, 297, 301, 302
reaction rate, 2
reactions, vii, 2, 8, 17, 39, 44, 50, 57, 253
real numbers, 182
real time, 264
realism, 197, 198
reality, 119, 197, 198, 276, 277, 279, 281
reasoning, vii, 61, 62, 63, 64
recall, 74, 175, 233, 237, 265, 291, 299
recession, 80
recognition, 305
recombination, 111, 252, 253, 254
rectification, 63
red shift, xi, 202, 243, 244, 249, 255
redistribution, 254
redshift, 218, 243, 244, 245, 248, 249, 251, 254, 277
reductionism, 198
redundancy, 65, 66, 70, 92, 96
reference frame, 270
reference system, 245, 304
regenerate, 250
reinforcement, 93
relativity, xi, xii, 94, 112, 139, 159, 165, 185, 186, 199, 215, 216, 243, 244, 245, 246, 247, 248, 253, 256, 261, 263, 271, 272, 273, 275, 276, 278, 280, 281, 282, 283
reproduction, 65
repulsion, 9, 83, 91, 92, 96, 222, 250
requirements, 21, 63
researchers, 186, 198, 201, 202, 291
residual error, 12, 32, 43
residuals, 55, 252
resistance, 70, 75, 86, 96
resolution, 90, 98
response, 211, 220, 304
restrictions, xi, 243
restructuring, 44, 50
rings, 71, 72, 73, 82, 83
root, 5, 20, 31, 40, 47, 48, 53, 122
root-mean-square, 5, 20, 31, 40, 47, 48, 53
roots, 122, 275
rotations, 263, 264, 268, 271, 305
rowing, 111, 118, 125
rules, 63, 230

S

SAP, 201
scalar field, viii, 103, 105, 119, 120, 121, 123, 129, 218, 261, 262
scattering, vii, 1, 3, 6, 7, 8, 9, 15, 16, 17, 18, 19, 20, 23, 24, 25, 26, 27, 28, 29, 30, 31, 32, 33, 34, 35, 36, 38, 39, 40, 41, 43, 44, 45, 46, 47, 48, 50, 51, 52, 53, 55, 56, 57, 251, 254
Schrödinger equation, 5, 6
science, 62, 63, 65, 67, 74, 84, 197, 200
scope, 247, 256, 286
Second World, 92
self-organization, 201
self-similarity, 65
sellers, 286
semi-microscopic approach, vii, 1
senses, 279, 280, 281
shape, 82, 83, 85, 90, 162, 164, 212, 234, 301
shock, 68, 79, 92
shock waves, 79, 92
showing, 82, 90, 287
simulation, 98
Singapore, 58, 227, 242, 308
smoothness, 65, 93
snaps, 94
solar system, 78, 82, 89, 246, 308
solution, 3, 5, 6, 7, 64, 98, 99, 107, 109, 110, 111, 114, 118, 122, 124, 132, 133, 134, 135, 136, 138, 139, 140, 142, 143, 144, 145, 151, 152, 153, 154, 155, 163, 187, 191, 192, 198, 203, 224, 230, 231, 234, 236, 237, 247, 248, 249, 256, 264, 281, 289, 291, 293, 294, 297, 300, 302, 304, 305
spacetime, xi, 141, 186, 208, 215, 216, 259, 260, 261, 262, 265, 266, 267, 268, 272, 274, 278, 279, 291
space-time, 113, 158, 173, 174, 175, 176, 178, 179, 180, 181, 182, 186, 197, 198, 199, 201
special relativity, 245, 263, 266, 267, 270, 273, 290
special theory of relativity, 215, 216
specialization, 65
species, 99, 232
speculation, 90, 97
speed of light, xii, 72, 81, 91, 95, 138, 203, 208, 221, 245, 248, 264, 285, 286, 295, 296, 297, 305
spin, 4, 6, 7, 8, 15, 25, 26, 31, 32, 33, 35, 39, 45, 46, 69, 71, 72, 73, 78, 79, 81, 82, 83, 84, 85, 88, 95, 96, 230, 232, 239, 281, 284, 295, 298, 305
St. Petersburg, 99

stability, 65, 89, 91, 96, 252
stars, vii, xi, 2, 8, 65, 71, 77, 78, 80, 81, 82, 83, 84, 85, 95, 96, 100, 104, 112, 164, 206, 207, 208, 243, 249, 250, 252, 253
state control, 174
sterile, 208
steroids, 84
storage, 75
stress, 139, 146, 259
stretching, ix, 72, 85, 104, 105, 149, 155, 158
string theory, 151, 182, 185, 186, 215, 216
strong force, 96
structure, vii, xi, 1, 2, 3, 6, 16, 28, 56, 63, 65, 66, 67, 68, 77, 104, 138, 186, 207, 215, 246, 249, 250, 252, 253, 254, 257, 259, 260, 263, 264, 277, 285, 295, 296, 304, 307, 308
structure formation, vii, 104, 207, 249
substitution, 151, 162, 221
substrate, viii, 103, 104, 105, 106, 107, 108, 109, 110, 111, 113, 114, 118, 119, 124, 129, 130, 137, 160
successive approximations, 150, 154
Sun, 8, 82, 85, 86, 87, 88, 89, 112, 166
superconductivity, 86
supermultiplet symmetry, vii, 1, 3
supernovae, 218, 254, 255
superstrings, 65, 67, 68, 69, 70, 72, 73, 74, 75, 76, 77, 78, 79, 81, 84, 86, 87, 89, 91, 92, 97
surface layer, 247
Sweden, 182
Switzerland, 166, 168, 170
symmetry, vii, 1, 3, 6, 45, 65, 72, 73, 89, 149, 230, 264, 273, 277
synchronization, 65, 70, 88

T

techniques, 62, 186, 210
technological progress, 197
technologies, 74
technology, 74, 77, 286
temperature, x, xi, 35, 65, 79, 81, 84, 92, 94, 111, 199, 205, 209, 214, 218, 219, 220, 221, 222, 223, 225, 226, 229, 230, 231, 235, 236, 250, 251, 254, 260, 301, 303
Temporal topos (t-topos), ix, 173
tension, ix, 104, 105, 149, 158, 216, 222
tensor field, 262
terminals, 75
textbooks, 229, 230, 246, 304
Theory of Everything, 309
thermal energy, 251
thermodynamic properties, 112

thermodynamics, viii, 61, 62, 64, 65, 74, 77, 199, 200, 230
thinning, 85
threshold level, 46, 51
throws, 82, 85, 89, 94
time periods, 176, 177, 178, 180, 181
tobacco, 281
topology, 174, 175
tornadoes, 88
torsion, 271, 287
torus, 68
total energy, xi, 210, 219, 220, 224, 229, 230, 231, 232, 253, 256, 277, 286, 287, 289, 290, 291, 292, 294, 300, 301, 304, 305
tracks, 216
trajectory, 78
transformation, 94, 95, 176, 248, 277
transformations, 117, 127, 148, 263, 264, 265, 268
translation, 264, 265, 270
transmission, 75, 76
transport, 135
traveling waves, 138
treatment, xii, 74, 175, 233, 285
trial, 62, 209, 210, 218
triggers, 35, 97
tunneling, ix, 181, 185, 186, 191, 195, 210, 220
tunneling effect, 210
turbulence, viii, 61, 63, 66, 67, 73, 74, 77, 79, 81, 84, 88, 93, 98
Turkey, 57

U

UK, 57, 59, 257
Ukraine, 185, 197
unification, 199
uniform, 67, 87, 110, 288
universality, 69, 72
universe, vii, viii, x, xi, xii, 61, 62, 63, 64, 66, 77, 78, 79, 80, 81, 82, 83, 84, 85, 86, 87, 89, 90, 92, 94, 97, 98, 99, 100, 104, 180, 185, 196, 197, 198, 199, 201, 205, 206, 207, 208, 209, 210, 211, 212, 213, 214, 215, 218, 219, 220, 221, 222, 223, 225, 226,227, 243, 244, 245, 246, 247, 248, 249, 251, 252, 253, 254, 255, 256, 264, 273, 275, 276, 278, 279, 280, 281, 282, 283
universes, 196, 230, 235, 291, 305
updating, 24
uranium, 72, 91, 92
USA, 173, 202, 227
Uzbekistan, 57

Index

V

vacuum, viii, ix, x, 65, 91, 103, 104, 105, 106, 107, 108, 112, 113, 114, 118, 123, 126, 149, 158, 159, 160, 161, 162, 164, 165, 186, 197, 198, 199, 201, 206, 207, 218, 224, 237, 245, 248, 252, 278, 279, 287, 297, 299
valence, 74, 93, 94
variables, 7, 108, 115, 127, 130, 174
variations, 100, 268, 296
vector, 94, 111, 116, 124, 128, 130, 140, 146, 160, 161, 270, 287, 290
velocity, ix, 90, 104, 105, 109, 111, 112, 144, 149, 156, 159, 162, 163, 164, 165, 202, 206, 207, 210, 211, 218, 220, 239, 259, 263, 265, 267, 268, 276, 277, 278
Venus, 212
vibration, xii, 65, 66, 87, 88, 90, 91, 93, 94, 95, 96, 150, 254, 275, 278, 279, 281
viruses, 281
viscosity, 69, 78, 79, 80, 81, 86, 88, 90, 235
vision, 281

W

Washington, 58
water, 66, 73, 84, 87, 88, 94, 280, 281, 282, 283
wave functions, vii, 1, 3, 4, 9, 50
wave number, 4, 5, 6, 7
wave vector, 111, 116, 124, 128
wavelengths, 87, 279
weak interaction, 207
weak-oscillating cosmic string, viii, 104, 105
Wisconsin, 99
worldview, 201

Y

yield, 62, 81, 218, 236, 277, 291, 303